Peripheral Nerve Sheath Tumors

Ignazio Gaspare Vetrano • Vittoria Nazzi
Editors

Peripheral Nerve Sheath Tumors

From Preclinical Studies to Advanced Therapies

Springer

Editors
Ignazio Gaspare Vetrano
Department of Neurosurgery
Oncological Neurosurgery Unit,
Fondazione IRCCS Istituto Neurologico
Carlo Besta
Milano, Italy

Vittoria Nazzi
Department of Neurosurgery
Functional Neurosurgery Unit,
Fondazione IRCCS Istituto Neurologico
Carlo Besta
Milano, Italy

ISBN 978-3-031-76381-6 ISBN 978-3-031-76382-3 (eBook)
https://doi.org/10.1007/978-3-031-76382-3

This Springer imprint is published by the registered company Springer Nature Switzerland AG
The registered company address is: Gewerbestrasse 11, 6330 Cham, Switzerland

I am grateful to my wife Elena, for her enduring support and encouragement. Her love is the reason for the greatest gift of my life, our son Alessandro Leone. Thank you for being my greatest motivation and for reminding me of what truly matters most.

Ignazio Gaspare Vetrano

I would like to express my sincere gratitude to all medical colleagues, supporting nursing staff in the operating theatres, on the intense care station and on the wards for their invaluable support and collaboration.

I extend my deepest appreciation to the patients who have entrusted me with their surgical care; your trust has been a constant source of purpose and dedication in my work.

Lastly, to my husband Stefano, thank you for your unwavering encouragement and understanding. To my daughters Claudia and Lucrezia, your love and inspiration have been my greatest motivation.

Vittoria Nazzi

Preface

Peripheral nerve tumors can range from benign to malignant and occur in any part of the body, often presenting significant challenges for diagnosis, treatment, and management. The complexity of these tumors lies in their interaction with critical nerve structures, which frequently necessitates a delicate balance between tumor resection and preservation of nerve function.

Peripheral nerve tumors can be classified into several categories, including schwannomas, neurofibromas, and malignant peripheral nerve sheath tumors (MPNSTs), each with distinct characteristics and clinical implications. The complex nature of these tumors requires a multidisciplinary approach, involving neurosurgeons, oncologists, radiologists, pathologists, and rehabilitation specialists, to optimize outcomes and improve the quality of life for patients.

The management of peripheral nerve tumors has traditionally relied heavily on surgical intervention, particularly for benign tumors where complete resection can often be curative. However, surgery carries inherent risks. Advances in microsurgical techniques and intraoperative monitoring or, more recently, the introduction of intraoperative fluorescent dyes, have improved surgical outcomes, but the potential for nerve damage and functional impairment remains a major concern.

Over the past few decades, significant progress has been made in the understanding of peripheral nerve tumors, particularly in terms of their molecular biology, genetics, pathology or, more recently, omics. All these factors have opened up new avenues for diagnosis and treatment, moving beyond traditional surgical approaches to include a range of novel therapeutic strategies. This book aims to provide an overview of the latest developments in the management of peripheral nerve tumors, highlighting innovative preclinical viewpoints and new treatment strategies, advances in surgical techniques, and emerging therapies.

This book also explores the role of imaging and diagnostics in the field of peripheral nerve tumors. Advances in imaging technologies, such as magnetic resonance imaging (MRI) and positron emission tomography (PET), have improved the ability to accurately diagnose and monitor these tumors, enabling more precise treatment planning. Furthermore, the introduction of oral selective inhibitors of MAPK kinase, or new surgical paradigms also for MPNST, are gradually changing the therapeutic management in case of multiple or malignant tumors.

In conclusion, the landscape of peripheral nerve tumor treatment is rapidly evolving, driven by advances in technology, molecular biology, and a deeper understanding of tumor pathophysiology. This book provides an in-depth examination of these developments, and it could offer a comprehensive resource for clinicians, researchers, and healthcare professionals involved in caring for patients with peripheral nerve tumors. By exploring both established and emerging treatment strategies, this book aims to contribute to the ongoing efforts to improve outcomes for patients with these challenging and often debilitating conditions.

Therefore, we wish to thank all the contributors to this book whose expertise in their specific field has provided and continues to provide new and solid foundations for peripheral nerve tumors.

Milan, Italy Ignazio Gaspare Vetrano
Milan, Italy Vittoria Nazzi

Contents

1 **Overview of Peripheral Nerve Sheath Tumors** 1
Anna Cecilia Lawson McLean

2 **Natural History and Clinical Evolution of Peripheral Nerve Sheath Tumors in Neurofibromatosis** 11
Brigitte C. Widemann, Eva Dombi, and Andrea M. Gross

3 **Advancement in Genetic and Molecular Landscape of Peripheral Nerve Sheath Tumors** 19
Suganth Suppiah, Chloe Gui, and Gelareh Zadeh

4 **Multi-Omic Analyses for Biomarker Discovery and Characterization of PNST** 25
Antonella Potenza, Tatiana Carrozzini, Gemma Gorla, Giuliana Pollaci, and Laura Gatti

5 **Preclinical NF1-PNST Models and Potential Drug Resistance Mechanisms** 37
Zhichao Wang, Wei Wang, and Yuehua Li

6 **MRI in the Diagnosis of PNST: State-of-the-Art and Recent Advancements** 53
Francesca Serpi, Domenico Albano, Salvatore Gitto, Carmelo Messina, and Luca Maria Sconfienza

7 **The Developing Role of Molecular Imaging PET in Peripheral Nerve Sheath Tumors** 71
Annachiara Arnone, Ignazio Gaspare Vetrano, and Pierpaolo Alongi

8 **A Clinical Summary of PNST in Adults** 83
Marica Eoli

9 **Clinical Assessment of Tumor Predisposition Syndromes: Related Nerve Sheath Tumors in Children** 93
Veronica Saletti, Rossana Bucci, Germana Lo Monaco, and Marco Moscatelli

10 **IOUS and CEUS in PNS Tumors** 107
Giovanni Carone, Massimiliano Del Bene, and Francesco Prada

11 Neurophysiological Evaluation and Intraoperative Monitoring in Peripheral Nervous System Tumors 113
Grazia Devigili and Valentina Catanzaro

12 Fluorescein-Guided Surgery of Peripheral Nerve Sheath Tumors 131
Jacopo Falco, Niccolò Innocenti, Ignazio Gaspare Vetrano, Marica Eoli, Morgan Broggi, Vittoria Nazzi, and Francesco Acerbi

13 Exoscope in PNS Tumor Surgery 139
Niccolò Innocenti, Jacopo Falco, Vittoria Cojazzi, Marica Eoli, Francesco Acerbi, Ignazio Gaspare Vetrano, and Vittoria Nazzi

14 Surgical Management of Peripheral Nerve Sheath Tumors 147
Maxwell T. Laws, Luis F. Castro, and Prashant Chittiboina

15 Surgical Management of Pediatric PNST 157
Julian Zipfel and Martin U. Schuhmann

16 Functional Status in Malignant Peripheral Nerve Sheath Tumors 167
Enrico Martin and J. Henk Coert

17 MEK Inhibitors and Other Medical Therapies for Benign Peripheral Nerve Sheath Tumors 175
Brigitte C. Widemann, Eva Dombi, and Andrea M. Gross

18 Systemic Treatment in Malignant Peripheral Nerve Sheath Tumors: Current Practice and Future Possibilities 183
Enrico Martin and Ingrid M. E. Desar

19 Proton and Carbon Ion Therapy for MPNST 195
Maria Rosaria Fiore, Agnieszka Chalaszczyk and Ester Orlandi

20 Radiomics and Machine Learning in PNST 211
Zhichao Wang, Chengjiang Wei, and Wei Wang

Overview of Peripheral Nerve Sheath Tumors

1

Anna Cecilia Lawson McLean

1.1 Introduction

Peripheral nerve sheath tumors (PNST) are by definition neoplasms that originate from peripheral nerve sheath cells. These include Schwann cells, the perineurium, and the epineurium (Fig. 1.1). PNST can occur sporadically or in the context of genetic disorders. The genetic disorders that are by definition associated with PNST are neurofibromatosis type 1 (NF1), nf2-related schwannomatosis (NF2), and non-nf2-schwannomatosis (SWN).

Over the recent years, we have gained a better understanding of PNST and new tumor entities have been detected through molecular analysis and whole-genome sequencing. This has led to new treatment strategies and advanced possibilities for targeted therapies. The 2021 edition of the World Health Organization (WHO) Classification of Tumors of the Central Nervous System lists six distinct entities of peripheral nerve sheath tumors under the section "cranial and paraspinal nerve tumors" (Table 1.1; [1]).

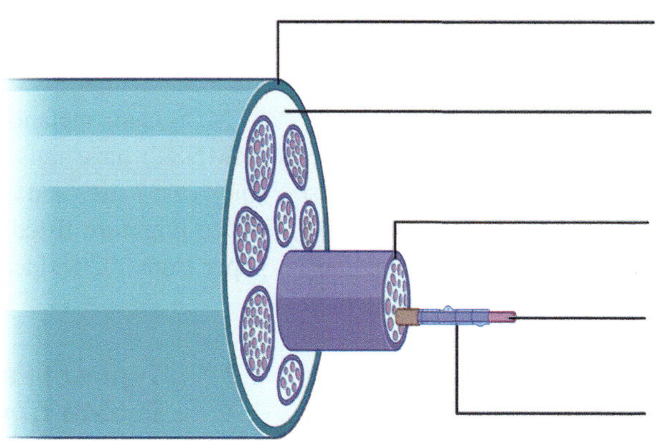

Fig. 1.1 Overview of the microscopic anatomy of peripheral nerves and the involved cell types. Tumors may originate from various locations. Peripheral nerve sheath tumors typically consist of neoplastic cells as well as nonneoplastic cell components that are shown in the illustration above (*Created with* BioRender.com)

A. C. Lawson McLean (✉)
Department of Neurosurgery, Jena University Hospital, Jena, Germany

I. G. Vetrano, V. Nazzi (eds.), *Peripheral Nerve Sheath Tumors*,
https://doi.org/10.1007/978-3-031-76382-3_1

Table 1.1 Peripheral nerve sheath tumors listed in the 2021 WHO Classification of Tumors of the Central Nervous System [1]

Tumor entity	Benign/malignant	Cells of origin
Schwannoma	Benign	Neoplastic Schwann cells
Neurofibroma	Benign (with few exceptions)	Neoplastic Schwann cells
Perineurioma	Benign	Neoplastic perineurial cells
HPNST	Benign	Neoplastic Schwann cells +- neoplastic perineurial cells
MPNST	Malignant	Neoplastic Schwann cells
MMNST	Malignant	Neoplastic Schwann precursor cells

Neurofibromas and schwannomas are the most common and best understood kinds of PNST. Over the recent years, perineuriomas and hybrid tumors have been characterized and added to the WHO classification. All of these entities are typically benign neoplasms. However, there are also two malignant kinds of peripheral nerve sheath tumors. The first type is blatantly called malignant peripheral nerve sheath tumor (MPNST) and represents a very aggressive type of sarcoma that affects the peripheral nerves. The second kind called malignant melanotic nerve sheath tumor (MMNST) an extremely rare entity that consists of Schwann cells with melanotic differentiation.

This chapter aims to give an overview of the different types of PNST and the associated tumor predisposition syndromes.

1.2 Epidemiology

Neurofibromas are usually regarded as the most common kind of PNST. However, there is no reliable on data on the epidemiology of solitary lesions. However, there is evidence that the incidence of NF1, which is by definition associated with the formation of neurofibromas, is 1:2500–1:3500 live births. NF1 is therefore the most common phakomatosis.

Schwannomas are also regarded as relatively common PNST. As with neurofibromas, there are no exact figures on the incidence of sporadic peripheral nerve schwannomas. The incidence of NF2 is 1:25,000 [2], and the incidence of SWN is 1:69,000 [3]. Both of which are typically associated with the formation of schwannomas.

Perineuriomas are very rare tumors that occur over a wide age range, without significant gender delectation [4]. There are no exact numbers on incidence. The available data is based on a limited case series. There is no known tumor-predisposition syndrome associated with perineurioma.

Similarly, the incidence of hybrid peripheral nerve sheath tumors (HPNST) is unknown and all available data is based on case series. Neurofibroma/schwannoma HPNST and schwannoma/perineurioma HPNST appear to commonly arise in the context of NF2 and SWN [5, 6].

Malignant peripheral nerve sheath tumors (MPNST) are very rare. The overall incidence is approximately 1.46 per million person-years, with an increased incidence among the elderly [7].

Malignant melanotic nerve sheath tumors (MMNST) have thus far only been reported in small case series and case reports. A 2020 review of the literature listed a total of 65 cases [8]. A paper from 2022 added another two cases [9].

1.3 Types of Peripheral Nerve Sheath Tumors

Peripheral nerve sheath tumors are commonly benign lesions of the peripheral nerves. Depending on their histological and molecular features, they can be grouped into six separate entities according to the recent fifth revision of the WHO Classification of Tumors of the Central Nervous System [1].

1.3.1 Neurofibromas

The most common type of peripheral nerve sheath tumors are neurofibromas. These tumors mainly consist of neoplastic Schwann cells but contain nonneoplastic components such fibroblasts, mast cells, perineurial-like cells, and residual axons in addition [10]. Histologically, they are composed of elongated cells intermixed with collagen and/or mucin bundles. Positivity for S-100 is low [11].

Neurofibromas are caused by a defective *nf1* gene, which encodes the protein neurofibromin. Neurofibromin exhibits its highest levels of expression in neurons, non-myelinating Schwann cells and oligodendrocytes [12]. Through its involvement in the Ras/MAPK, Akt/mTOR, ROCK/LIMK/cofilin, and cAMP/PKA pathways neurofibromin regulates many fundamental cellular processes, such as proliferation, cell growth, and survival [13].

Solitary or localized neurofibromas of the larger peripheral nerves are often difficult to distinguish from schwannomas based on their radiographic and macroscopical appearance. They may appear sporadically but can also be associated with NF1. Depending on their location, surgical resection is most likely the treatment option of choice. In contrast to schwannomas, neurofibromas are typically more difficult to dissect from the carrier nerve and several nerve fascicles may be affected. Thus, there is a higher risk of postoperative neurological deficits.

Cutaneous neurofibromas are typically associated with NF1. In most cases, they are a cosmetic issue and do not cause neurological deficits. Malignant transformation is very rare. However, the psychological effect of these tumors, which also occur in the facial area should not be underestimated [14]. Treatment options include laser ablation and surgical resection.

Plexiform neurofibromas are also associated with NF1. They develop in 20–50% of patients with NF1 and can cause substantial complications including pain, motor impairment, and disfigurement [15]. By definition, plexiform neurofibromas affect several nerve fascicles and can spread through several layers of tissues, including the cutis. Due to their multinodular enlargement, the terms "bag of worms" or "string of onions" have been coined to describe their appearance [10]. There is a high rate of malignant transformation (8–13% in NF1 patients; [16]), and treatment is often difficult for these tumors. Surgical resection is associated with a high risk of neurological deficits and complete resection can be difficult to achieve due to the extent of the tumors and the involvement of several layers of tissues. The MEK inhibitor selumetinib has been shown to reduce tumor volume of plexiform neurofibromas in approximately 70% of patients in preclinical studies that led to its approval for both children and adults with plexiform neurofibromas in the context of NF1 [15, 17]. While selumetinib is currently the only approved MEK inhibitor, evidence shows that other MEK inhibitors such as trametinib may be similarly effective in the pediatric cohort [18].

Atypical neurofibromatous neoplasms of uncertain biological potential (ANNUBP) are the most recent addition to the WHO Classification of Tumors of the Central Nervous System within the neurofibroma category [1]. They represent a subgroup of tumors with worrisome histological features but not clearly deserving a diagnosis of malignancy. Diagnostic criteria for ANNUBP include neurofibromas with at least two of the following four features: histological atypia, loss of the typical neurofibroma architecture, hypercellularity, and increased mitotic activity [>1/50 but <3 mitotic figures per 10 High power fields (HPFs) or <1.5 mitoses/mm^2] [10]. On a genetic level, the progression from plexiform neurofibroma to ANNUBP seems to be driven by the deletion of *CDKN2A/B*. Further malignant transformation to MPNST likely involves chromosomal rearrangements and inactivation of the

PRC2 genes, loss of DNA repair genes, and copy-number increase of signal transduction and cell-cycle genes [19].

1.3.2 Schwannomas

Schwannomas are another common type of peripheral nerve sheath tumors. They consist of neoplastic Schwann cells and may occur sporadically or in the context of genetic disorders. Histologically, they are composed of a mixture of cellular Antoni type A areas and more loosely organized less cellular Antoni type B areas. Antoni type A areas consist predominantly of entangled Schwann cell processes in a sparse stroma containing basement membrane substance and collagen fibrils. Antoni type B areas are sparsely cellular and edematous and feature a variable macrophage infiltrate. The tumor capsule consists of multiple layers of collagen fibers. In contrast to neurofibromas, there are no residual axons within the tumorous tissue. On immunohistochemistry, Schwannomas are positive for mature Schwann cell markers SOX10 and S-100 protein [10, 20].

So far, defects of the *nf2*, *SMARCB1*, and *LZTR1* genes have been shown to cause the formation of schwannomas [21–23]. In most cases, the *nf2* gene, which encodes the protein merlin, is affected. Loss of merlin leads to Ras and Rac activation, which stimulates cell growth and proliferation through several signaling pathways, such as the Hippo, WNT, and TGF pathways [24, 25].

Schwannomas may occur sporadically or in the context of genetic disorders. They can affect both peripheral and cranial nerves. As most peripheral nerve schwannomas have a collagenous capsule, complete surgical excision can usually be performed without injury of the carrier nerve. Nevertheless, electrophysiological monitoring is highly recommended during the procedure to avoid postoperative deficits.

Intracranially, schwannomas most commonly arise from the VIII cranial nerve. According to US American data, vestibular schwannomas have an incidence of 1.14:10,00,000 [26]. Vestibular schwannomas can be treated by microsurgical resection/decompression or stereotactic radiosurgery [27].

1.3.3 Perineurioma

Perineuriomas are rare nerve sheath tumors that are composed of neoplastic perineural cells. They come as soft tissue (extraneural) or intraneural variants. Soft tissue perineuriomas occur in the subcutaneous tissues of the trunk and limbs as a painless, solitary nodule or mass. Intraneural perineuriomas, on the other hand, are restricted to the boundaries of a nerve [28]. They primarily affect the extremities of young adults and children. Histologically, the perineurial cells show predominantly storiform growth patterns. The stroma is collagenous, with about 20% of cases containing myxoid matrix. Mitotic activity is usually low. Immunohistochemical staining is positive for CD34 in about 50% of cases. The staining for S100 and GFAP is negative [4].

Patients with intraneural perineuriomas usually present with atrophy and motor dysfunction. Complete resection is not recommended. Rather, a decompression in the manner of a long-distance epineurotomy should be attempted [29].

1.3.4 Hybrid Tumors

Hybrid peripheral nerve sheath tumors (HPNST) were first included in the 4th edition of the World Health Organization (WHO) Classification of Tumors of Soft Tissue and Bone in 2013. They typically are hybrids of neurofibromas/schwannomas or schwannomas/perineuriomas. Hybrids of neurofibromas/perineuriomas are very rare. HPNST has histological and immunohistochemical features of both merged tumor entities.

Schwannoma/perineurioma HPNST seem to occur sporadically and to not associated with genetic syndromes [5]. Neurofibroma/schwannoma HPNST, on the other hand, are linked to NF2 and SWN [30, 31]. Next-generation sequencing has identified that *ERBB2* mutations in these types of HPNST could be a treatable target and preliminary trials with the ERBB inhibitor lapatinib have shown promising results [6].

1.3.5 Malignant Peripheral Nerve Sheath Tumors

MPNST are very rare tumors, which account for approximately 2% of sarcomas (Ng 2013). They can arise from neuroectodermal cells of peripheral or cranial nerves, and most commonly arise in peripheral nerves of the extremities. The sciatic nerve is the predominantly affected nerve (Sobscuk 2020), [4]. Approximately half of them occur in the context of NF1. MPNST in the context of NF1 appear to affect younger patients and to have a worse prognosis than sporadic MPNST [32]. On a molecular level, similarly to schwannomas and neurofibromas, MPNST show an over-activation of RAS-mediated pathways. In addition, alterations in p53, p16, and epidermal growth factor receptor (EGFR) have been shown to be associated with the formation of MPNST [33].

The gold standard in the treatment of MPNST is surgery with wide negative margins [34–36]. Still, the local recurrence is about 50% and distant metastatic disease occurs in 30–60% of patients [4]. Adjuvant radiation therapy is typically used to reduce local recurrence. However, radiation itself is a risk factor for the development of MPNST in NF1, so risk and benefit should be assessed carefully in each case [37]. With regard to systemic therapy, anthracycline-based chemotherapy has been shown to improve overall survival. According to an EORTC soft tissue and bone sarcoma group study published in 2011, a combination of doxorubicin and ifosfamide had the best response rates [38]. Several targeted therapies have been proposed and evaluated in preclinical and clinical trials (up to phase II). Based on the molecular alterations found in MPNST, mTOR inhibitors such as everolimus and MEK inhibitors such as sirolimus have been proposed as potentially effective agents. However, no single agents have proven to improve overall survival thus far. Currently, several trials investigating combination therapies are ongoing ([39], NCT03433183).

1.3.6 Malignant Melanocytic Nerve Sheath Tumors

To date, only 67 cases of MMNST, formerly known as melanocytic schwannomas, have been described in the literature [8, 9]. They are composed of neoplastic Schwann cells that produce melanin and therefore appear dark in color. Schwann cells and melanocytes both originate from Schwann cell precursors in the neural crest, hence their potential ability to produce melanin. MMNST can be divided into psammomatous and non-psammomatous subtypes [8]. In approximately half of the cases, they are associated with the so-called Carney complex, which is an autosomal dominantly inherited familial syndrome. The Carney complex is characterized by myxomas (cardiac, cutaneous, and mammary), spotty pigmentation, and endocrine overactivity (Cushing's syndrome and acromegaly) (Alexiev 2018). Based on the little data available, surgical resection is the standard of care. Local recurrence and distal metastasis were reported in up 35% and 44% of cases, respectively (Torres-Mora 2014).

1.3.7 Association with Genetic Disorders

There are three genetic disorders that are associated with the formation of peripheral nerve sheath tumors: neurofibromatosis type 1 (NF1), nf2-schwannomatosis (NF2), and non-nf2-schwannomatosis (SWN).

1.3.7.1 Neurofibromatosis Type 1

NF1 is an autosomal-dominant tumor predisposition syndrome also known as von Recklinghausen disease. The incidence is 1:2500–1:3500 neonates, regardless of ethnicity or racial background

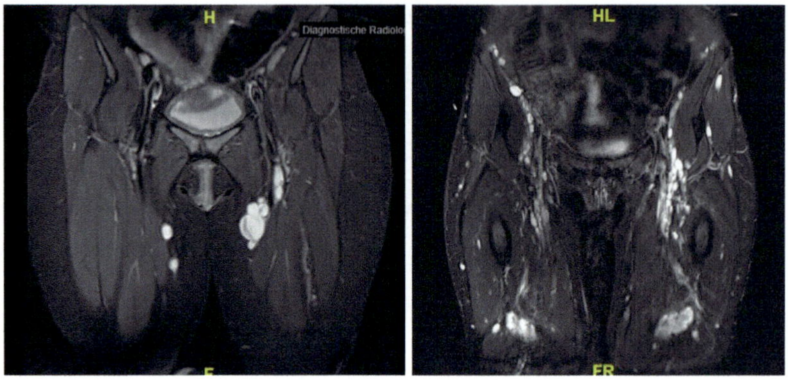

Fig. 1.2 Comparison of schwannomas of the femoral nerve in a patient with nf2-related schwannomatosis (left) and plexiform neurofibromas of the femoral nerve in a patient with neurofibromatosis type 1 (right) (*Images courtesy of PD Dr. Lobsien and Prof. Klisch, Erfurt*)

[40, 41]. It is caused by a mutation of the *nf1* gene, which is located on chromosome 17q11.2 and which encodes the protein neurofibromin. Deficiency of neurofibromin leads to an over-activation of RAS, resulting in cellular proliferation [42]. NF1 is associated with the formation of multiple cutaneous neurofibromas that commonly first appear in adolescence and increase in number in adulthood. Patients may also develop singular neurofibromas of peripheral nerves as well as plexiform neurofibromas (Fig. 1.2). In 8–13% of patients, plexiform neurofibromas degenerate into MPNST, which are the main reason for a lowered life expectancy in patients with NF1 [16]. Other hallmarks of NF1 include ≥6 café au lait spots, optic nerve glioma, and ≥2 Lisch nodules (iris hamartomas) [43].

1.3.7.2 Nf2-Related Schwannomatosis

NF2 is also an autosomal-dominant hereditary tumor predisposition syndrome, but it is significantly less common than NF1. The incidence is 1:25,000 live births [2]. Defects of the *nf2* gene, which is located on 22q11.2, lead to a deficiency in the proto-oncogene merlin. Merlin deficiency, similar to loss of neurofibromin, leads to RAS over-activation resulting in cellular proliferation [24]. Bilateral vestibular schwannomas are hallmarks of NF2. However, patients are prone to develop various types of tumors of the central and peripheral nervous systems, including schwannomas of the peripheral nerves, meningiomas, and intramedullary spinal ependymomas (Fig. 1.2). Non-tumorous manifestations of NF2 include subcapsular cataracts and polyneuropathy [44, 45].

1.3.7.3 Non-nf2-Schwannomatosis

SWN is a very rare disease with an incidence of 1:69,000 neonates [3]. Patients typically present with multiple schwannomas of the peripheral nerves that are similar to those seen in sporadic cases or NF2. In contrast to NF2, the leading symptom is neuropathic chronic pain. Patients also lack other hallmarks of NF2, such as bilateral vestibular schwannomas or subcapsular cataract. Typically, mutations in the *SMARCB1* or the *LZTR1* genes lead to SWN [46, 47]. More recent research revealed that in some cases of SWN, *ERBB2* mutations give rise to segmental neurofibroma/schwannoma hybrid tumors [6].

References

1. Louis DN, Perry A, Wesseling P, Brat DJ, Cree IA, Figarella-Branger D, Hawkins C, Ng HK, Pfister SM, Reifenberger G, Soffietti R, von Deimling A, Ellison DW. The 2021 WHO classification of tumors of the central nervous system: a summary. Neuro Oncol. 2021;23(8):1231–51. https://doi.org/10.1093/neuonc/noab106. PMID: 34185076; PMCID: PMC8328013.
2. Evans DG, Moran A, King A, Saeed S, Gurusinghe N, Ramsden R. Incidence of vestibular schwannoma and neurofibromatosis 2 in the North West of England over a 10-year period: higher incidence than previously thought. Otol Neurotol. 2005;26(1):93–7. https://doi.org/10.1097/00129492-200501000-00016.
3. Evans DG, Bowers NL, Tobi S, Hartley C, Wallace AJ, King AT, Lloyd SKW, Rutherford SA, Hammerbeck-Ward C, Pathmanaban ON, Freeman

SR, Ealing J, Kellett M, Laitt R, Thomas O, Halliday D, Ferner R, Taylor A, Duff C, Harkness EF, Smith MJ. Schwannomatosis: a genetic and epidemiological study. J Neurol Neurosurg Psychiatry. 2018;89(11):1215–9. https://doi.org/10.1136/jnnp-2018-318538. Epub 2018 Jun 16.
4. Mocellin S. Perineurioma. In: Soft tissue tumors. Cham: Springer; 2021. https://doi.org/10.1007/978-3-030-58710-9_199.
5. Hornick JL, Bundock EA, Fletcher CD. Hybrid schwannoma/perineurioma: clinicopathologic analysis of 42 distinctive benign nerve sheath tumors. Am J Surg Pathol. 2009;33(10):1554–61. https://doi.org/10.1097/PAS.0b013e3181accc6c.
6. Ronellenfitsch MW, Harter PN, Kirchner M, Heining C, Hutter B, Gieldon L, Schittenhelm J, Schuhmann MU, Tatagiba M, Marquardt G, Wagner M, Endris V, Brandts CH, Mautner VF, Schröck E, Weichert W, Brors B, von Deimling A, Mittelbronn M, Steinbach JP, Reuss DE, Glimm H, Stenzinger A, Fröhling S. Targetable ERBB2 mutations identified in neurofibroma/schwannoma hybrid nerve sheath tumors. J Clin Invest. 2020;130(5):2488–95. https://doi.org/10.1172/JCI130787. PMID: 32017710; PMCID: PMC7190903.
7. Bates JE, Peterson CR, Dhakal S, Giampoli EJ, Constine LS. Malignant peripheral nerve sheath tumors (MPNST): a SEER analysis of incidence across the age spectrum and therapeutic interventions in the pediatric population. Pediatr Blood Cancer. 2014;61(11):1955–60. https://doi.org/10.1002/pbc.25149. Epub 2014 Aug 17.
8. Solomou G, Dulanka Silva AH, Wong A, Pohl U, Tzerakis N. Extramedullary malignant melanotic schwannoma of the spine: case report and an up to date systematic review of the literature. Ann Med Surg (Lond). 2020;59:217–23. https://doi.org/10.1016/j.amsu.2020.10.003. PMID: 33088497; PMCID: PMC7559563.
9. Yeom JA, Song YS, Lee IS, Han IH, Choi KU. Malignant melanotic nerve sheath tumors in the spinal canal of psammomatous and non-psammomatous type: two case reports. World J Clin Cases. 2022;10(24):8735–41. https://doi.org/10.12998/wjcc.v10.i24.8735. PMID: 36157803; PMCID: PMC9453363.
10. Belakhoua SM, Rodriguez FJ. Diagnostic pathology of tumors of peripheral nerve. Neurosurgery. 2021;88(3):443–56. https://doi.org/10.1093/neuros/nyab021. PMID: 33588442; PMCID: PMC7884141.
11. Woertler K. Tumors and tumor-like lesions of peripheral nerves. Semin Musculoskelet Radiol. 2010;14(5):547–58. https://doi.org/10.1055/s--0030-1268073. Epub 2010 Nov 11.
12. Daston MM, Scrable H, Nordlund M, Sturbaum AK, Nissen LM, Ratner N. The protein product of the neurofibromatosis type 1 gene is expressed at highest abundance in neurons, Schwann cells, and oligodendrocytes. Neuron. 1992;8(3):415–28. https://doi.org/10.1016/0896-6273(92)90270-n.
13. Bergoug M, Doudeau M, Godin F, Mosrin C, Vallée B, Bénédetti H. Neurofibromin structure, functions and regulation. Cells. 2020;9(11):2365. https://doi.org/10.3390/cells9112365. PMID: 33121128; PMCID: PMC7692384.
14. Leidger A, Vosschulte M, Nieder TO, Mautner VF. Sexual self-esteem and psychological burden of adults with neurofibromatosis type 1. Front Psychol. 2022;13:883019. https://doi.org/10.3389/fpsyg.2022.883019. PMID: 35756263; PMCID: PMC9220094.
15. Dombi E, Baldwin A, Marcus LJ, Fisher MJ, Weiss B, Kim A, Whitcomb P, Martin S, Aschbacher-Smith LE, Rizvi TA, Wu J, Ershler R, Wolters P, Therrien J, Glod J, Belasco JB, Schorry E, Brofferio A, Starosta AJ, Gillespie A, Doyle AL, Ratner N, Widemann BC. Activity of selumetinib in neurofibromatosis type 1-related plexiform neurofibromas. N Engl J Med. 2016;375(26):2550–60. https://doi.org/10.1056/NEJMoa1605943. PMID: 28029918; PMCID: PMC5508592.
16. Evans DG, Baser ME, McGaughran J, Sharif S, Howard E, Moran A. Malignant peripheral nerve sheath tumours in neurofibromatosis 1. J Med Genet. 2002;39(5):311–4. https://doi.org/10.1136/jmg.39.5.311. PMID: 12011145; PMCID: PMC1735122.
17. Gross AM, Wolters PL, Dombi E, Baldwin A, Whitcomb P, Fisher MJ, Weiss B, Kim A, Bornhorst M, Shah AC, Martin S, Roderick MC, Pichard DC, Carbonell A, Paul SM, Therrien J, Kapustina O, Heisey K, Clapp DW, Zhang C, Peer CJ, Figg WD, Smith M, Glod J, Blakeley JO, Steinberg SM, Venzon DJ, Doyle LA, Widemann BC. Selumetinib in children with inoperable plexiform neurofibromas [Erratum in: N Engl J Med. 2020;383(13):1290.]. N Engl J Med. 2020;382(15):1430–42. https://doi.org/10.1056/NEJMoa1912735. Epub 2020 Mar 18. PMID: 32187457; PMCID: PMC7305659.
18. McCowage GB, Mueller S, Pratilas CA, et al. Trametinib in pediatric patients with neurofibromatosis type 1 (NF-1)-associated plexiform neurofibroma: a phase I/IIa study. J Clin Oncol. 2018;36(15):10504.
19. Pemov A, Hansen NF, Sindiri S, Patidar R, Higham CS, Dombi E, Miettinen MM, Fetsch P, Brems H, Chandrasekharappa SC, Jones K, Zhu B, Wei JS, National Intramural Sequencing Center (NISC) Comparative Sequencing Program; National Cancer Institute (NCI) Division of Cancer Epidemiology and Genetics (DCEG) Cancer Genomics Research Laboratory, Mullikin JC, Wallace MR, Khan J, Legius E, Widemann BC, Stewart DR. Low mutation burden and frequent loss of CDKN2A/B and SMARCA2, but not PRC2, define premalignant neurofibromatosis type 1-associated atypical neurofibromas. Neuro Oncol. 2019;21(8):981–92. https://doi.org/10.1093/neuonc/noz028. PMID: 30722027; PMCID: PMC6682216.
20. Erlandson RA. Peripheral nerve sheath tumors. Ultrastruct Pathol. 1985;9(1–2):113–22. https://doi.org/10.3109/01913128509055493.

21. Piotrowski A, Xie J, Liu YF, Poplawski AB, Gomes AR, Madanecki P, Fu C, Crowley MR, Crossman DK, Armstrong L, Babovic-Vuksanovic D, Bergner A, Blakeley JO, Blumenthal AL, Daniels MS, Feit H, Gardner K, Hurst S, Kobelka C, Lee C, Nagy R, Rauen KA, Slopis JM, Suwannarat P, Westman JA, Zanko A, Korf BR, Messiaen LM. Germline loss-of-function mutations in LZTR1 predispose to an inherited disorder of multiple schwannomas. Nat Genet. 2014;46(2):182–7. https://doi.org/10.1038/ng.2855. Epub 2013 Dec 22. PMID: 24362817; PMCID: PMC4352302.
22. Rouleau GA, et al. Genetic linkage of bilateral acoustic neurofibromatosis to a DNA marker on chromosome 22. Nature. 1987;329(6136):246–8.
23. Sestini R, Bacci C, Provenzano A, Genuardi M, Papi L. Evidence of a four-hit mechanism involving SMARCB1 and NF2 in schwannomatosis-associated schwannomas. Hum Mutat. 2008;29(2):227–31. https://doi.org/10.1002/humu.20679.
24. Morrison H, Sperka T, Manent J, Giovannini M, Ponta H, Herrlich P. Merlin/neurofibromatosis type 2 suppresses growth by inhibiting the activation of Ras and Rac. Cancer Res. 2007;67(2):520–7. https://doi.org/10.1158/0008-5472.CAN-06-1608.
25. Mota M, Shevde LA. Merlin regulates signaling events at the nexus of development and cancer. Cell Commun Signal. 2020;18(1):63. https://doi.org/10.1186/s12964-020-00544-7. PMID: 32299434; PMCID: PMC7164249.
26. Cioffi G, Yeboa DN, Kelly M, Patil N, Manzoor N, Greppin K, Takaoka K, Waite K, Kruchko C, Barnholtz-Sloan JS. Epidemiology of vestibular schwannoma in the United States, 2004–2016. Neurooncol Adv. 2020;2(1):vdaa135. https://doi.org/10.1093/noajnl/vdaa135. PMID: 33241216; PMCID: PMC7672330.
27. Yakkala VK, Mammi M, Lamba N, Kandikatla R, Paliwal B, Elshibiny H, Corrales CE, Smith TR, Mekary RA. Audiovestibular symptoms and facial nerve function comparing microsurgery versus SRS for vestibular schwannomas: a systematic review and meta-analysis. Acta Neurochir (Wien). 2022;164(12):3221–33. https://doi.org/10.1007/s00701-022-05338-z. Epub 2022 Aug 13.
28. Macarenco RS, Ellinger F, Oliveira AM. Perineurioma: a distinctive and underrecognized peripheral nerve sheath neoplasm. Arch Pathol Lab Med. 2007;131(4):625–36. https://doi.org/10.5858/2007-131-625-PADAUP.
29. Brand C, Pedro MT, Pala A, Heinen C, Scheuerle A, Braun M, Antoniadis G. Perineurioma: a rare entity of peripheral nerve sheath tumors. J Neurol Surg A Cent Eur Neurosurg. 2022;83(1):1–5. https://doi.org/10.1055/s-0041-1726110. Epub 2021 May 24.
30. Feany MB, Anthony DC, Fletcher CD. Nerve sheath tumours with hybrid features of neurofibroma and schwannoma: a conceptual challenge. Histopathology. 1998;32(5):405–10. https://doi.org/10.1046/j.1365-2559.1998.00419.x.
31. Harder A, Wesemann M, Hagel C, Schittenhelm J, Fischer S, Tatagiba M, Nagel C, Jeibmann A, Bohring A, Mautner VF, Paulus W. Hybrid neurofibroma/schwannoma is overrepresented among schwannomatosis and neurofibromatosis patients. Am J Surg Pathol. 2012;36(5):702–9. https://doi.org/10.1097/PAS.0b013e31824d3155.
32. Widemann BC. Current status of sporadic and neurofibromatosis type 1-associated malignant peripheral nerve sheath tumors. Curr Oncol Rep. 2009;11(4):322–8. https://doi.org/10.1007/s11912-009-0045-z. PMID: 19508838; PMCID: PMC6689400.
33. Perry A, Kunz SN, Fuller CE, Banerjee R, Marley EF, Liapis H, Watson MA, Gutmann DH. Differential NF1, p16, and EGFR patterns by interphase cytogenetics (FISH) in malignant peripheral nerve sheath tumor (MPNST) and morphologically similar spindle cell neoplasms. J Neuropathol Exp Neurol. 2002;61(8):702–9. https://doi.org/10.1093/jnen/61.8.702.
34. Abbas JS, Holyoke ED, Moore R, Karakousis CP. The surgical treatment and outcome of soft-tissue sarcoma. Arch Surg. 1981;116(6):765–9. https://doi.org/10.1001/archsurg.1981.01380180025006.
35. Dunn GP, Spiliopoulos K, Plotkin SR, Hornicek FJ, Harmon DC, Delaney TF, Williams Z. Role of resection of malignant peripheral nerve sheath tumors in patients with neurofibromatosis type 1. J Neurosurg. 2013;118(1):142–8. https://doi.org/10.3171/2012.9.JNS101610. Epub 2012 Oct 26.
36. Scaife CL, Pisters PW. Combined-modality treatment of localized soft tissue sarcomas of the extremities. Surg Oncol Clin N Am. 2003;12(2):355–68. https://doi.org/10.1016/s1055-3207(03)00003-6.
37. Yamanaka R, Hayano A. Radiation-induced malignant peripheral nerve sheath tumors: a systematic review. World Neurosurg. 2017;105:961–970.e8. https://doi.org/10.1016/j.wneu.2017.06.010. Epub 2017 Jun 8.
38. Kroep JR, Ouali M, Gelderblom H, Le Cesne A, Dekker TJA, Van Glabbeke M, Hogendoorn PCW, Hohenberger P. First-line chemotherapy for malignant peripheral nerve sheath tumor (MPNST) versus other histological soft tissue sarcoma subtypes and as a prognostic factor for MPNST: an EORTC soft tissue and bone sarcoma group study. Ann Oncol. 2011;22(1):207–14. https://doi.org/10.1093/annonc/mdq338. Epub 2010 Jul 23. PMID: 20656792; PMCID: PMC3003614.
39. Hassan A, Pestana RC, Parkes A. Systemic options for malignant peripheral nerve sheath tumors. Curr Treat Options Oncol. 2021;22(4):33. https://doi.org/10.1007/s11864-021-00830-7.
40. Evans DG, Howard E, Giblin C, Clancy T, Spencer H, Huson SM, Lalloo F. Birth incidence and prevalence of tumor-prone syndromes: estimates from a UK family genetic register service. Am J Med Genet A. 2010;152A(2):327–32. https://doi.org/10.1002/ajmg.a.33139.

41. Huson SM, Compston DA, Clark P, Harper PS. A genetic study of von Recklinghausen neurofibromatosis in South East Wales. I. Prevalence, fitness, mutation rate, and effect of parental transmission on severity. J Med Genet. 1989;26(11):704–11. https://doi.org/10.1136/jmg.26.11.704. PMID: 2511318; PMCID: PMC1015740.
42. Upadhyaya M, Shaw DJ, Harper PS. Molecular basis of neurofibromatosis type 1 (NF1): mutation analysis and polymorphisms in the NF1 gene. Hum Mutat. 1994;4(2):83–101. https://doi.org/10.1002/humu.1380040202.
43. NIH Consensus Statement. National Institutes of Health Consensus Development Conference Statement: neurofibromatosis. Bethesda, MD, USA, July 13–15, 1987. Neurofibromatosis. 1988;1(3):172–8.
44. McLaughlin ME, Pepin SM, Maccollin M, Choopong P, Lessell S. Ocular pathologic findings of neurofibromatosis type 2. Arch Ophthalmol. 2007;125(3):389–94. https://doi.org/10.1001/archopht.125.3.389.
45. Sperfeld AD, Hein C, Schröder JM, Ludolph AC, Hanemann CO. Occurrence and characterization of peripheral nerve involvement in neurofibromatosis type 2. Brain. 2002;125(Pt 5):996–1004. https://doi.org/10.1093/brain/awf115.
46. Hulsebos TJ, Plomp AS, Wolterman RA, Robanus-Maandag EC, Baas F, Wesseling P. Germline mutation of INI1/SMARCB1 in familial schwannomatosis. Am J Hum Genet. 2007;80(4):805–10. https://doi.org/10.1086/513207. Epub 2007 Feb 16. PMID: 17357086; PMCID: PMC1852715.
47. Kehrer-Sawatzki H, Farschtschi S, Mautner VF, Cooper DN. The molecular pathogenesis of schwannomatosis, a paradigm for the co-involvement of multiple tumour suppressor genes in tumorigenesis. Hum Genet. 2017;136(2):129–48. https://doi.org/10.1007/s00439-016-1753-8. Epub 2016 Dec 5. PMID: 27921248; PMCID: PMC5258795.

Natural History and Clinical Evolution of Peripheral Nerve Sheath Tumors in Neurofibromatosis

2

Brigitte C. Widemann, Eva Dombi, and Andrea M. Gross

2.1 Introduction

Neurofibromatosis type 1 (NF1) is an autosomal dominant, genetic condition characterized by the development of a variety of manifestations including histologically benign and malignant peripheral nerve sheath tumors (PNST) [1, 2]. Mutations in the *NF1* tumor suppressor result in RAS pathway activation, which is critical to the pathogenesis of PNST [3]. Much has been learned about the natural history and clinical evolution of these tumors during the past two decades [4–6]. This increased understanding has been instrumental to the development of prevention and management strategies, interventional clinical trials, and the first regulatory approval of a medical treatment for PNST. In this chapter, we will review the current knowledge of the natural history of PNST and its implications for clinical management.

The development of PNST is a hallmark feature of NF1. Table 2.1 summarizes key aspects of the genetic changes, presentation, and management of cutaneous neurofibromas (CNF), plexiform neurofibromas (PNF), atypical neurofibromas (ANF), and malignant peripheral nerve sheath tumors (MPNST).

B. C. Widemann (✉) · E. Dombi · A. M. Gross
Pediatric Oncology Branch, Center for Cancer Research, National Cancer Institute, Bethesda, MD, USA
e-mail: widemanb@mail.nih.gov; dombie@mail.nih.gov; andrea.gross@nih.gov

2.1.1 Cutaneous Neurofibromas (CNF)

Cutaneous neurofibromas (CNF) are the most frequent PNST in NF1 and almost every individual with NF1 will develop at least one CNF during their lifetime [7]. However, the number of CNF varies greatly among individuals, ranging from very few to thousands of tumors. CNF consist of multiple cell types including Schwann cells, fibroblasts, mast cells, and macrophages. As with all NF1-associated PNST, CNF are characterized by biallelic loss of *NF1* in the tumorigenic Schwann cells. These tumors typically develop in later childhood, adolescence, and throughout life. While CNF never undergo malignant transformation, they can cause substantial morbidity including disfigurement, pruritus, and social, emotional, and psychological distress [8, 9]. To date, there are no approved medical therapies for CNF. In 2018, a comprehensive research strategy was published based on a meeting of experts with the goal of advancing the development of effective therapies for NF1 CNF [10]. The development of CNF early in childhood has been associated with the development of a larger number of CNF in later life. However, many aspects of the natural history of CNF

I. G. Vetrano, V. Nazzi (eds.), *Peripheral Nerve Sheath Tumors*,
https://doi.org/10.1007/978-3-031-76382-3_2

Table 2.1 Characteristics of NF1-associated peripheral nerve sheath tumors

Characteristic	Cutaneous neurofibroma (CNF)	Plexiform neurofibroma (PNF)	Atypical neurofibroma (ANF)	Malignant peripheral nerve sheath tumor (MPNST)
Incidence (%)	>95	30–50	Unknown	8–15.8
Median age at diagnosis (year)	Adolescent, young adult	Young child	27 (7.6–60)	13–36
Development	Peripheral nerve	Peripheral nerve	PNF/nerve	PNF/ANF/nerve
Tumor genetics	Biallelic *NF1* loss	Biallelic *NF1* loss	+ Loss *CDKN2A/B*	+ Frequently loss of *polycomb repressive complex: EED, SUZ12,* p53…
Histopathology	Benign	Benign	Benign—borderline	Malignant: low, intermediate, high grade
Clinical symptoms	Disfigurement, pruritus, social emotional distress	Large slow growth, pain, functional loss, disfigurement	Firm nodular tumor, faster growth compared to PNF, pain, functional loss	Rapidly enlarging tumor, worsening/ severe pain, functional loss
Imaging	Photography, ultrasound, small, frequently multiple cutaneous lesions	MRI: frequently large mass, complex shape, central dot sign, slow growth	MRI: frequently distinct nodular lesion, loss of central dot sign, faster growth compared to PNF, FDG-PET: increased uptake	MRI: rapidly growing mass with heterogeneous contrast uptake, diffusion restriction FDG-PET: increased uptake CT chest/abdomen/ pelvis: possible distant metastases
Treatment	Surgical approaches; investigational trials	Complete surgical resection frequently not feasible MEK inhibitor selumetinib approved for children with inoperable, symptomatic PNF; investigational trials	Marginal surgical resection for lesions considered to be at increased risk for malignant transformation; investigational trials for unresectable ANF	Complete surgical resection with wide negative margins essential for cure of high grade MPNST. Radiation and chemotherapy as indicated; investigational trials for unresectable, recurrent, metastatic MPNST

remain unpredictable, which can cause substantial distress to individuals with NF1 and their family members. There are very few genotype–phenotype correlations that predict CNF burden in individuals with NF1. In addition, even within families carrying the same germline *NF1* mutation the number of CNF that develop can vary greatly among individuals. Other factors that may contribute to CNF heterogeneity include hormonal factors, spatial and temporal loss of heterozygosity of the normal *NF1* allele, gene modifiers, and the tumor microenvironment [11].

Several observational natural history studies have been conducted and are ongoing to better characterize the development of CNF throughout life and to develop better methods of detection and measurement of CNF. Methods used in interventional and natural history studies to quantitate CNF burden include calipers, volume photography, ultrasound, and whole-body photography [12]. Additional studies are ongoing to develop a detailed understanding of the clinical burden of CNF and to develop and validate patient-reported outcome (PRO) instruments, such as the Skindex

PRO, that will allow measuring clinical benefit in interventional trials targeting CNF [13]. Knowledge gained from these studies will assist in the development of clinical trials with meaningful endpoints to measure the potential clinical benefit of preventing CNF development or growth of these lesions in addition to achieving shrinkage. Many efforts at increasing the understanding of CNF are ongoing, including the Neurofibromatosis Therapeutic Acceleration Program (NTAP) [14] and the Response Evaluation in Neurofibromatosis and Schwannomatosis (REiNS) [8, 15] international working group, which aim to increase the understanding of the pathogenesis, biology, and clinical course of CNF in addition to developing meaningful endpoints and outcome measures for interventional clinical trials. Importantly, interventional clinical trials directed at CNF are ongoing and include oral as well as topical mitogen-activated protein kinase kinase (MEK) inhibitors.

2.1.2 Plexiform Neurofibromas (PNF)

Plexiform neurofibromas (PNF) are histologically benign PNST, which develop in approximately 50% of individuals with NF1 based on whole-body magnetic resonance imaging (WB-MRI) studies [5]. Like CNF, PNF consist of multiple cell types, but in contrast to discrete CNF, PNF involve multiple fascicles and branches of nerves. Several definitions of PNF with variable appearances and clinical characteristics have been proposed in the past. A recent consensus review has provided a way to classify PNF based on clinically relevant imaging characteristics on MRI, including the morphology and internal structure as well as the depth and the relationship to adjacent tissues [5]. This classification system may be helpful for the understanding of PNF responses in interventional trials. Figure 2.1 provides examples of several types of PNF.

Most PNF develop in early childhood or may be congenital, and the development of new superficial PNF later in life has only been documented once to date in studies using WB-MRI to monitor PNF changes over time [4, 16]. A review of WB-MRI conducted on the National Institutes of Health (NIH) natural history study concluded that no new bulky PN developed in this cohort [4]. The anatomic location of a PNF impacts when it is detected and what symptoms it causes. Internal thoracic or abdominal PNF tend to be detected later in life unless they are impinging on vital structures such as the airway. In contrast, orbital or facial PNF, which often have a superficial component and are close to critical structures or in confined anatomical regions, frequently are detected earlier in life. Unlike CNF, PNF are at risk to undergo malignant transformation to MPNST with a lifetime incidence of MPNST in individuals with NF1 of 8–15.8% [17, 18]. The natural history of PNF is thus in stark contrast to that of CNF even though PNF are also only genomically characterized by biallelic loss of NF1 in the tumorigenic Schwann cells. It is unclear to date why these two genomically similar tumors manifest at different ages and have such clinically different behaviors. Recent preclinical studies suggest that in addition to genetic changes, the tumor microenvironment contributes to the pathogenesis of PNST and may at least in part explain the differences between these tumors [19].

Depending on their location, PNF can cause a variety of morbidities including disfigurement, pain, vision loss, motor, or bowel and bladder dysfunction as well as airway compromise. A retrospective review of patients enrolled in an NIH natural history study demonstrated that many of the children had clinically significant morbidity early in life at the time of their first presentation to NIH and that this morbidity did not improve over time [20].

Few studies have utilized WB-MRI to comprehensively characterize PNF burden. These studies demonstrate that individuals with NF1 can have more than one PNF and that many also present with a total body tumor burden that is characterized by the involvement of most nerves. The development of volumetric MRI analysis to sensitively and reproducibly measure the volume of PNF has greatly aided the understanding of the

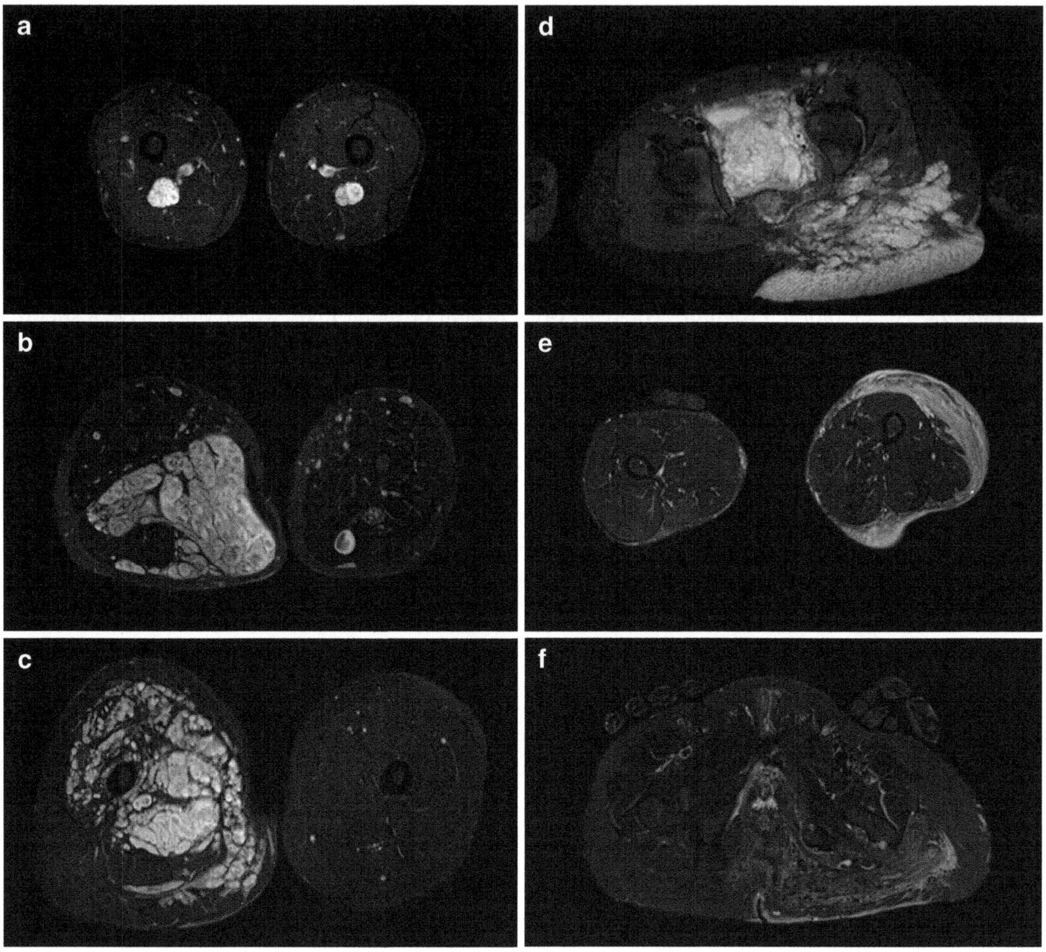

Fig. 2.1 Examples of plexiform neurofibromas (PNF) on axial STIR MRI demonstrate the structural diversity of these tumors: (**a**) bilateral enlargement of the sciatic nerves, (**b**) well-demarcated expansive mass in the posterior compartment if the right thigh, (**c**) anterior thigh PNF intertwined with muscle tissue, (**d**) pelvic and gluteal PNF extending into the dermal and subcutaneous tissue layers, (**e**) diffuse superficial PNF of the lateral thigh, and (**f**) streaky, ill-defined PNF of the left gluteal region

natural history of these tumors [4, 21, 22]. Longitudinal natural history studies have demonstrated that PNF grow most rapidly in early childhood where up to a 200% increase in PNF volume can be seen annually. A slower growth rate is seen in adolescence and very limited or no growth in most adults. Unlike CNF, there is no evidence that PNF growth increases with puberty [23]. In contrast to the growth behavior of PNF in children and young adults [4], a recently published study of 10-year follow-up with WB-MRI and volumetric MRI analysis in older adults with NF1 demonstrated slow, spontaneous shrinkage of the majority of internal neurofibromas [24]. PNF growth rates have not been shown to differ between anatomic locations and, importantly, PNF growth is not erratic as previously suggested. While the growth rate of PNF varies between people, the growth rate remains fairly constant within individuals during early childhood. The transition of more rapid growth to decreased growth rate, cessation of growth and slow tumor shrinkage, also varies between individuals, and factors predicting these changes have not been identified to date [4].

A better understanding of the growth of PNF through childhood and adulthood through natural history studies has helped the development of meaningful clinical trial endpoints for studies of targeted therapies directed at PNF. For example, clinical trials for young children can meaningfully assess the impact of an interventional agent using a change in time to progression as an endpoint, whereas in clinical trials directed at PNF in adults, which typically do not grow and may demonstrate slow shrinkage, this would not be a relevant endpoint, and assessing PNF tumor shrinkage is more appropriate.

Volumetric MRI analysis has become the standard method to evaluate response in clinical trials for PNF and was used to assess response defined as PNF shrinkage ≥20% compared to baseline in the phase 2 registration trial of the MEK inhibitor selumetinib for children with inoperable symptomatic PNF, also called SPRINT [25, 26]. This study resulted in the FDA approval of selumetinib for children with NF1 and inoperable symptomatic PNF in 2020.

2.1.3 Malignant Transformation of PNF

PNF are at risk for transformation to MPNST, and while MPNST are typically diagnosed in adults, these tumors have infrequently been described in young children as well. The diagnosis of an MPNST can be challenging as symptoms experienced in PNF can overlap with those of MPNST. Changes in clinical characteristics such as an increase in pain the development of new pain as well as changes in imaging characteristics can aid in this diagnosis [5]. Most MPNST develop in preexisting PNF, and until fairly recently, it was thought that these tumors arise directly from PNF. However, over the past decade, ANF, which frequently develop in preexisting PNF, have been identified as borderline malignant lesions and, at least in a subset of tumors, the more direct precursor to MPNST [27]. Volumetric MRI analysis of PNST has helped identification of premalignant ANF or MPNST. For example, growth of a PNF by more than 20% per year in an adult should raise concern for the possibility of an ANF or MPNST. Very few genotype phenotype correlations exist in NF1, however whole gene deletion predicts for the development of more severe manifestations and individuals with whole gene deletion are also at greater risk for development of MPNST [5].

2.1.4 Atypical Neurofibromas (ANF)

Atypical neurofibromas are characterized histologically based on loss of neurofibroma architecture, increased cellularity, infrequent mitosis, and/or cellular atypia. The presence of two or more of these features has been described as atypical neurofibromatous neoplasm of uncertain biological potential (ANNUBP), and these tumors are considered to be at increased risk for transformation to MPNST [27]. In addition, ANF have been characterized by the frequent loss of *CDKN2A/B*. This genetic change has not been described in PNF and therefore likely is the first step toward malignant transformation [28–30]. Importantly, during the past decade, the clinical characteristics of ANF have been described. The NIH NF1 Natural History study using longitudinal WB-MRI in children and young adults with NF1 first identified the presence of lesions called distinct nodular lesions (DNL) [4, 27, 31]. These lesions are defined by a distinct nodular appearance with a lack of the central dot sign and a diameter of at least 3 cm. These lesions are typically more avid on fluorodeoxyglucose positron emission tomography (FDG-PET) when compared to surrounding PNF. DNL also have a faster growth rate than PNF in the same individual, and unlike PNF, which tend to slowdown in growth during adulthood, DNL growth appears to be independent of age. Unlike PNF, new DNL often develop throughout life. Figure 2.2 shows the development of a DNL in an individual with NF1. In the NIH Natural History study at the time

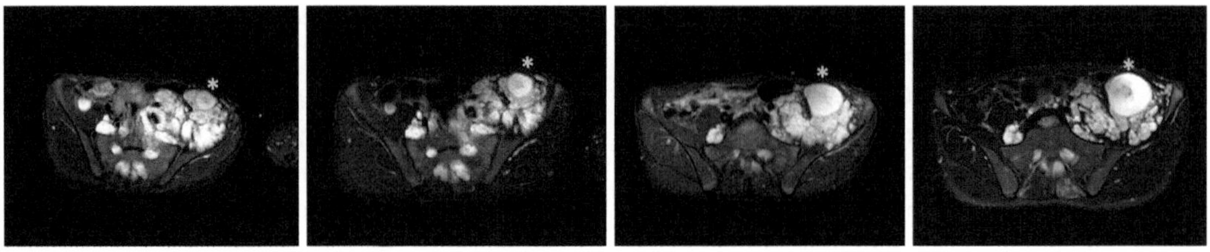

Fig. 2.2 Axial STIR MRI demonstrating the development of a distinct nodular lesion (DNL) (*) in a preexisting left pelvic plexiform neurofibroma (PNF) in an individual with neurofibromatosis type 1 (NF1)

of the first MRI, 58 of 122 patients had at least 1 DNL [median age at first MRI 9.4 years (3.1, 20)]. In contrast to PNF, the development of 83 new DNL was observed on follow-up MRIs at a median age of 17 years (10, 23.9). In addition, spontaneous regression of DNL has been described with a greater incidence of regression in adults compared to children [4, 24]. These findings suggest that DNL have a different biology than PNF. While DNL can be asymptomatic, pain has been described as the most frequent symptom of these lesions. Given the more rapid growth, FDG-PET avidity and potential presence of pain as clinical symptoms DNL have frequently been biopsied to rule out malignant transformation. Many of these lesions have been found to be ANF on histological exam, some appeared as benign neurofibromas, and some of the lesions have been identified to have *CDKN2A/B* loss. However, it is unknown, what proportion of DNL are ANF. In addition, the incidence of *CDKN2A/B* loss in DNL and in ANF is not known presently. A retrospective review of 63 patients with 76 biopsy-proven ANF showed that these lesions are at increased risk for malignant transformation [27]. An NCI consensus conference on PNST and malignant transformation identified that marginal resection of ANF may allow for safe surgical resection of a potentially premalignant lesion and thus prevention of transformation to highly aggressive MPNST [32]. Several case series have been reported confirming that safe marginal resection of ANF is feasible and that these lesions do not recur after resection [33, 34]. In the absence of effective treatment strategies for unresectable high-grade MPNST, marginal resection of ANNUBP is thus a potential strategy to prevent the development of MPNST. Additional studies are needed to better characterize the malignant potential of ANF and to develop tailored evidence-based management approaches.

2.1.5 Malignant Peripheral Nerve Sheath Tumors (MPNST)

Malignant peripheral nerve sheath tumors (MPNST) are aggressive sarcomas with a high propensity for metastasis and recurrence after surgical resection [35]. Most MPNST are high grade based on histological evaluation characterized by high mitotic rate and the presence of necrosis. Approximately 50% of MPNST develops in individuals with NF1, and MPNST is the highest contributor to the death rate of people with NF1. The only potentially curative treatment approach for MPNST to date remains complete surgical resection with wide negative margins. The presence of an unresectable or metastatic MPNST currently does not have realistic curative treatment options. For this reason, it is critical to detect MPNST early at a time when surgical resection may still be feasible.

In a patient with NF1, the presence of an enlarging palpable or visible mass and development of new pain or a change in pain that persists as well as a new neurologic compromise should raise the concern for the potential presence of an MPNST and result in the appropriate workup including a detailed clinical exam and imaging typically including an MRI. In contrast to PNF and ANF, MPNST have a heterogeneous appearance on MRI, including heterogeneous contrast enhancement. In addition, MRI with evaluation of diffusion-weighted sequences allows to assess the apparent diffusion coefficient, which frequently shows diffusion restriction in an MPNST. Most MPNST also demonstrate increased glucose uptake on FDG-PET [36]. The knowledge that at least some MPNST arise from preexisting ANF provides a potential opportunity to prevent MPNST. In particular, patients who may be at greater risk for the development of MPNST may benefit from surveillance strategies, which allow for early detection of PNST at risk for malignant transformation. This approach will be piloted in an NF1 surveillance study at the NIH (NCT06222203) and is also undergoing evaluation in an ongoing study at Johns Hopkins Hospital, Baltimore, MD, USA (NCT05677594). In these studies, patients who have potential risk factors for MPNST, such as prior radiation treatment, presence of a DNL, history of a prior MPNST, or large PNF burden, will undergo surveillance imaging with WB-MRI at regular intervals, in addition to clinical evaluation and possibly additional imaging, and biopsy and surgical interventions, as clinically indicated. These studies may provide a framework for interventional surgical strategies to prevent MPNST development. While WB-MRI allows for the detection of changes in PNST, these studies often cannot be performed frequently, as they are costly and currently mostly available in the research setting. The development of liquid biomarkers of malignant transformation therefore is an exciting development. Recent studies have demonstrated that cell-free DNA (cfDNA) can separate benign PNF from MPNST, and work is in progress to refine these strategies to also identify ANF [37]. While this is approach is currently limited to research studies, there is thus hope that in the future liquid biomarkers may become available as a clinical test to separate different types of PNST. This would allow for more frequent monitoring without the need to travel in patients at greater risk for transformation and positively impact the monitoring of the natural history of PNST in people with NF.

In summary, substantial progress has been made in the understanding of the natural history of PNST. Importantly, this knowledge has improved the management of PNST, enabled the development of meaningful interventional clinical trials, and allowed for the development of strategies to prevent transformation to MPNST.

References

1. Ly KI, Blakeley JO. The diagnosis and management of neurofibromatosis type 1. Med Clin N Am. 2019;103(6):1035–54.
2. Ferner RE. Neurofibromatosis 1 and neurofibromatosis 2: a twenty first century perspective. Lancet Neurol. 2007;6(4):340–51.
3. Ratner N, Miller SJ. A RASopathy gene commonly mutated in cancer: the neurofibromatosis type 1 tumour suppressor. Nat Rev Cancer. 2015;15(5):290–301.
4. Akshintala S, Baldwin A, Liewehr DJ, Goodwin A, Blakeley JO, Gross AM, et al. Longitudinal evaluation of peripheral nerve sheath tumors in neurofibromatosis type 1: growth analysis of plexiform neurofibromas and distinct nodular lesions. Neuro Oncol. 2020;22(9):1368–78.
5. Fisher MJ, Blakeley JO, Weiss BD, Dombi E, Ahlawat S, Akshintala S, et al. Management of neurofibromatosis type 1-associated plexiform neurofibromas. Neuro Oncol. 2022;24(11):1827–44.
6. Korf BR. Plexiform neurofibromas. Am J Med Genet. 1999;89(1):31–7.
7. Ortonne N, Wolkenstein P, Blakeley JO, Korf B, Plotkin SR, Riccardi VM, et al. Cutaneous neurofibromas: Current clinical and pathologic issues. Neurology. 2018;91(2 Suppl 1):S5–S13.
8. Cannon A, Pichard DC, Wolters PL, Adsit S, Erickson G, Lessing AJ, et al. Perspective of adults with neurofibromatosis 1 and cutaneous neurofibromas: implications for clinical trials. Neurology. 2021;97(7 Suppl 1):S15–24.
9. Cannon A, Sarin KY, Petersen AK, Pichard DC, Wolters PL, Erickson G, et al. Perspectives of adolescents with neurofibromatosis 1 and cutaneous neurofibromas: implications for clinical trials. Clin Trials. 2023;21:67–72.
10. Blakeley JO, Wolkenstein P, Widemann BC, Lee J, Le LQ, Jackson R, et al. Creating a comprehensive research strategy for cutaneous neurofibromas. Neurology. 2018;91(2 Supplement 1):S1–4.
11. Jiang C, McKay RM, Lee SY, Romo CG, Blakeley JO, Haniffa M, et al. Cutaneous neurofibroma heterogeneity: factors that influence tumor burden in neurofibromatosis type 1. J Invest Dermatol. 2023;143(8):1369–77.
12. Li Y, Blakeley JO, Ly I, Berman Y, Lau J, Wolkenstein P, et al. Current and emerging imaging techniques for neurofibromatosis type 1-associated cutaneous neurofibromas. J Invest Dermatol. 2023;143(8):1397–405.
13. Fertitta L, Sarin KY, Bergqvist C, Patel E, Peiffer B, Moryousef S, et al. cNF-Skindex in adults living with

neurofibromatosis 1: severity strata in France and validation in United States adults. J Invest Dermatol. 2023;143(11):2226–32.
14. Blakeley JO, Le LQ, Lee SY, Ly I, Rhodes SD, Romo CG, et al. A call for discovery and therapeutic development for cutaneous neurofibromas. J Invest Dermatol. 2023;143(8):1351–7.
15. Maguiness S, Berman Y, Rubin N, Dodds M, Plotkin SR, Wong C, et al. Measuring the effect of cutaneous neurofibromas on quality of life in neurofibromatosis type 1. Neurology. 2021;97(7 Suppl 1):S25–31.
16. Nguyen R, Dombi E, Widemann BC, Solomon J, Fuensterer C, Kluwe L, et al. Growth dynamics of plexiform neurofibromas: a retrospective cohort study of 201 patients with neurofibromatosis 1. Orphanet J Rare Dis. 2012;7:75.
17. Evans DG, Baser ME, McGaughran J, Sharif S, Howard E, Moran A. Malignant peripheral nerve sheath tumours in neurofibromatosis 1. J Med Genet. 2002;39(5):311–4.
18. Uusitalo E, Rantanen M, Kallionpaa RA, Poyhonen M, Leppavirta J, Yla-Outinen H, et al. Distinctive cancer associations in patients with neurofibromatosis type 1. J Clin Oncol. 2016;34(17):1978–86.
19. Yang FC, Ingram DA, Chen S, Zhu Y, Yuan J, Li X, et al. Nf1-dependent tumors require a microenvironment containing Nf1+/− and c-kit-dependent bone marrow. Cell. 2008;135(3):437–48.
20. Gross AM, Singh G, Akshintala S, Baldwin A, Dombi E, Ukwuani S, et al. Association of plexiform neurofibroma volume changes and development of clinical morbidities in neurofibromatosis 1. Neuro Oncol. 2018;20(12):1643–51.
21. Cai W, Kassarjian A, Bredella MA, Harris GJ, Yoshida H, Mautner VF, et al. Tumor burden in patients with neurofibromatosis types 1 and 2 and schwannomatosis: determination on whole-body MR images. Radiology. 2009;250(3):665–73.
22. Dombi E, Solomon J, Gillespie AJ, Fox E, Balis FM, Patronas N, et al. NF1 plexiform neurofibroma growth rate by volumetric MRI: relationship to age and body weight. Neurology. 2007;68(9):643–7.
23. Kotch C, Dombi E, Shah AC, Smith K, Brown S, Li Y, et al. Retrospective cohort analysis of the impact of puberty on plexiform neurofibroma growth in patients with neurofibromatosis type 1. J Pediatr. 2023;260:113513.
24. Ly KI, Merker VL, Cai W, Bredella MA, Muzikansky A, Thalheimer RD, et al. Ten-year follow-up of internal neurofibroma growth behavior in adult patients with neurofibromatosis type 1 using whole-body MRI. Neurology. 2023;100(7):e661–e70.
25. Dombi E, Ardern-Holmes SL, Babovic-Vuksanovic D, Barker FG, Connor S, Evans DG, et al. Recommendations for imaging tumor response in neurofibromatosis clinical trials. Neurology. 2013;81(21 Suppl 1):S33–40.
26. Gross AM, Wolters PL, Dombi E, Baldwin A, Whitcomb P, Fisher MJ, et al. Selumetinib in children with inoperable plexiform neurofibromas. N Engl J Med. 2020;382(15):1430–42.
27. Higham CS, Dombi E, Rogiers A, Bhaumik S, Pans S, Connor SEJ, et al. The characteristics of 76 atypical neurofibromas as precursors to neurofibromatosis 1 associated malignant peripheral nerve sheath tumors. Neuro Oncol. 2018;20(6):818–25.
28. Beert E, Brems H, Daniels B, De Wever I, Van Calenbergh F, Schoenaers J, et al. Atypical neurofibromas in neurofibromatosis type 1 are premalignant tumors. Genes Chromosomes Cancer. 2011;50(12):1021–32.
29. Pemov A, Hansen NF, Sindiri S, Patidar R, Higham CS, Dombi E, et al. Low mutation burden and frequent loss of CDKN2A/B and SMARCA2, but not PRC2, define pre-malignant neurofibromatosis type 1-associated atypical neurofibromas. Neuro Oncol. 2019;21(8):981–22.
30. Pemov A, Li H, Patidar R, Hansen NF, Sindiri S, Hartley SW, et al. The primacy of NF1 loss as the driver of tumorigenesis in neurofibromatosis type 1-associated plexiform neurofibromas. Oncogene. 2017;36(22):3168–77.
31. Meany H, Dombi E, Reynolds J, Whatley M, Kurwa A, Tsokos M, et al. 18-Fluorodeoxyglucose-positron emission tomography (FDG-PET) evaluation of nodular lesions in patients with neurofibromatosis type 1 and plexiform neurofibromas (PN) or malignant peripheral nerve sheath tumors (MPNST). Pediatr Blood Cancer. 2013;60(1):59–64.
32. Reilly KM, Kim A, Blakely J, Ferner RE, Gutmann DH, Legius E, et al. Neurofibromatosis type 1-associated MPNST state of the science: outlining a research agenda for the future. J Natl Cancer Inst. 2017;109(8):djx124.
33. Bernthal NM, Putnam A, Jones KB, Viskochil D, Randall RL. The effect of surgical margins on outcomes for low grade MPNSTs and atypical neurofibroma. J Surg Oncol. 2014;110(7):813–6.
34. Nelson CN, Dombi E, Rosenblum JS, Miettinen MM, Lehky TJ, Whitcomb PO, et al. Safe marginal resection of atypical neurofibromas in neurofibromatosis type 1. J Neurosurg. 2019;133:1516–26.
35. Prudner BC, Ball T, Rathore R, Hirbe AC. Diagnosis and management of malignant peripheral nerve sheath tumors: current practice and future perspectives. Neurooncol Adv. 2020;2(Suppl 1):i40–i9.
36. Ahlawat S, Blakeley JO, Rodriguez FJ, Fayad LM. Imaging biomarkers for malignant peripheral nerve sheath tumors in neurofibromatosis type 1. Neurology. 2019;93(11):e1076–e84.
37. Szymanski JJ, Sundby RT, Jones PA, Srihari D, Earland N, Harris PK, et al. Cell-free DNA ultra-low-pass whole genome sequencing to distinguish malignant peripheral nerve sheath tumor (MPNST) from its benign precursor lesion: a cross-sectional study. PLoS Med. 2021;18(8):e1003734.

3 Advancement in Genetic and Molecular Landscape of Peripheral Nerve Sheath Tumors

Suganth Suppiah, Chloe Gui, and Gelareh Zadeh

3.1 Introduction

Peripheral nerve sheath tumors (PNSTs) are a relatively uncommon soft tissue neoplasm that arises throughout the central and peripheral nervous system. These tumors fall within a spectrum ranging from benign tumors, such as schwannomas and cutaneous neurofibromas, to high-grade malignant tumors referred to as malignant peripheral nerve sheath tumors (MPNSTs). Although most benign PNSTs have negligible malignant potential, individuals can still suffer from intractable pain, neurological dysfunction, and, in the case of genetic tumor predisposition syndromes, cosmetic disfigurement. PNSTs pose a major challenge in diagnosis and management, often requiring a multidisciplinary team of surgeons, radiologists, oncologists, and pathologists with expertise in the field.

Magnetic resonance imaging (MRI), the gold standard in investigating PNSTs, can identify the presence of a PNST but cannot reliably distinguish between the subtypes of benign PNSTs (schwannoma vs. neurofibroma) or detect malignant transformation into an MPNST. Therefore, physicians often rely on a combination of clinical gestalt and imaging characteristics to determine the optimal management plans for individuals with PNSTs. Currently, surgery remains the only therapeutic option for benign PNSTs. In the case of MPNSTs, gross total resection with a negative margin is recommended, although not always possible due to the surrounding vital neurovascular structures. In addition, adjuvant radiation and chemotherapy for MPNSTs have demonstrated minimal effect on overall survival.

The lack of effective medical treatment options for both benign and malignant PNSTs has stimulated interest in exploring targeted therapies. There has also been a movement in pathology to adopt novel tumor classifying schemes combining histopathology and molecular biomarkers to better predict tumor behavior, as witnessed in the new 2021 World Health Organization (WHO) central nervous system tumor classification. However, our understanding of the molecular drivers of oncogenesis in PNSTs is limited. Genetic tumor-predisposing syndromes, such as the neurofibromatosis family of disorders, have provided the most insight into the molecular landscape of these tumors and identified *NF1*, *NF2*, *SMARCB1*, and *LZTR1* genes as major

S. Suppiah · C. Gui
Division of Neurosurgery, Department of Surgery, University of Toronto, Toronto Western Hospital, Toronto, ON, Canada
e-mail: suganth.suppiah@mail.utoronto.ca; Chloe.gui@mail.utoronto.ca

G. Zadeh (✉)
Division of Neurosurgery, Department of Surgery, University of Toronto, Toronto Western Hospital, Toronto, ON, Canada

Department of Neurosurgery, Mayo Clinic, Rochester, Minnesota, USA
e-mail: gelareh.zadeh@uhn.ca

I. G. Vetrano, V. Nazzi (eds.), *Peripheral Nerve Sheath Tumors*,
https://doi.org/10.1007/978-3-031-76382-3_3

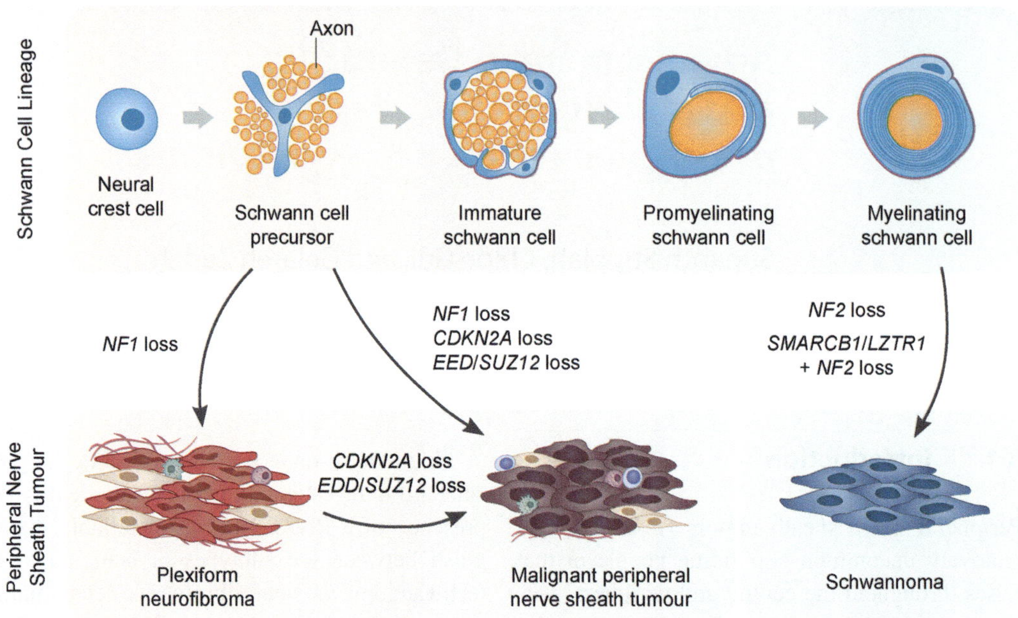

Fig. 3.1 Summary of the molecular alterations that have been identified in peripheral nerve sheath tumors

players in tumor development. Newer genomic studies have identified other genetic alterations, such as *SH3PX2DA-HTRA1* fusions and *KIR2DL5* mutations, in a subset of sporadic tumors. Here we review the current state of knowledge on the genetic and epigenetic landscape of peripheral nerve sheath tumors (Fig. 3.1).

3.2 Tumor Predisposition Syndromes

PNSTs are a major feature of tumor predisposition syndromes such as neurofibromatosis type 1 (NF-1), neurofibromatosis type 2 (NF-2), and schwannomatosis (SWN). Much of our initial understanding of nerve sheath tumor biology was based on the genetics of the neurofibromatosis family of syndromes. Friedrich von Recklinghausen, a German pathologist, first described neurofibromas associated with NF-1 in 1882. NF-1 is an autosomal dominant neurocutaneous disorder affecting 1 in 2500 births. Clinically, it is characterized by the development of multiple neurofibromas and cutaneous pigmentary abnormalities. Individuals with NF-1 harbor a germline mutation in the *neurofibromatosis 1 (NF1)* gene located on chromosome 17q11.2, which encodes neurofibromin, a GTPase-activating protein that functions as a tumor suppressor through its negative regulation of the RAS pathway. Similarly, NF-2 is an autosomal dominant genetic disorder caused by a mutation in *NF2* that affects 1 in 25,000 individuals. Although NF-2 is a hereditary disorder, 50–60% of cases are caused by de novo mutations, and as many as one-third of cases are due to somatic mosaicism. The development of bilateral vestibular schwannomas is pathognomonic for NF-2. In addition, multiple spinal and peripheral nerve schwannomas can arise in NF-2 individuals. Finally, SWN is another rare genetic disorder, affecting 1 in 40,000 individuals, that is, characterized by the development of multiple cutaneous, peripheral nerve schwannomas. Molecular studies indicate that SWN is a heterogeneous disorder, with a subset of patients harboring mutations in *SMARCB1* or *LZTR1*, both found on chromosome 22. Understanding the germline mutations in these genetic disorders provided the initial insight into the underlying molecular drivers of oncogenesis in PNSTs.

3.2.1 Neurofibroma

In NF-1-associated peripheral nerve neurofibromas, loss of *NF1* gene plays a major role in oncogenesis. *NF1* is a relatively large gene that consists of 60 exons and is known to generate several isoforms via alternative splicing. Most mutations in NF1 are loss-of-function mutations through various mechanisms such as point mutations, splice site mutations, frameshift, insertions, and large deletions. Neurofibromin 1 negatively regulates the RAS pathway by converting RAS-GTP to the inactive RAS-GDP, leading to the inhibition of downstream signaling. The RAS/MAPK pathway regulates cell proliferation, differentiation, and senescence.

NF1 loss of heterozygosity (LOH) is exclusively in S-100-positive Schwann cells and absent in the perineural cells and fibroblasts in NF-1-associated peripheral nerve neurofibromas [1]. This suggests that biallelic *NF1* inactivation is only required in the Schwann cells for tumor development. Interestingly, Schwann cells with both homozygous and heterozygous loss of *NF1* coexist within neurofibromas, which may be explained by a second hit mutation occurring secondarily in a tumor that already developed polyclonally or, alternatively, a tumor arising through a two-hit mechanism in a stem/progenitor cell and stimulating the infiltration of nonneoplastic cells such as heterozygous Schwann cells, perineural cells, and fibroblasts [2]. In NF-1 mouse models, loss of *NF1* in Schwann cell lineage is a sufficient genetic alteration to generate tumors [1, 3, 4]. However, *NF1* haploinsufficiency is required in the non-Schwannian cells within the microenvironment for the tumor to develop [3, 5]. In contrast, the molecular landscape of sporadic neurofibromas is much less understood, although biallelic *NF1* inactivation in Schwann cells has been previously reported [6].

Although cutaneous neurofibromas are difficult to differentiate from peripheral nerve neurofibromas on histopathology, there is growing evidence that cutaneous neurofibromas are a separate tumor entity. Skin-derived precursors (SKPs), stem cells residing in the dermis, have been implicated as the cell of origin in cutaneous neurofibromas [2, 7]. Biallelic loss of *NF1* in SKPs is enough to induce the development of neurofibromas [7]. In addition, these SKPs can differentiate into neurons, Schwann cells, smooth muscle cells, and adipocytes under specific conditions [2, 8, 9]. SKPs with biallelic *NF1* loss implanted next to sciatic nerves have also been shown to develop into plexiform neurofibromas, thereby demonstrating its pluripotency. Again, tumor microenvironment is of importance in cutaneous neurofibromas formation as SKPs with *NF1* loss implanted in mice developed only in pregnant female mice [7]. This is consistent with the clinical observation that most cutaneous neurofibromas develop during puberty and increase in numbers during pregnancy and with the use of oral contraceptive pill [10].

Beyond the role of *NF1* in cutaneous and peripheral nerve neurofibromas, little is known about the molecular drivers in tumorigenesis. Runt-related transcription factor-1 (*RUNX1* or *AML1*) overexpression has been observed in human plexiform neurofibromas and in mouse Schwann cell progenitors, with inhibition by short hairpin RNA or by pharmacological agents reducing mouse neurofibroma sphere numbers in vitro [11]. *RUNX1* is known to be important for hematopoiesis, angiogenesis, differentiation of T and B cells, neuronal development, and glial cell differentiation [12–15]. Targeted deletion of *RUNX1* in Schwann cells and Schwann cell progenitor cells delayed neurofibroma formation in vivo [11]. These findings suggest that RUNX1 has a role in neurofibroma initiation. In addition, *KIR2DL5* mutations have been observed in a subset of sporadic cutaneous neurofibromas [16]. *KIR2DL5* encodes a transmembrane glycoprotein receptor expressed on NK cells and some T lymphocytes, but its role outside of the immune system is unknown. Anastaski et al. (2017), also, demonstrated that *KIR2DL5* knockdown with short hairpin RNA decreased expression of the receptor and increased Schwann cell proliferation, suggesting that it regulated cell growth [16].

3.2.2 Schwannoma

Similarly, schwannomas arise from the transformation of Schwann cells. At the molecular level, loss of *NF2* occurs in all NF-2-associated schwannomas and in 60% of sporadic tumors. *NF2* is found on chromosome 22q11.2 and encodes merlin, a cytoskeleton protein thought to have a tumor-suppressive function. Merlin-deficient cells can proliferate uncontrolled due to a lack of contact-mediated tumor suppression. Recurrent mutations in *ARID1A, ARID1B*, and *DDR1* have also been reported in schwannomas [17]. In addition, a recurrent in-frame *SH3PXD2A-HTRA1* fusion, resulting from a balanced 19-Mb chromosomal inversion on chromosome 10q, has been observed in 10% of schwannomas [17]. The fusion resulted in elevated phosphorylated ERK, increased proliferation, and increased invasion in vivo, with pharmaceuticals targeting the MEK-ERK pathway demonstrating therapeutic efficacy [17].

Since SWN individuals harbor germline *SMARCB1* or *LZTR1* (leucine-zipper-like transcriptional regulator 1) mutations, these genes have been suggested to play a role in schwannoma tumorigenesis. *SMARCB1* is part of the SWI/SNF (switch/sucrose nonfermentable) chromatin remodeling complex that relieves repressed chromatin structures, allowing for transcription. This tumor suppressor gene is found on chromosome 22q and has been shown to interact with *ARID1A*, which also encodes for a component of the SWI/SNF complex [18]. Complete loss-of-function mutations in *SMARCB1* result in the development of rhabdoid tumors, while SWN involves hypomorphic *SMARCB1* mutations giving rise to mutant proteins with residual function [19]. Similarly, *LZTR1* is a transcriptional regulator and may play a role in stabilizing the Golgi complex. *LZTR1* is also found on chromosome 22q11. However, biallelic inactivation of SMARCB1 or LZTR1 is insufficient for schwannoma development. In this subset of schwannomas, a four-hit/three-step model has been proposed, where the first hit is the germline SMARCB1 or LZTR1 mutation. The second hit is chromosome 22q LOH through a contiguous deletion that spans the *NF2*, *SMARCB1*, and *LZTR1* genes. This causes the loss of one *NF2* allele and the remaining *SMARCB1* or *LZTR1* allele, resulting in the biallelic inactivation of *SMARCB1* or *LZTR1*. Finally, a mutation occurs in the remaining *NF2*, leading to biallelic inactivation of *NF2* as well [19]. In the case of SWN schwannomas, it appears that the loss of two tumor suppressors is required for tumor development.

3.2.3 Malignant Peripheral Nerve Sheath Tumors

About half of all MPNSTs arise in the setting of NF-1 and in many cases are a result of malignant transformation of an underlying plexiform neurofibroma. The majority of MPNSTs, both NF-1-associated and sporadic, harbor *NF1* mutations [20, 21]. However, the tumor needs to accumulate a series of other genetic alterations to reach its malignant potential. Somatic alterations in *CDKN2A* (cyclin-dependent kinase inhibitor 2A), a tumor suppressor found on chromosome 9, are found in 80% of all MPNSTs [20]. The most common alteration is homozygous deletion of *CDKN2A* in MPNST. However, heterozygous deletion of *CDKN2A* is observed in atypical neurofibromas, thought to be a premalignant variant, which suggests that loss of *CDKN2A* occurs early in malignant transformation [22].

Recently, loss-of-function alterations of the PRC2 (polycomb repressive complex 2) components, such as EED or SUZ12, have been identified in up to 90% of MPNSTs [20, 21]. The PRC2 complex has histone methyltransferase activity and silences chromatin by trimethylating H3K27me3 (histone H3 on lysine 27). In most cases, there is a compensatory increase in H3K27ac [23]. Loss of PRC2 amplifies RAS-driven transcription and potentiates the effects of NF1 mutation [23]. As a result, loss of H3K27me3 is a sensitive marker for MPNSTs [22, 24–26]. Reconstitution of SUZ12 in MPNST cell lines has shown to increase H3K27me3 and decrease in H3K27ac [23]. Mutations in the epigenetic regulators play a major role in MPNST development.

3.3 Conclusion

Further work is needed to understand the biological drivers of the spectrum of nerve sheath tumors to guide the development of targeted therapies. Furthermore, better molecular subclassification of these tumors will lead to advancements in precision medicine.

References

1. Perry A, Roth KA, Banerjee R, Fuller CE, Gutmann DH. NF1 deletions in S-100 protein-positive and negative cells of sporadic and neurofibromatosis 1 (NF1)-associated plexiform neurofibromas and malignant peripheral nerve sheath tumors. Am J Pathol. 2001;159(1):57–61.
2. Gutiérrez-Rivera A, Iribar H, Tuneu A, Izeta A. Skin-derived precursor cells as an in vitro modelling tool for the study of type 1 neurofibromatosis. Stem Cells Int. 2012;2012:646725. https://www-hindawi-com.myaccess.library.utoronto.ca/journals/sci/2012/646725/
3. Zhu Y, Ghosh P, Charnay P, Burns DK, Parada LF. Neurofibromas in NF1: Schwann cell origin and role of tumor environment. Science. 2002;296(5569):920–2.
4. Carroll SL, Nancy R. How does the Schwann cell lineage form tumors in NF1? Glia. 2008;56(14):1590–605.
5. Yang FC, Ingram DA, Chen S, Zhu Y, Yuan J, Li X, et al. Nf1-dependent tumors require a microenvironment containing Nf1+/− and c-kit-dependent bone marrow. Cell. 2008;135(3):437–48.
6. Tiziana SC, Vult VSF, Domanski HA, Nils M, Fredrik M. Biallelic somatic inactivation of the NF1 gene through chromosomal translocations in a sporadic neurofibroma. Int J Cancer. 2005;117(6):1055–7.
7. Le LQ, Shipman T, Burns DK, Parada LF. Cell of origin and microenvironment contribution for NF1-associated dermal neurofibromas. Cell Stem Cell. 2009;4(5):453–63.
8. Toma JG, Akhavan M, Fernandes KJL, Barnabé-Heider F, Sadikot A, Kaplan DR, et al. Isolation of multipotent adult stem cells from the dermis of mammalian skin. Nat Cell Biol. 2001;3(9):778–84.
9. Fernandes KJL, McKenzie IA, Mill P, Smith KM, Akhavan M, Barnabé-Heider F, et al. A dermal niche for multipotent adult skin-derived precursor cells. Nat Cell Biol. 2004;6(11):1082–93.
10. Friedman JM. Review article: Neurofibromatosis 1: clinical manifestations and diagnostic criteria. J Child Neurol. 2002;17(8):548–54.
11. Li H, Zhao X, Yan X, Jessen WJ, Kim MO, Dombi E, et al. Runx1 contributes to neurofibromatosis type 1 neurofibroma formation. Oncogene. 2016;35(11):1468–74.
12. Inoue K, Shiga T, Ito Y. Runx transcription factors in neuronal development. Neural Dev. 2008;3:20.
13. Aldskogius H, Berens C, Kanaykina N, Liakhovitskaia A, Medvinsky A, Sandelin M, et al. Regulation of boundary cap neural crest stem cell differentiation after transplantation. Stem Cells. 2009;27(7):1592–603.
14. Nagamachi A, Htun PW, Ma F, Miyazaki K, Yamasaki N, Kanno M, et al. A 5′ untranslated region containing the IRES element in the Runx1 gene is required for angiogenesis, hematopoiesis and leukemogenesis in a knock-in mouse model. Dev Biol. 2010;345(2):226–36.
15. Ichikawa M, Asai T, Saito T, Seo S, Yamazaki I, Yamagata T, et al. AML-1 is required for megakaryocytic maturation and lymphocytic differentiation, but not for maintenance of hematopoietic stem cells in adult hematopoiesis. Nat Med. 2004;10(3):299–304.
16. Anastasaki C, Dahiya S, Gutmann DH. KIR2DL5 mutation and loss underlies sporadic dermal neurofibroma pathogenesis and growth. Oncotarget. 2017;8(29):47574–85.
17. Agnihotri S, Jalali S, Wilson MR, Danesh A, Li M, Klironomos G, et al. The genomic landscape of schwannoma. Nat Genet. 2016;48(11):1339–48.
18. Shain AH, Pollack JR. The spectrum of SWI/SNF mutations, ubiquitous in human cancers. PLoS One. 2013;8(1):e55119.
19. Kehrer-Sawatzki H, Farschtschi S, Mautner VF, Cooper DN. The molecular pathogenesis of schwannomatosis, a paradigm for the co-involvement of multiple tumour suppressor genes in tumorigenesis. Hum Genet. 2017;136(2):129–48.
20. Lee W, Teckie S, Wiesner T, Ran L, Granada CNP, Lin M, et al. PRC2 is recurrently inactivated through *EED* or *SUZ12* loss in malignant peripheral nerve sheath tumors. Nat Genet. 2014;46(11):1227.
21. Zhang M, Wang Y, Jones S, Sausen M, McMahon K, Sharma R, et al. Somatic mutations of *SUZ12* in malignant peripheral nerve sheath tumors. Nat Genet. 2014;46(11):1170.
22. Röhrich M, Koelsche C, Schrimpf D, Capper D, Sahm F, Kratz A, et al. Methylation-based classification of benign and malignant peripheral nerve sheath tumors. Acta Neuropathol (Berl). 2016;131(6):877–87.
23. De Raedt T, Beert E, Pasmant E, Luscan A, Brems H, Ortonne N, et al. PRC2 loss amplifies Ras-driven transcription and confers sensitivity to BRD4-based therapies. Nature. 2014;514(7521):247–51.
24. Cleven AHG, Sannaa GAA, de Bruijn IB, Ingram DR, van de Rijn M, Rubin BP, et al. Loss of H3K27 tri-methylation is a diagnostic marker for malignant peripheral nerve sheath tumors and an indicator for an inferior survival. Mod Pathol. 2016;29(6):582.
25. Prieto-Granada CN, Wiesner T, Messina JL, Jungbluth AA, Chi P, Antonescu CR. Loss of H3K27me3 expression is a highly sensitive marker for sporadic and radiation-induced MPNST. Am J Surg Pathol. 2016;40(4):479–89.
26. Schaefer IM, Fletcher CD, Hornick JL. Loss of H3K27 trimethylation distinguishes malignant peripheral nerve sheath tumors from histologic mimics. Mod Pathol. 2016;29(1):4.

4 Multi-Omic Analyses for Biomarker Discovery and Characterization of PNST

Antonella Potenza, Tatiana Carrozzini, Gemma Gorla, Giuliana Pollaci, and Laura Gatti

4.1 Biomarker Definition and Characterization

The term "biomarker" is a portmanteau of "biological marker" commonly used to refer to a wide range of medical signs that can be measured accurately and reproducibly. In 2001, the National Institutes of Health Biomarkers Definitions Working Group indicated that the correct denomination of a biomarker is a specific characteristic that is objectively measured and evaluated as an indicator of physiological and pathogenic processes or pharmacological responses to a therapeutic intervention [1].

A few years later, the International Program on Chemical Safety, led by the World Health Organization and in coordination with the United Nations and the International Labor Organization, broadened the concept of biomarker by defining it as "any substance, structure or process and also products that can be measured in the body and can influence or predict the incidence and outcome of disease." Additionally, an even wider definition of a biomarker takes into account the effects of treatments, interventions, and unintended environmental exposures, including blood pressure, imaging markers, specific analytes found in biofluids and tissues. Since medical signs exert an important role in clinical practice, biomarkers may be considered the most objective and quantifiable medical signs that can be reproducibly measured through modern scientific laboratory approaches [2]. To reach clinical practice, each biomarker needs to undergo (1) validation and (2) qualification processes.

1. The validation assessment evaluates a biomarker and its specific measurement performance, to set the range of conditions under which the marker will provide repeatable and accurate data [3]. A biomarker is considered valid when it can reliably track patient outcomes or characteristics. Thus, the validity of a biomarker is closely linked to the function and to the ability to discriminate between different clinical features.

   The validation step estimates the determination of specific biomarker criteria such as:

T. Carrozzini · G. Gorla · L. Gatti (✉)
Laboratory of Neurobiology and UCV, Neurology IX Unit, Fondazione IRCCS Istituto Neurologico Carlo Besta, Milan, Italy
e-mail: antonella.potenza@istituto-besta.it; tatiana.carrozzini@istituto-besta.it; gemma.gorla@istituto-besta.it; laura.gatti@istituto-besta.it

A. Potenza · G. Pollaci
Laboratory of Neurobiology and UCV, Neurology IX Unit, Fondazione IRCCS Istituto Neurologico Carlo Besta, Milan, Italy

Department of Pharmacological and Biomolecular Sciences, Università di Milano, Milan, Italy
e-mail: giuliana.pollaci@istituto-besta.it

I. G. Vetrano, V. Nazzi (eds.), *Peripheral Nerve Sheath Tumors*,
https://doi.org/10.1007/978-3-031-76382-3_4

- Sensitivity: ability of an appropriate biomarker to be measured with adequate precision in a biological sample
- Specificity: ability of a biomarker to distinguish patients who are responders to an intervention from those who are non-responders in terms of changes in clinical endpoints (measure of how a patient feels, functions, or survives)

2. The qualification process associates a specific biomarker with one or more specific biological processes and clinical endpoints. Thus, a biomarker is subjected to a formal regulatory procedure that ensures its reliability, leading to a precise interpretation and application in medical product development.

   Several biomarkers have been differently defined according to their putative applications. Considering their functions and connections to diverse aspects of the disease, biomarkers can be divided into different categories (Fig. 4.1):

   – Diagnostic biomarker: it detects or confirms the presence of a disease or condition of interest. Such biomarkers may be also subcategorized as "early detection marker," allowing a more timely intervention, and "disease classification," not only to identify a disorder but also to redefine the stratification of the disease.
   – Monitoring biomarker: it can be measured several times to assess the status of a disease, to monitor its natural progression, or to detect an effect of a medical product or an environmental agent.
   – Predictive biomarker: it could predict response or future probable adverse events. The presence or change in a predictive biomarker could forecast an individual or group of individuals more likely to experience a favorable or unfavorable effect from exposure to a medical product or environmental agent.
   – Prognostic/outcomes biomarker: it is used to identify the likelihood of a clinical event, disease recurrence, or to estimate the progression of a disorder. Although this distinction is not uniformly accepted, the BEST working group concluded that prognostic biomarkers should be differentiated from susceptibility/risk biomarkers, which deal with the transition from a healthy state to disease. Furthermore, they are distinguished from predictive biomarkers that identify factors associated with the effect of intervention or exposure.

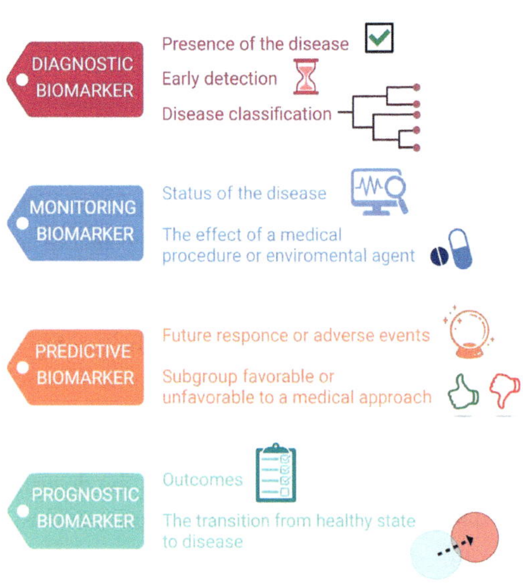

Fig. 4.1 Different types of biomarkers

A good biomarker must be stable to remain unaffected by sampling and storage procedures. Furthermore, the ideal biological samples have to be collected in a noninvasive way, giving the possibility to clinicians to gather them several times over a period of time without any consequences for the patient. In addition, putative biomarkers have to be measured accurately and reproducibly in order to carry out comparisons between patients and healthy donors (HD). Currently, the most frequently used biofluid is blood, but urine, cerebrospinal fluid, tissue biopsies, saliva, milk, seminal fluid, tears, pleural fluid, and others may be employed, taking into account their different availability and the procedure difficulties correlated to their collection (Fig. 4.2). Specifically, blood-based biomarkers have the advantage to be the most standardized and frequently repeated. Furthermore, other biofluids may be more repre-

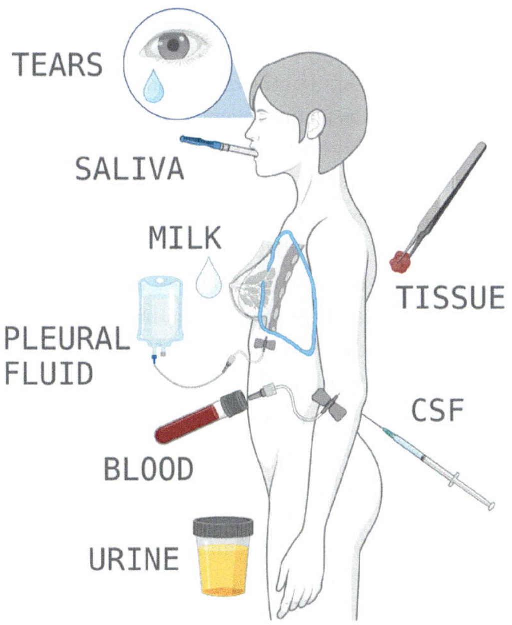

Fig. 4.2 Main biofluids used to measure biomarkers

sentative and informative than peripheral blood since a biomarker will give more indications if identified in close proximity to the site where damage and/or the origin of the disease is presumed. However, once the biomarker has been detected, its association with disease and its role in the clinic setting need to be determined.

4.2 Multi-Omic Analyses

Biological matrices can be investigated through traditional techniques, including the antibody-based ELISA tests as well as more recent and sensitive immunoassays. Since these techniques look for a single or a limited number of analytes, a significant period of time and a large amount of samples are needed. With the advent of omic technologies able to study the entire proteome/lipidome/metabolome of a particular matrix, the amount of biomarkers analyzed in a single experiment has become even greater, allowing for the production of much more data in reduced time spam. Omic techniques include genomics, transcriptomics, proteomics, metabolomics, and lipidomics, which correspond to global analyses of genes, RNA, proteins, metabolites, lipids, and all the changes in DNA that do not involve alterations in the sequence (epigenomics). The main goal of an omic study is to obtain a comprehensive understanding of the biological system under investigation with the final aim to associate, for example, the omics-based molecular measurements with clinical information like outcomes, risk of recurrence, or response to therapy. Integration of multi-omics data providing information on biomolecules from different layers seems to be a promising approach to better understand the biology complexity. Thus, omics-based measurements could provide for a predictive or prognostic model of a particular condition or disease in a more accurate and faster way than using standard clinical approaches.

4.2.1 Genomic and Transcriptomic Analysis

Genomic science is the identification, measurement, or comparison of genomic features such as DNA sequence, structural variation, copy number variation, or regulatory and functional element observation at a genomic scale. Methods for genomic analysis typically require high-throughput sequencing or microarray hybridization and bioinformatics.

Within the cancer research field, circulating cell-free nucleic acids (cf-NA), such as DNA, mRNA, and microRNA (miRNA) reached an increasing interest during past decades, since substantial amounts were found in the blood of cancer patients [4]. Notably, the release of nucleic acids into the bloodstream has been associated with apoptotic and necrotic processes and possibly also to active secretion of cancer cells in the tumor environment [5–7]. The tumor size and status as well as the clearance, degradation, and physiological filtering processes of the blood and lymphatic circulation strongly impact the amount of cell-free DNA (cfDNA) released from tumor cells [8]. Notably, the presence of cfDNA molecules has been determined not only in blood samples but also in urine, cerebrospinal fluid, pleural fluid, and saliva.

Thus, the concept of liquid biopsy has been widely associated with the analysis of circulating tumor DNA (ctDNA), including both cell-free genomic and mitochondrial DNA. It is interesting to note that ctDNA assays constitute useful applications in cancer-related studies, from early diagnosis and prognosis to the evaluation of residual diseases, the selection of therapeutic approaches in accordance with specific factors, and the disease surveillance [9]. Actually, detection and quantification of circulating nucleic acids represent a challenge for clinical purposes, since the determination is mostly hampered by their short length and low concentration in body fluids. However, a wide range of experimental strategies have been established to detect them properly [4]. Indeed, in order to identify tumor-related genetic and epigenetic alterations, which are closely associated with cancer development, progression, and resistance to therapy, initial studies focused on quantitative PCR assays [10–12]. Subsequently, to target individual patient-specific point mutations, PCR-based assays moved to personalized approaches [13]. Then, the rising interest in next-generation sequencing (NGS) approaches has allowed for the research of parallel multiple digital sequence alterations in plasma cfDNA molecules, leading to a targeted sequencing strategy of regions recurrently mutated in a specific cancer subtype [14]. Anyway, this second generation of liquid biopsy is limited by insufficient mutant molecules being sampled. Indeed, in a low-volume tumor, few or none mutant molecules may be identified in a sample using digital PCR, thus hampering the sensitivity of both first- and second-generation liquid biopsies. Then, due to the high number of mutations detected so far, liquid biopsy moved to a third generation, which guided the conversion of digital signals into analogical ones [15]. In particular, this last generation of liquid biopsies referring to whole-exome sequencing and whole-genome sequencing (WES, WGS) may be divided into tumor-derived and non-tumor-derived (microenvironment) approaches. For instance, *fragmentomics* is an emerging field based on the identification of differences in fragmentation patterns between tumor-derived and normal cfDNA. Studies carried out in the last years have revealed the relative abundance of short ctDNA fragments in cancer patients, and, as a consequence, size-based selection emerged to be a logical next step to enrich tumor signals [16].

Furthermore, based on existing differences in methylation profile between tumor and normal tissues, plasma *methylomics* arose as a novel tool to represent and study differential methylation patterns in the bloodstream. Indeed, plasma methylation profiles allow for a highly sensitive categorization of samples (in both early and advanced disease stages), to determine how different tissues may contribute to the circulating DNA pool (DNA methylation patterns are considered tissue-specific) and to examine the expression of cancer-related genes [17].

Alternatively, *nucleosomics* made it possible to determine nucleosome positioning and occupancy based on sequencing depths across the genome. Patterns of nucleosome occupancy differ between normal and cancer cells and have been coupled with distinct gene expression patterns [18].

On the other hand, non-tumor-derived approaches provide insights into tumor-to-host or microenvironmental relationships which can be exploited by aligning unpaired readings with microbial (microbiomic) genomes and by studying platelet RNA (transcriptomic) profile [17, 19].

Besides the quantification of cfDNA, valuable insights originated from transcriptomic approaches, which represent the study of the complete set of all the ribonucleic acid (RNA) molecules expressed by a specific cell, tissue, or organism [20, 21]. Indeed, transcriptomic analyses cover all types of transcripts, including messenger RNAs (mRNAs), microRNAs (miRNAs), and different types of small- (miRNA, siRNA, piRNA, snRNA, snoRNA, YRNA) or long-noncoding RNAs (pseudogene RNA, telomeric RNA, and circRNA). It is suggested that RNAs could be implied in cell-to-cell communication and act as hormones [4], and it has been proved that they can be detected also in extracellular fluids [22]. In particular, as far as cancer patients, circulating RNA molecules are found to be surprisingly stable in the bloodstream despite the boosted amount of RNases observed among these subjects. Intriguingly, though the presence of circulating coding mRNA has been remarked in oncological patients, a wide number of studies focused on cell-free non-coding RNAs whose different expression between patients and healthy subjects suggested them as suitable biomarker candidates for diagnostic tests [23–28].

In particular, miRNAs are assessed as useful cancer biomarker candidates, since their altered expression may affect tumor microenvironment, supporting cell growth and invasion.

Dealing with modern transcriptomic strategies, high-throughput methods are able to analyze the expression of multiple transcripts in different physiological or pathological conditions [29]. The most sensitive technique used for miRNA, lncRNA, and circRNA identification and quantification is RNA-sequencing (RNA-seq). On the other hand, a helpful alternative approach is represented by the application of microarrays for nucleic acid detection and analysis [30–32]. Moreover, the method still addressed as the "gold standard" for diagnostic aims is the RT-qPCR [33]. Indeed, adjustments of the standard RT-qPCR techniques have been established for the detection of short RNA molecules (including miRNA) [34]. Therefore, RT-qPCR is considered the most appealing method for routine evaluation in clinical laboratories due to its undoubted properties: high sensitivity, specificity, and reproducibility.

4.2.2 Proteomic Analysis

The proteome is the complete set of proteins expressed by a cell, tissue, or organism. The study of the proteome is quite complex because proteins undergo posttranslational modifications, have different spatial configurations and intracellular localizations, and interact with further proteins as well as with other molecules. One of the major goals for proteomic investigations is the discovery of panels of protein markers that could serve for the prediction, confirmation, and monitoring of a disease state. The most used techniques for proteomic analyses are protein microarrays and mass spectrometry. Protein microarray analysis is a high-throughput method used to track the expression levels, interactions, and activities of proteins on a large scale belonging to a complex sample. The chip consists of a support such as a glass slide, nitrocellulose membrane, bead, or microtiter plate with a chemically modified surface, to which an array of capture proteins is bound. Probe molecules, typically labeled with a fluorescent dye, are added to the array. Any reaction between the probe and the immobilized protein emits a fluorescent signal which is read by a laser scanner. Although protein microarrays are not quantitative approaches, they represent valuable strategies since they are rapid, automated, economical, and highly sensitive, consuming small quantities of samples and reagents. There are three classes of protein microarray: (1) analytical protein arrays used for proteome screening and biomarker discovery, (2) functional protein arrays to study protein interaction and activity, and (3) reverse phase protein arrays that are useful tools for biomarker validation. It is important to state that the three different tools differ depending on the type of analyte adsorbed to the surface (antibody, peptide, antigen, protein, etc.).

Mass spectrometry (MS) is the most comprehensive approach for the quantitative profiling of proteins as well as their interactions and modifications. It is considered a fruitful technique to study protein expression, posttranslational modifications, and interactions, thus providing for a wide range of possible analyses of protein expression and functions. Proteomic studies require simplification of the protein mixtures into less complex components that are more amenable to analysis. For this purpose, proteins are usually separated either by 1-dimensional (1D) or 2D gel electrophoresis or by a chromatographic technique. The main types (most used) of MS techniques are electrospray ionization and matrix-assisted laser desorption/ionization. These two strategies differ in the way proteins or peptides are ionized for mass determination. Modern mass spectrometers allow for the identification of peptides and proteins at subfemtomolar (<10−15 mol) levels, thus potentially enabling the study of low-abundant proteins, such as those proteins involved in signal-transduction processes.

4.2.3 Lipidomic Analysis

When thinking about lipids for clinical use, cardiovascular risk biomarkers (like cholesterol, LDL, HDL, etc.) arise, but they are just a very small part of one of the four principal categories of biological macromolecules. Indeed, as a fundamental component of organisms, lipids are structural and functional molecules that play important biological functions, including their role in cellular barriers, membrane matrices, energy storage, and signaling. Lipids participate in a variety of cellular processes, because of the great diversity of structures and physiochemical properties that characterize them. Lipids are generally classified as hydrophobic or amphiphilic molecules based on their biosynthesis and chemical structures.

By the term "lipidomics," we refer to the large-scale analysis of the complete set of lipids (the lipidome) in any biological system. Lipidomic research covers all aspects of individual lipids and lipid-derived species, including their structure, function, and interactions with other cellular constituents such as proteins, other lipids, and metabolites in both health and disease conditions. Altered pathways of lipid metabolism play a key role in the prediction, prognosis, and diagnosis of many common diseases. Most bioactive mediators arise from the cleavage of lipid constituents from cell membranes due to the activity of phospholipases. Changes in a specific lipidomic pathway that occur over a long period of time result in a stress condition for the organism. As a consequence, an allostatic load usually begins to accumulate (the initial stage of a clinical condition), until the organism recovers from the load or fails to adapt. If there is a failure, a disruption of allostatic adaptation can be observed as an overproduction or reduction of some biomarkers of the metabolic pathway. Abnormalities in lipid metabolism may be a defining marker of the stage of a lipidome-related disease or, alternatively, correlations may be detected when comparing patients to HD.

The ultimate goal of lipidomics is to explore the underlying mechanisms of lipid metabolism and their changes under pathological conditions, which should lead to the discovery of biomarkers that can be used for disease diagnosis or treatment, as drug targets, for guidance in precision/personalized medicine and dietary intervention.

Due to the inherent chemical complexity of the lipidome and the resulting challenges associated with its analysis, advances in the field of lipidomics have lagged far behind advances made in other omic disciplines.

The most common method of separating analytes according to the strength of their interaction between the solid stationary phase and the liquid mobile phase is high-performance liquid chromatography (HPLC). Another technique that can be used is reversed-phase high-performance liquid chromatography (RP-HPLC), which allows the separation of molecules on the basis of hydrophobicity. The separation depends on the hydrophobic binding of the solute molecule from the mobile phase to the immobilized hydrophobic ligands attached to the stationary phase. However, different matrix effects and different solvent compositions can influence their ionization, leading to inaccurate lipid quantification [35].

For this reason, hydrophilic interaction chromatography (HILIC) was devised to perform a specific separation of polar lipids by distribution, adsorption, ion exchange, and exclusion chromatography [36]. Moreover, ultra-high-performance liquid chromatography (UHPLC) represents an alternative rapid chromatography, in which nanocolumns are used. The latter technique has a good lipid separation efficiency, and it is very fast.

In addition to these techniques, the most appropriate analytical method for the analysis of lipid molecules or large assemblies of them (the lipidome) is liquid chromatography coupled with mass spectrometry (MS).

Liquid chromatography is a separation method that is usually combined with MS to achieve more selective detection and more accurate identification of lipids. Specifically, these two techniques are used in a coupled way because lipid compounds tend to overlap in MS techniques in the absence of analytical separation, especially in complex matrices.

MS-based lipidomics can be divided into two subfields: non-targeted and targeted lipidomics. In particular, non-targeted lipidomics, also known as the global lipidomic approach, aims to identify and quantify the entire asset of lipids in a system. In contrast, targeted lipidomics focuses on the quantitative determination of specific lipid classes. Non-targeted lipidomics is mostly used to obtain an overview of lipid molecules that can be detected in a system, while targeted lipidomics is more applicable to solving specific and complex biological problems (issues). Routine procedures for MS-based lipidomics consist of biological sample collection, lipid extraction, sample pretreatment, MS data acquisition, post-acquisition data processing, and physiological significance clarification.

The combination of proteomics, lipidomics, genomics, and transcriptomics represents a useful approach to provide comprehensive and significant data. Undoubtedly, this combined method is expected to evolve in the future, giving us increasingly complex information. In particular, the cross-linked multi-omic research will supply new insights into deeply understanding the pathogenic mechanisms of diseases, thus representing a helpful source for biomarker discovery.

4.3 Insights Into Biomarker Discovery in Peripheral Nerve Sheath Tumors

The identification of peripheral nerve sheath tumor (PNST) biomarkers and the subsequent development of investigative tools will allow an early diagnosis, thereby highlighting the transition of neurofibromas (NF)/plexiform neurofibromas (pNF) into highly aggressive tumors.

Biomarkers could be exploited together with gene delivery technologies to control and contrast the growth of tumors, as well as being a potential diagnostic tool. To date, several blood biomarkers have already been studied in patients with PNST, including those considered predictors of malignant peripheral nerve sheath tumor (MPNST) development and progression. These biomarkers have been so far evaluated through different biochemical approaches.

4.3.1 Nucleic Acids as Biomarker in PNST

Plasma cfDNA concentrations in healthy subjects typically fluctuate between 0 and 100 ng/mL, while increased blood levels of cf-DNA reflect pathological processes, including malignant and benign lesions, inflammatory diseases, stroke, trauma, and sepsis [8, 10]. Currently, it has not yet been elucidated whether a disease state such as neurofibromatosis type 1 (NF1) may affect the individual's cfDNA concentration. Thus, the opportunity to examine how NF1 may refer to cfDNA concentration emerged as crucial for optimizing the use of cfDNA assays in treatment outcome evaluation as well as in planning screening for NF1-related malignancies. For this purpose, Kalliopaa and colleagues had recently carried out a study aimed at exploring the effects of NF1 on plasma cfDNA release [37]. The authors did not find a statistically significant difference between individuals with NF1 and control subjects. Age at sampling was marginally significant since older age was associated with lower cfDNA concentration among individuals with NF1 but not in controls. Intriguingly, patients with six or more subcutaneous neurofibromas (scNF) seemed to have higher normalized cfDNA concentration than individuals with less than 6 scNF. The authors stated that the current sample size and the lack of volumetric information on tumor burden may have prevented the supposed association between pNF/scNF and higher cfDNA concentration among NF1 patients. Indeed, benign NF coupled with NF1 may form a substantial tumor burden, which could affect and modify cfDNA concentrations, thus reflecting a NF1-specific process. Additionally, it has been suggested that age may exert a crucial role in intratumoral inflammation changes, leading to impaired cfDNA concentrations in NF1.

Furthermore, a clinical trial carried out by Rishi Lulla and colleagues was performed to assess the utility of microRNAs in the blood of NF1 patients as an early predictor of cancer as well as a marker of therapy response. In particular, the research group proposed an observational case-only prospective study to evaluate the presence of microRNAs in archived tumor tissue and blood of child and young adult patients with NF-1 (2–21 years). Their study aimed to evaluate miRNA expression patterns both in (1) tissue of low-grade gliomas and (2) between patients with and without imaging findings of gliomas [38].

Very recently, a multicentric study has been carried out by two large NF1 referral centers (National Cancer Institute and Washington University in St. Louis), where plasma cfDNA was analyzed using ultralow-pass whole-genome sequencing (ULP-WGS). The authors found out that cfDNA belonging to MPNST patients is characterized by a shorter fragmentation profile if compared to pNF patients or HD. The authors, due to the quantification of genome-wide copy number alterations (CNAs) in cfDNA, were able to evaluate plasma cfDNA fraction directly derived from tumor. Interestingly, the latter fraction revealed a real difference between MPNST and pNF (86% accuracy). Plasma cfDNA from MPNST and pNF patients, differently from HD, harbored focal copy number loss of NF1. Moreover, MPNST patient cfDNA was found to be associated with greater tumor genomic instability than pNF patients. In conclusion, cfDNA fragment analysis followed by ULP-WGS may reasonably be used as a biomarker for treatment response and screening identification for early detection of MPNST. Thus, this research has to be considered extremely relevant since it is the first evidence revealing the ability of liquid biopsy to distinguish between benign and malignant tumors in a heritable cancer predisposition syndrome [39].

Lastly, a whole exome sequencing (WES) approach has allowed to elucidate the genetic events of clonal progression leading to MPNSTs, by identifying the molecular events critical for MPNST spread. Mutations in the TRIM family of proteins, a class of E3-Ubiquitin ligases, were present at higher rates in metastatic MPNST samples. TRIM proteins are ubiquitously expressed in normal tissue. Specifically, TRIM23 seems to have a central role in regulating metastatic behavior. Therefore, the opportunity to target this clinically relevant population of cancer cells can lead to the identification and development of disease progression biomarkers [40].

4.3.2 Proteins as Biomarkers in PNST

A considerable amount of protein biomarkers has been studied in serum and plasma samples of patients affected by PNST using traditional techniques (e.g., ELISA). Among these biomarkers, the most relevant is the adrenomedullin. Adrenomedullin is a vasodilator peptide hormone, able to stimulate the growth of new blood vessels and to increase the tolerance of cells to oxidative stress and hypoxic injury. It is currently well established that adrenomedullin is implicated in the pathogenesis of many tumors (such as glioblastoma, lung, breast, and ovarian carcinoma) as well as in the proliferation and migration of tumor cells. Over the years, it has been observed that adrenomedullin is highly expressed in MPNST patients if compared to NF-1 patients, thus suggesting its potential role in malignant transformation [41].

Another possible diagnostic biomarker for NF-1 [42] is melanoma inhibitory activity (MIA), which has been found elevated in the plasma of NF-1 patients. MIA is a small secreted protein, normally expressed by cartilage, overexpressed by melanoma, and implicated in metastasis. Indeed, MIA is involved in in vitro inhibition of cell adhesion to the extracellular matrix, thus promoting metastasis formation [43].

The AXL protein is a member of the TAM family of receptor tyrosine kinases (RTKs), widely expressed in normal cells and tissues, which regulates cell survival. The soluble fraction of the extracellular domain of AXL (sAXL) has been demonstrated to be elevated in several cancer types. Furthermore, plasma levels of sAXL have been found to correlate with various tumor size and volume. Thus, sAXL may be used to better stratify PNST patients. Indeed, patients with pNF showed significantly higher levels of sAXL if compared to patients without pNF, highlighting an important correlation in male rather than in female patients. Therefore, the plasma level of sAXL may provide a useful biomarker for monitoring tumor growth (progression) and treatment efficacy in NF1 patients with a pNF or MPNST [44].

As it is widely known, NF-1 is also characterized by chronic inflammation. Repetitive inflammation has been generally recognized as a stimulating factor for tumor onset, primarily due to the reactive oxygen species-induced DNA mutations and because of an increased level of released cytokines and growth factors from infiltrating inflammatory cells. In fact, patients with NF-1 have high levels of pro-inflammatory factors, such as interleukin 6 (IL-6), gamma interferon (INFγ), and tumor necrosis factor-alpha (TNF). There are some factors, such as insulin-like growth factor binding protein1 (IGFBP-1) and the chemokine RANTES, that seem to be more expressed in MPNST patients than NF-1 patients and that can therefore be used to discriminate between the two subtypes. Moreover, it has been registered a higher level of IGFBP-1 and chemokine RANTES in MPNST patients in comparison with people without MPNST. Serum IGFBP1 concentration is also correlated with internal tumor load, thus being a potential risk marker for malignant transformation, differently from pro-inflammatory cytokines whose levels did not depend on tumor load, but increased systematically in NF-1 patients. Instead, RANTES is an inflammatory chemokine known to mediate chemotactic activity in immune cells, such as T cells and monocytes. RANTES is expressed from tumor cells and is closely associated with an advanced stage of the disease, suggesting a role in cancer progression. Increased serum levels of RANTES and IGFBP1 may be the result of high secretion by the cancer cells themselves or by immune cells in response to the neoplastic process, or by both mechanisms. A further stratification of NF-1 patients into different clinical groups (cutaneous NF, pNF, and MPNST) might be carried out according to the differential expression of IGFBP1 and RANTES.

As mentioned above, despite the progress in proteomic techniques, plasma analyses are not currently available for PNSTs. Thus, preliminary results may reasonably pave the way for future omic analyses, which will hopefully lead to a complete and more representative picture of the disease.

4.3.3 Lipids as Biomarkers in PNST

Tumors are characterized by a deregulation of lipid metabolism and stimulation of lipid synthesis, which affects both cholesterol and fatty acids and could be the result of activation of oncogenic pathways. Therefore, lipids appear to play an important role in tumor progression.

The lipidomic analysis in PNSTs has only recently been introduced. So far, the study carried out by Vetrano et al. [45] is the only one in which lipidomic approaches have been employed in PNST characterization, to understand the cross-talk between lipid profile and tumor characteristics (schwannomas, SCH; NFs, pNFs). This is a very innovative and exploratory approach, since little is known about the possible imbalance of the lipid profile in the PNST. In particular, the untargeted and targeted lipidomic analyses performed on PNST surgical samples revealed peculiar lipid profiles in several benign tumors, regardless of the common and shared biological origin. Specifically, the targeted lipidomic was carried out on sphingolipids, a class of lipid molecules which represents one of the main components of the plasma membrane, which participate in the immune response and can initiate and/or support inflammation and cellular apoptosis in hereditary, fibrotic, and progressive diseases. The results showed that SCH are characterized by an increase in coenzyme Q (CoQ), a mitochondrial factor with powerful antioxidant properties, which can be suggestive of a defense mechanism to counteract possible damage, due to oxidative events. Additionally, high levels of ether lysophospholipid have been detected, and this may indicate an inflammatory hallmark in these tumors. An inflammatory state in SCH is also confirmed by increased levels of ceramide, a key mediator belonging to sphingolipids. Levels of sphingomyelin and plasmalogens were found decreased, an evidence that could be associated with a local demyelination process capable of creating a possible disturbance in the physiological development of the nerves, considering the cytoarchitecture of the SCH. In patients with NF, on the other hand, the untargeted analysis identified an elevated ceramide level, which was also confirmed by the targeted analysis. Therefore, NF appears to have a greater inflammatory feature than other types of PNST. Furthermore, the increase in ceramide is also related to the increase in vitamin A ester, which is the intracellular storage form of retinol. Retinyl esters are hydrolyzed to release retinol, which in turn is oxidized to retinaldehyde and retinoic acid that promotes differentiation and increasing ceramide levels. Furthermore, vitamin A is a strong antioxidant agent, which may explain the decrease in oxidized phospholipids and triglycerides present in these tumors. The results also showed that pNFs have a lower global lipid content than other types of PNST, probably linked to an increase in extracellular components as well as to the peculiar collagen deposition.

Finally, the study pointed out that different PNSTs show specific alterations in lipid profile, which determines precise histopathological features that may influence the more or less aggressive tumor behavior [45].

4.4 Conclusion

The need for translational research on circulating biomarkers for cancer patients' stratification is urgent, to guide individual treatment decisions and provide information to patients and relatives. In order to improve individualized treatment and effective secondary prevention, novel biomarkers enabling assessment of patients' specific outcome are needed. Ideally, biomarkers should assist in the diagnosis, prognosis, and treatment of diseases. Liquid biopsy appears to be a game changer that may revolutionize current cancer patient management. Unlike the invasive and costly conventional tissue biopsies which only provide a snapshot of gene/protein expression at a given time and location, liquid biopsy may be used for early diagnosis, monitoring tumor dynamics/progression over time, and detecting metastasis development and treatment responses. Liquid biopsy is quickly moving from research into clinical practice being noninvasive and easily repeatable. Directing future research toward the identification of biomarkers that comprise a specific molecular fingerprint, as compared to more nonspecific markers of general physiological processes, will be essential. Determining the relationships between any given measurable biomarker and relevant clinical endpoints is mandatory to decipher the complex disease pathogenesis of several pathological conditions, rare and not rare, including PNSTs. The rarity of PNST and the gap of knowledge about its pathogenic mechanisms and disease progression could take advantage from such novel cutting-edge techniques. The reported omics-based approaches are aimed at identifying potential biomarker candidates for diagnostic/prognostic/predictive purposes. The discovery of putative biomarkers reflecting specific pathophysiologic aspects of PNSTs will enable a better management of the disease, thus providing a more personalized patients' care (Fig. 4.3).

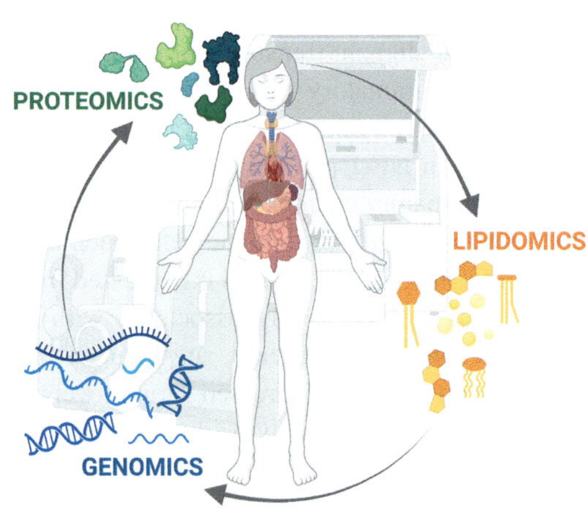

Fig. 4.3 Omics approaches

References

1. Biomarkers Definitions Working Group. Biomarkers and surrogate endpoints: preferred definitions and conceptual framework. Clin Pharmacol Ther. 2001;69(3):89–95. https://doi.org/10.1067/mcp.2001.113989.
2. Strimbu K, Tavel JA. What are biomarkers? Curr Opin HIV AIDS. 2010;5(6):463–6. https://doi.org/10.1097/COH.0b013e32833ed177.
3. Hunter DJ, Losina E, Guermazi A, Burstein D, Lassere MN, Kraus V. A pathway and approach to biomarker validation and qualification for osteoarthritis clinical trials. Curr Drug Targets. 2010;11(5):536–45. https://doi.org/10.2174/138945010791011947.
4. Szilágyi M, Pös O, Márton É, et al. Circulating cell-free nucleic acids: main characteristics and clinical application. Int J Mol Sci. 2020;21(18):6827. https://doi.org/10.3390/ijms21186827.
5. Choi JJ, Reich CF 3rd, Pisetsky DS. The role of macrophages in the in vitro generation of extracellular DNA from apoptotic and necrotic cells. Immunology. 2005;115(1):55–62. https://doi.org/10.1111/j.1365-2567.2005.02130.x.
6. Stroun M, Maurice P, Vasioukhin V, et al. The origin and mechanism of circulating DNA. Ann N Y Acad Sci. 2000;906:161–8. https://doi.org/10.1111/j.1749-6632.2000.tb06608.x.
7. Gahan PB, Swaminathan R. Circulating nucleic acids in plasma and serum. Recent developments. Ann N Y Acad Sci. 2008;137:1–6. https://doi.org/10.1196/annals.1448.050.
8. Schwarzenbach H, Hoon DS, Pantel K. Cell-free nucleic acids as biomarkers in cancer patients. Nat Rev Cancer. 2011;11(6):426–37. https://doi.org/10.1038/nrc3066.
9. Im YR, Tsui DWY, Diaz LA Jr, Wan JCM. Next-generation liquid biopsies: embracing data science in oncology. Trends Cancer. 2021;7(4):283–92. https://doi.org/10.1016/j.trecan.2020.11.001.
10. Wan JCM, Massie C, Garcia-Corbacho J, et al. Liquid biopsies come of age: towards implementation of circulating tumour DNA. Nat Rev Cancer. 2017;17(4):223–38. https://doi.org/10.1038/nrc.2017.7.
11. Diehl F, Li M, Dressman D, et al. Detection and quantification of mutations in the plasma of patients with colorectal tumors. Proc Natl Acad Sci U S A. 2005;102(45):16368–73. https://doi.org/10.1073/pnas.0507904102.
12. Vogelstein B, Papadopoulos N, Velculescu VE, Zhou S, Diaz LA Jr, Kinzler KW. Cancer genome landscapes. Science. 2013;339(6127):1546–58. https://doi.org/10.1126/science.1235122.
13. Heitzer E, Haque IS, Roberts CES, Speicher MR. Current and future perspectives of liquid biopsies in genomics-driven oncology. Nat Rev Genet. 2019;20(2):71–88. https://doi.org/10.1038/s41576-018-0071-5.
14. Newman AM, Bratman SV, To J, et al. An ultrasensitive method for quantitating circulating tumor DNA with broad patient coverage. Nat Med. 2014;20(5):548–54. https://doi.org/10.1038/nm.3519.
15. Newman AM, Lovejoy AF, Klass DM, et al. Integrated digital error suppression for improved detection of circulating tumor DNA. Nat Biotechnol. 2016;34(5):547–55. https://doi.org/10.1038/nbt.3520.
16. Jiang P, Lo YMD. The long and short of circulating cell-free DNA and the ins and outs of molecular diagnostics. Trends Genet. 2016;32(6):360–71. https://doi.org/10.1016/j.tig.2016.03.009.
17. Sun K, Jiang P, Chan KC, et al. Plasma DNA tissue mapping by genome-wide methylation sequencing for noninvasive prenatal, cancer, and transplantation assessments. Proc Natl Acad Sci U S A. 2015;112(40):E5503–12. https://doi.org/10.1073/pnas.1508736112.
18. Liu Y, Liu T-Y, Weinberg DE, White BW, De La Torre CJ, Tan CL, Schmitt AD, Selvaraj S, Tran V, Laurent LC, Cabel L, Bidard F-C, Putcha G, Haque IS. Spatial co-fragmentation pattern of cell-free DNA recapitulates in vivo chromatin organization and identifies tissue-of-origin. BiorXiv. 2019;2019:564773. https://doi.org/10.1101/564773.
19. Poore GD, Kopylova E, Zhu Q, et al. Microbiome analyses of blood and tissues suggest cancer diagnostic approach. Nature. 2020;579(7800):567–74. https://doi.org/10.1038/s41586-020-2095-1.
20. Morozova O, Hirst M, Marra MA. Applications of new sequencing technologies for transcriptome analysis. Annu Rev Genomics Hum Genet. 2009;10:135–51. https://doi.org/10.1146/annurev-genom-082908-145957.
21. Wolf JB. Principles of transcriptome analysis and gene expression quantification: an RNA-seq tutorial. Mol Ecol Resour. 2013;13(4):559–72. https://doi.org/10.1111/1755-0998.12109.
22. Zhao C, Sun X, Li L. Biogenesis and function of extracellular miRNAs. ExRNA. 2019;1:38. https://doi.org/10.1186/s41544-019-0039-4.
23. Larrea E, Sole C, Manterola L, et al. New concepts in cancer biomarkers: circulating miRNAs in liquid biopsies. Int J Mol Sci. 2016;17(5):627. https://doi.org/10.3390/ijms17050627.
24. Penyige A, Márton É, Soltész B, et al. Circulating miRNA profiling in plasma samples of ovarian cancer patients. Int J Mol Sci. 2019;20(18):4533. https://doi.org/10.3390/ijms20184533.
25. Márton É, Lukács J, Penyige A, et al. Circulating epithelial-mesenchymal transition-associated miRNAs are promising biomarkers in ovarian cancer. J Biotechnol. 2019;297:58–65. https://doi.org/10.1016/j.jbiotec.2019.04.003.
26. Koirala P, Zou DH, Mo YY. Long non-coding RNAs as key regulators of cancer metastasis. J Cancer Metastasis Treat. 2016;2:1–10. https://doi.org/10.4103/2394-4722.171829.
27. Boon RA, Jaé N, Holdt L, Dimmeler S. Long non-coding RNAs: from clinical genetics to therapeutic

targets? J Am Coll Cardiol. 2016;67(10):1214–26. https://doi.org/10.1016/j.jacc.2015.12.051.
28. Lu M. Circular RNA: functions, applications and prospects. ExRNA. 2020;2:1. https://doi.org/10.1186/s41544-019-0046-5.
29. Milward EA, Shahandeh A, Heidari M, Johnstone DM, Daneshi N, Hondermarck H. Transcriptomics. In: Bradshaw RA, Stahl PD, editors. Encyclopedia of cell biology. Academic Press; 2016. p. 160–5.
30. Ziemann M, Kaspi A, El-Osta A. Evaluation of microRNA alignment techniques. RNA. 2016;22(8):1120–38. https://doi.org/10.1261/rna.055509.115.
31. Amanullah M, Yu M, Sun X, et al. MDEHT: a multivariate approach for detecting differential expression of microRNA isoform data in RNA-sequencing studies. Bioinformatics. 2020;36(9):2657–64. https://doi.org/10.1093/bioinformatics/btaa015.
32. Williams Z, Ben-Dov IZ, Elias R, et al. Comprehensive profiling of circulating microRNA via small RNA sequencing of cDNA libraries reveals biomarker potential and limitations. Proc Natl Acad Sci U S A. 2013;110(11):4255–60. https://doi.org/10.1073/pnas.1214046110.
33. Čapková M, Šáchová J, Strnad H, et al. Microarray analysis of serum mRNA in patients with head and neck squamous cell carcinoma at whole-genome scale. Biomed Res Int. 2014;2014:408683. https://doi.org/10.1155/2014/408683.
34. Forero DA, González-Giraldo Y, Castro-Vega LJ, Barreto GE. qPCR-based methods for expression analysis of miRNAs. Biotechniques. 2019;67(4):192–9. https://doi.org/10.2144/btn-2019-0065.
35. Avela HF, Sirén H. Advances in lipidomics. Clin Chim Acta. 2020;510:123–41. https://doi.org/10.1016/j.cca.2020.06.049.
36. Hu T, Zhang JL. Mass-spectrometry-based lipidomics. J Sep Sci. 2018;41(1):351–72. https://doi.org/10.1002/jssc.201700709.
37. Kallionpää RA, Ahramo K, Aaltonen M, Pennanen P, Peltonen J, Peltonen S. Circulating free DNA in the plasma of individuals with neurofibromatosis type 1. Am J Med Genet A. 2021;185(4):1098–104. https://doi.org/10.1002/ajmg.a.62081.
38. https://clinicaltrials.gov/ct2/show/NCT01541800.
39. Szymanski JJ, Sundby RT, Jones PA, et al. Cell-free DNA ultra-low-pass whole genome sequencing to distinguish malignant peripheral nerve sheath tumor (MPNST) from its benign precursor lesion: a cross-sectional study. PLoS Med. 2021;18(8):e1003734. https://doi.org/10.1371/journal.pmed.1003734.
40. Godec A, Jayasinghe R, Chrisinger JSA, et al. Whole exome sequencing reveals the maintained polyclonal nature from primary to metastatic malignant peripheral nerve sheath tumor in two patients with NF1. Neurooncol Adv. 2019;2(Suppl 1):i75–84. https://doi.org/10.1093/noajnl/vdz026.
41. Hummel TR, Jessen WJ, Miller SJ, et al. Gene expression analysis identifies potential biomarkers of neurofibromatosis type 1 including adrenomedullin. Clin Cancer Res. 2010;16(20):5048–57. https://doi.org/10.1158/1078-0432.CCR-10-0613.
42. Kolanczyk M, Mautner V, Kossler N, et al. MIA is a potential biomarker for tumour load in neurofibromatosis type 1. BMC Med. 2011;9:82. https://doi.org/10.1186/1741-7015-9-82.
43. Yoshida Y, Furumura M, Tahira M, Horie T, Yamamoto O. Serum biomarker in neurofibromatosis type 1. J Dermatol Sci. 2012;67(2):155–8. https://doi.org/10.1016/j.jdermsci.2012.04.010.
44. Johansson G, Peng PC, Huang PY, et al. Soluble AXL: a possible circulating biomarker for neurofibromatosis type 1 related tumor burden [published correction appears in PLoS One. 2015;10(3):e0119975]. PLoS One. 2014;9(12):e115916. https://doi.org/10.1371/journal.pone.0115916.
45. Vetrano IG, Dei Cas M, Nazzi V, et al. The lipid asset is unbalanced in peripheral nerve sheath tumors. Int J Mol Sci. 2021;23(1):61. https://doi.org/10.3390/ijms23010061.

5 Preclinical NF1-PNST Models and Potential Drug Resistance Mechanisms

Zhichao Wang, Wei Wang, and Yuehua Li

The establishment of appropriate models for preclinical studies and further advanced therapies are very important. In this chapter, various in vitro or in vivo models of plexiform neurofibromas (PNFs) and malignant peripheral nerve sheath tumors (MPNSTs) are separately represented as models of benign and malignant peripheral nerve sheath tumors (PNSTs). Models of other kinds of PNSTs could use these models as references. This chapter introduces the cell and animal models of PNST, in particular PNF and MPNST, to increase the understanding of diseases. By discussing the modeling methods of cells and animals, we intended to identify the common origin of different types of neurofibroma cells and analyze the developmental process of PNST models. Furthermore, a full understanding of these models can be helpful to study the molecular and cellular mechanisms underlying neurofibroma development, select potential effective therapeutic agents, and ultimately decrease the morbidity and mortality rates of patients in the clinic [1]. Available PNST models are shown in Fig. 5.1.

Cell models can be further divided into 2D and 3D models based on their growth pattern. Both traditional 2D cell culture methods and 3D cell culture with the application of Matrigel or other matrices can be used as preclinical cell models to evaluate cell functions and gene and protein conditions in various microenvironmental changes. However, there are some differences between these two models. 2D cell culture, including tissue-extracted primary cells and cell lines, is regarded as a simple, highly effective, and repeatable method to study tumor biology, molecular changes, and therapeutic strategies. It is also a common and fundamental method for preclinical drug screening. Moreover, the use of tumor cell monocultures can avoid the risk of contamination of other cells and provide a relatively pure space to understand the specific changes that occur within tumor cells compared with other components in the microenvironment of neurofibroma in response to a given condition [2]. By using coculture methods with Transwell plates or in various cell ratios, tumor cells can be cultured with other cell types for evaluation of the cell-to-cell indirect influence or direct influence. 2D and 3D cell culture PNST models, similar to cell models of other diseases, show similar trends in the response of different agents. Therapies that show little effect on the viability of PNST cells often do not have higher efficacy when grown in more realistic 3D cell models [3].

Z. Wang (✉) · W. Wang · Y. Li
Department of Plastic and Reconstructive Surgery, Shanghai Ninth People's Hospital, Shanghai Jiao Tong University School of Medicine, Shanghai, China
e-mail: wangzhichao@sh9hospital.org.cn

I. G. Vetrano, V. Nazzi (eds.), *Peripheral Nerve Sheath Tumors*,
https://doi.org/10.1007/978-3-031-76382-3_5

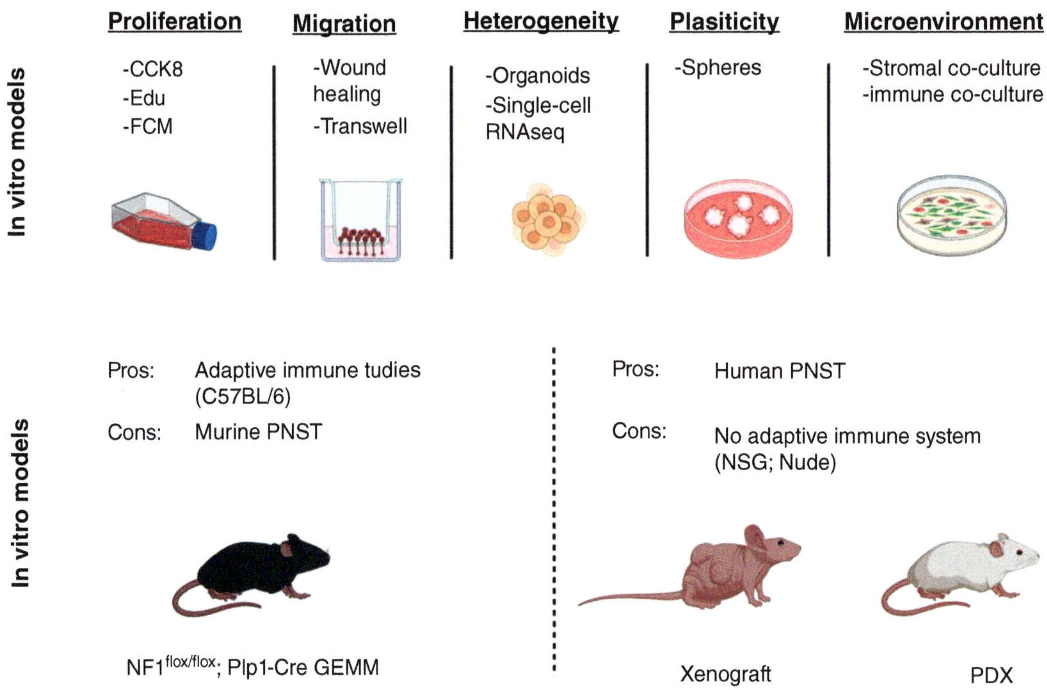

Fig. 5.1 Available PNST in vitro and in vivo models

However, compared to 3D cell culture, there are still some disadvantages to 2D cell culture, including (1) inevitable genotypic and transcriptome changes that deviate from the original tumor after several passages in serum medium [4]; (2) limitation in mimicking the actual tumor environment, such as oxygen levels, nutrient levels, and pH gradients [5]; (3) limitation in studying interactions with the microenvironment [6]; and (4) lack of influence by the main metabolic organs [7]. In other words, the 3D cell system has several advantages: (1) better-mimicking tumor in vivo morphology and spatial arrangement and its interaction with the extracellular matrix, (2) slower growth in 3D that is in accordance with in vivo growth rate, (3) increased energy production and different gene expression profiles, and (4) differences in therapeutic sensitivity. Detail comparisons between 2D and 3D cell cultures are shown in Fig. 5.2. Various 3D models have been established, including spheroids, and the recent expanding investigation field of iPSCs and organoids [2].

In vivo animal models can be used to solve complex questions about cell–cell interactions or tumor microenvironmental influences to study the mechanism underlying tumor development. Similar to other tumor types, the most commonly used animal models of PNSTs are genetically engineered mouse models (GEMMs) and patient-derived xenografts (PDXs).

In addition to in vitro cell models, representative in vivo models such as patient-derived xenografts (PDXs) and engineered animal models also help investigate the characteristics or disease mechanisms. On the other hand, patient-derived xenograft (PDX) models resemble tumors in cellular complexity and can retain tumor heterogeneity. Immunodeficient mice are often used for the implantation of PDXs, which makes immunity-related studies of autologous immune cells difficult. Moreover, the low reimplantation

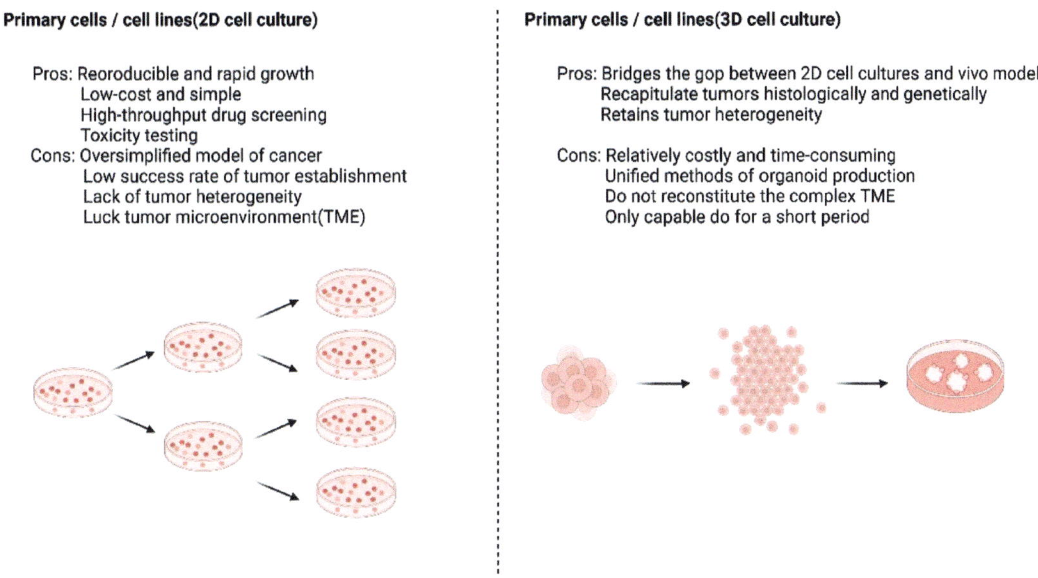

Fig. 5.2 The comparison between 2D cell culture and 3D cell culture

Table 5.1 GEMMs generated by Cre-Lox system and the NF1-related phenotypes

Genotype	Cell type	Phenotype
GFAP-Cre; Nf1$^{flox/mut}$	Astrocytes	Optic pathway glioma
P0A-Cre;Nf1$^{flox/-}$	Nonmyelinating SCs	Plexiform neurofibroma
Plp-CreER;Nf1$^{flox/flox}$	Schwann cell precursors (SCPs)	Plexiform neurofibroma
Krox20-Cre;Nf1$^{flox/-}$	SCPs	Plexiform neurofibroma
Dhh-Cre;Nf1$^{flox/flox}$	SCPs	Plexiform/cutaneous neurofibroma
Dhh-Cre;Nf1$^{flox/flox}$; Ink4a/Arf$^{+/-}$; Ink4a/Arf$^{-/-}$; Ink4a/Arf$^{flox/flox}$	SCPs	Atypical neurofibroma, MPNST
CMV-CreER T2;Rosa25;Nf1$^{flox/-}$	Skin-derived precursors (SKPs)	Cutaneous neurofibroma
Hoxb7-Cre;Nf1$^{flox/flox}$;LacZ	SKPs	Cutaneous/plexiform neurofibroma
Prss56-Cre;Rosa26-tdTom;Nf1flox/flox	Boundary cap cells	Cutaneous/plexiform neurofibroma
Postn-Cre;Nf1$^{flox/flox}$;Arf$^{flox/+}$	SCPs	Plexiform neurofibroma, atypical neurofibromatous neoplasms of uncertain biologic potential (ANNUBP), MPNST
Mx1-Cre; Nf1$^{flox/flox}$	Bone marrow	JMML
CAD5-Cre^{ERT2}/Rosa26-LSL-td-tomato; Nf1$^{flox/flox}$	Vascular endothelium	Altered proliferation and morphogenesis
Col2α1-Cre; Nf1$^{flox/flox}$	Chondrocytes	Skeletal dysplasia

rates, relatively large expense, and construction time are other major limitations of PDX [8] (Table 5.1). However, cancer is a highly heterogeneous disease. Preclinical models that fully recapitulate tumor pathophysiology are critical for the accurate assessment of drug efficacy and toxicity [9]. In general, the combination of in vivo and in vitro models will be especially useful for a comprehensive study to better represent the biological characteristics and therapeutic strategies of tumors.

5.1 Benign PNST-PNF

5.1.1 2D Cell Model: Primary PNF Cells

Primary human PNF cell cultures are commonly derived from freshly resected neurofibromas [10–12]. According to the protocol, the tumor mass is cut into pieces and fully digested by collagenase and dispase in 37 °C oscillators. Suspensions of multiple cell types are dispersed in a single cell and can better display the heterogeneity of neurofibroma than cell lines. By culturing these cells in selected cell culture medium supplemented with certain neuron growth factors or using immune-magnetic beads specific for tumor biomarkers and fluorescence-activated cell sorting, these cell suspensions can be further screened and the characteristics of primary PNST cells can be maintained [13].

One of the outstanding problems of primary cell models is that after several passages [13], cells do not proliferate well and show genotypic and transcriptome changes [2, 4, 14], which limits their application in further molecular and cellular analyses [15]. Moreover, although primary cell cultures can represent individual differences in tumors better than cell lines, they still have disadvantages when used as oversimplified models of PNF compared to 3D cell culture models, including difficulties in imitating the various cell types and the complex tumor environment [2].

5.1.2 2D Cell Model: PNF Cell Lines

To solve the problems of limited passages in primary cells, several PNF-related immortalized cell lines have been established, such as ipNF05.5, ipNF05.5mix, ipNF95.6, and ipNF9511.bc, and they have been widely used for basic research on neurofibroma in recent decades [16]. By using lentiviral transduction of TERT human and Cdk4 murine transgenes, immortalized human Schwann cells can be generated, which prolongs the proliferation times of primary PNF cells. In particular, different genotypes of Schwann cells with and without NF1 mutations were obtained from normal nerves of healthy people, from non-neoplastic nerves of patients with NF1, or from neoplastic plexiform neurofibromas of patients with NF1, and these cells were generated by Dr. Margaret R. Wallace from the University of Florida. $NF1^{-/-}$ Schwann cells derived from PNFs provide a convenient and stable cell model and can be further used as a platform for the study of MPNST progression [16, 17]. Moreover, PNF cell lines growing in a matrix with physical support and mechanical cues can be further used to establish a 3D cell model to recapitulate a more realistic SC microenvironment [1, 13]. The NF1 research group of Shanghai Ninth People's Hospital also established immortalized PNF cell lines with genetic information from Chinese patients and used them for in vitro cell models [18].

However, the disadvantages of cell line culture are obvious. The most notable problem is the evitable change in cellular biological status in tumor heterogeneity and genetic diversity during the process of immortalization. Passaging cell lines in culture medium for a long time may also cause genetic drift, which may further increase genetic variability and reduce genetic reproducibility. In addition, unlike malignant cell lines, most benign PNST cells are unable to form tumor masses regardless of the transplanted location, such as the inguinal region, in situ sciatic nerve, or renal capsule of immunodeficient mice, or the number of cells, which seriously limits their application for in vivo evaluation. While the PNF cell lines ipNF04.4 and ipNF03.3 can form tumor masses in the in situ sciatic nerves of immunocompromised mice, the relatively difficult process to generate the model and uncertain drug delivery efficacy may limit their application.

Taken together, cell lines, as another common 2D cell model, also have the disadvantage of being too simplistic for thorough studies. 3D cell models can, to some certain, optimize this disadvantage and imitate the in vivo tumor microenvironment better for preclinical studies [9].

5.1.3 3D Cell Model: Spheroids

Spheroids can be established in cancer research as in vitro 3D cell microanalogs and made by growing cells in low attachment dishes with appropriate centrifugal velocity for sphere formation, which can be used for the investigation of in vitro drug screening assays [19].

5.1.4 3D Cell Model: Organoids

Organoids use stem cells or iPSCs cultured with certain tissue growth factors and Matrigel and develop by self-organization, proliferation, and differentiation in vitro. These cells can finally develop into 3D structures, reflect the characteristics of digested organs, and can be established on a large scale [20]. The cells that are derived to establish organoids can be extracted from freshly resected normal tissue or tumors [9]. In addition, the different components of the tumor mass, e.g., PNSTs, such as pathological Schwann cells, fibroblasts, endothelial cells, and immune cells, can grow into the matrix together and form tumors [1, 21, 22].

Thus, organoids show the preponderance of a variety of cell types and can recapitulate the interactions between tumor cells and other cell components in vitro with more valuable results [2]. By a standardized establishment process, organoid biobanks and microfluidic organs-on-a-chip can both be built that keep the genetic information from different patients for drug screening and personalized optimized treatment strategies. More importantly, the organoids can be directly xenografted in immunodeficient mice and used for in vivo research with clinical heterogeneity in genetic and phenotypic characteristics, and they can be considered the link between in vitro cell models and in vivo PDX models. Although there are currently no reports on the use of PNF-related organoids in studies, human cerebral organoids are already widely used as excellent platforms for disease research, such as glioma [2] and glioblastoma [20], with good repeatability. These existing studies of organoids can inspire future exploration directions and accelerate the study of PNSTs [9]. However, some challenges with this technology remain. Organoids are composed of various cell types in complex 3D structures; however, it is difficult to thoroughly predict or identify the cell types and corresponding ratio, which indicates the difficulty in determining the accuracy of native organ composition or architecture. Another shortcoming, particularly in iPSC-derived organoids, is the limitation of tissue maturation that restricts the growth and development of organoids. Moreover, the overall morphogenesis and environmental factors may still be different compared to those of real organs in vivo due to the established protocol, which involves the addition of matrices such as Matrigel to mimic the in vivo environment. With the use of animal-derived culture agents, the risks of additional immunogens and pathogens must be overcome for potential clinical applications [23].

5.1.5 3D Cell Model-Induced Pluripotent Stem Cell (iPSC) PNF Model

Induced pluripotent stem cells (iPSCs) are another valuable method for recapitulating genetic heterogeneity and studying tumorigenic processes. They can be generated from tumor-associated fibroblasts [24, 25] or directly from tumor cells to acquire isogenic *NF1*$^{+/-}$ or *NF1*$^{-/-}$ iPSC lines [26]. The process to build this model is displayed in Fig. 5.3. According to previous studies, *NF1*$^{-/-}$ iPSCs have already been made from PNF cells from NF1 patients [26]. After their differentiation from neural crest cells into Schwann cells, several changes occur, including acceleration in proliferation, disability in myelination, and the trend toward 3D spheroids compared with wild-type iPSCs; thus, they can be regarded as a valuable PNF cell model [17]. In a recent study, neurofibroma spheres built with a mixture of *NF1*$^{-/-}$ Schwann cells induced from PNF-derived iPSCs and primary *NF1*$^{+/-}$ fibroblasts derived from PNFs finally generated PNF-like masses. The researchers grafted cell mixtures in situ into the sciatic nerve of nude mice and generated a model specifically to study the inter-

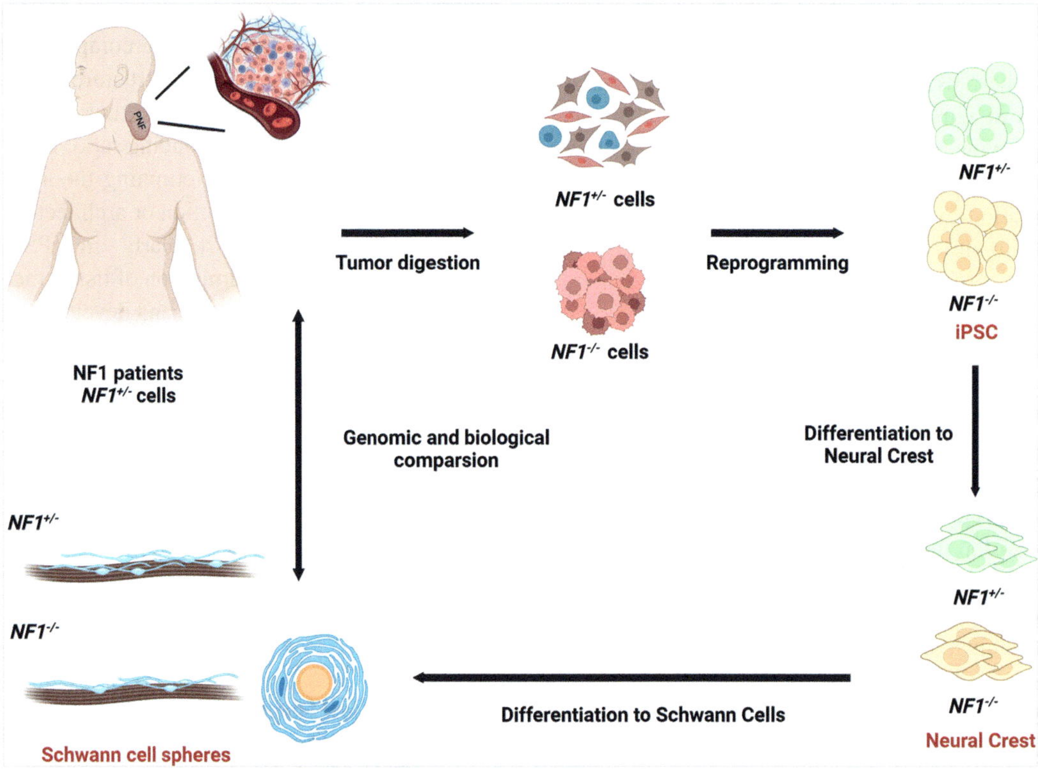

Fig. 5.3 The process to establish the iPSC model

actions between Schwann cells themselves and Schwann cells and fibroblasts [27]. iPSCs can also be a proper cell model to study the genetic and epigenetic changes for tumor progression and can combine the technique of CRISPR-Cas9 or other DNA-edited tools [26] to have better tractable platforms for tumorigenesis or future preclinical therapeutic screening strategies [28].

There are also some limitations of iPSC models. The difficulties in accelerating remodeling efficiency, such as how to optimize the methods to change the chromosomal and genomic composition or epigenetic state to reprogram cancer cells, limit technical applications. In addition to the study of diseases with late onset points, it is also difficult to use the iPSC model for the immature embryonic or fetal cell state. Another concern is how to avoid the risk of tumorigenicity, such as variants in copy number or loss of heterozygosity from iPSCs [29], particularly when needed for culture for prolonged periods of time [30, 31]. Genetic heterogeneity of iPSCs may also affect the results, as genetic variation may confound the model in certain diseases without appropriately designed controls [32].

5.1.6 Animal Model: Patient-Derived Xenograft (PDX)-PNF Models

A patient-derived xenograft (PDX) model is constructed by grafting freshly surgically resected tumor tissues into the rich-blood armpit or inguinal region, called heterotopic subcutaneous engraftment, or orthotopic engraftment with a more physiological site, for instance, the sciatic nerve for PNF, of immunodeficient mice such as the NOD/SCID/IL2Rγnull (NSG) strain [19]. The process to establish this animal model is shown in Fig. 5.4. Compared with other animal models, PDX models can correlate directly with human cancers, retain the original structure of the tumor and cell type proportion, and retain the

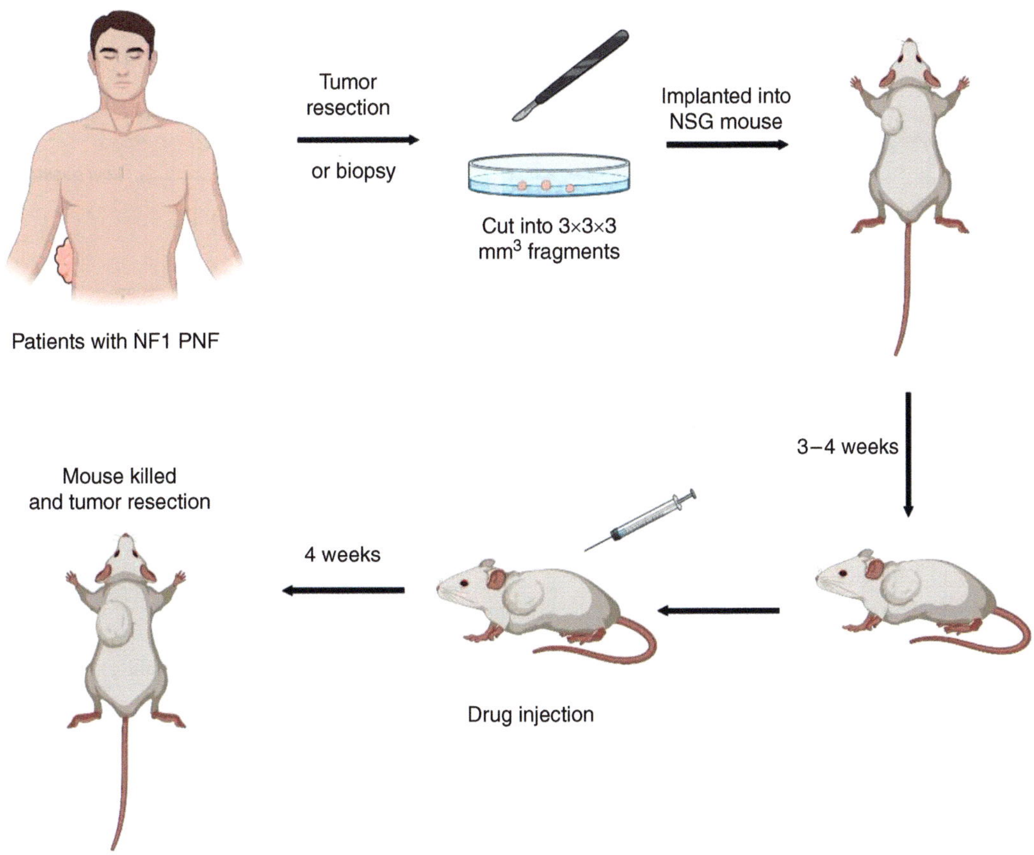

Fig. 5.4 The process to establish the PDX model

main molecular components and histological characteristics of human PNFs to a large extent for a period of time. Similar to the advantages of organoids, PDX models are also useful to study the interactions among various cell types and the tumor microenvironmental components. Further comparisons between PDX models and organoids are shown in Fig. 5.5.

However, there are limitations to PDX models. Due to the direct engraftment transfer of resected tumors, there is no PDX model available for studying the tumorigenesis of PNFs [2]. Another limitation is related to the complicated experimental procedures, including the long operation process to construct the PDX model, immediate experimental operations, variability in growth ability, and relatively high expense [2, 33, 34]. Moreover, the small number of xenografts from resected tumors cannot fully represent the complete heterogeneous information of tumors. Therefore, inevitable genotypic and transcriptome differences should be considered when evaluating the metabolism and pharmacokinetics of new therapeutic methods [2]. The PDX model also faces the risk of human tumor tissues being subsequently replaced by murine stroma and the occurrence of genetic deviation after several passages of PDX. In addition, the difficulties in investigating the tumor microenvironment cannot be ignored when using immune-deficient mice to generate models [33]. There are also some problems related to moral or ethical issues in PDX models [2]. One solution is to graft multiregion selection of the tumor mass to obtain relatively comprehensive genetic characteristics and use the early passages of xenografts to maintain therapeutic accuracy in drug screening [34].

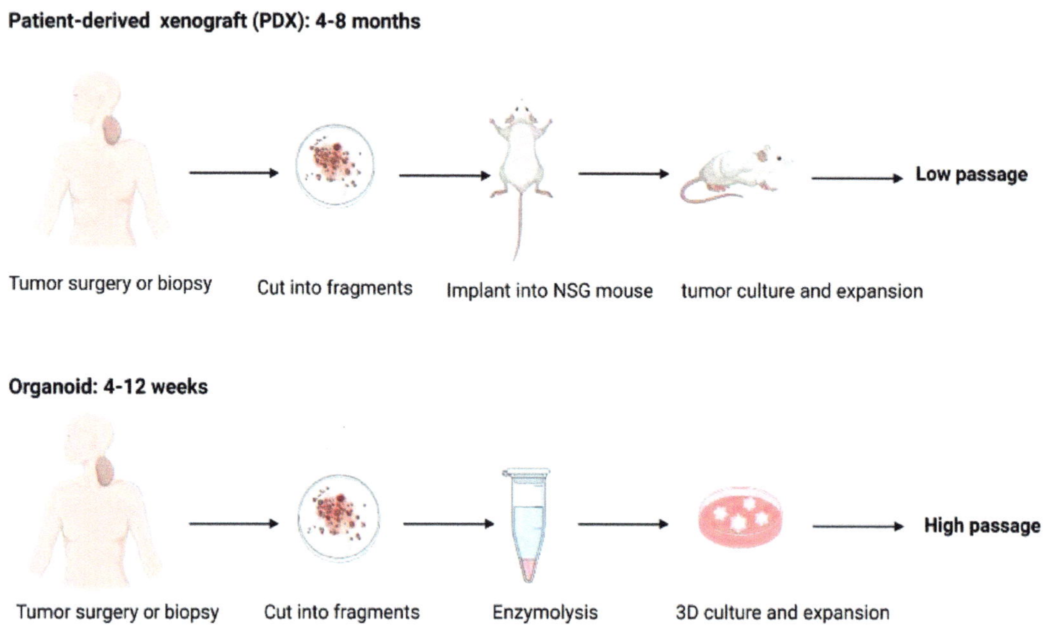

	PDX	Organoid
Generation efficiency	+	++
Genetic manipulation	-/+	++
Genetic stability	++	++
Scalability	+	++
Integratable immune system	-/+	-/+
Intra-tumor heterogeneity	++	+
Tumour-stroma interaction	+	-/+
Low-throughput drug screens	+	++
High-throughput drug screens	-	++
Prediction of clinical response	++	+=
Ease of use	+	++

Fig. 5.5 Summaries of the advantages and disadvantages between PDX and organoids

5.1.7 Animal Model: Genetically Engineered Mouse Model (GEMM) of PNF

GEMMs combine the tissue-specific expression of oncogenes or the tissue-specific deletion of tumor suppressors to drive the growth of primary tumor cells. The first GEMM of NF1 was described in 1994, and the germline inactivation of *Nf1*$^{+/-}$ mice led to a nonfunctional mutation without classical manifestations [35]. Mice with homozygous *Nf1* exon 31 mutations died in the embryonic stage due to cardiac defects. Although heterozygous *Nf1*$^{\Delta 31/+}$ mice showed no NF1 phenotypes, these initial studies have accumulated experience for the development of more advanced modeling strategies. For GEMMs, the conditional knockout strategies, i.e., Cre-loxP systems have been widely used by deleting one or both *Nf1* alleles in specific cell stages in Schwann cells. For instance, the specific *Nf1* loss in Schwann cells may further develop into PNF-like lesions in chimeric *Nf1*$^{flox/-}$: Krox20-Cre mice [35, 36]. It is noteworthy that the application of the Cre-lox system in GEMMs is able to induce mutations in a tissue-restrictive manner and then regulate the location and opportunity of gene expression. This system helps achieve establish mouse models that develop specific manifestations in NF1 patients, including neurofibromas, optic glioma, and other phenotypes (Table 5.1, [35]). Furthermore, in addition to direct approaches to the ablation of the *Nf1* gene, modified genes or other nonsense suppression or exon-skipping modifications of *Nf1* can also be used for the promotion of neurofibroma-related phenotypes [37]. For instance, mouse models with a common nonsense mutation in humans (nmNF1, exon 18 c.2041C>T; p.Arg681*) can significantly reduce the expression of neurofibromin and give rise to spinal cord-originated PNFs [37]. The use of GEMMs has also proven to have a similar origin in both CNFs and PNFs, as *Nf1*$^{-/-}$ Schwann cells with the loss of either Hoxb7 [38] or Prss56 [39] can both generate characteristic neurofibroma.

GEMMs compensate for some limitations of other models, as they retain the genetic heterogeneity of PNF. This model can be used to study tumorigenesis, interactions between different cell types in the tumor microenvironment, and the malignant or metastatic progression of benign PNFs [36], especially to prove the role of gene changes in specific cell types in the process of disease. GEMMs display typical phenotypes, such as neurofibroma formation, optic gliomas, learning difficulties or bone lesions, and histopathological biomarkers, such as high expression of S100 and sox-10, as in human PNFs [40]. Moreover, GEMMs enable spontaneous tumor growth instead of traditional tumor implantation. This growth mode is more similar to the actual human tumor growth pattern and more suitable for biological mechanism research. However, these mouse models lack other common clinical NF1 phenotypes, such as café au lait macules, Lisch nodules, and cutaneous neurofibromas. They also show limitations in mimicking the disease process or the complicated background of NF1 patients in the clinic. The relatively short lifespan of mice limits the study of disease progression. Therefore, despite the irreplaceable advantages of GEMMs in tumor studies, they are still not the best platform to completely mimic the tumor heterogeneity of humans.

5.1.8 Animal Model: Genetic Engineering Pig Models

In recent years, studies of genetically engineered pig models have shown promise in presenting the clinical features of NF1 and can be regarded as complementary to GEMMs in some ways. Compared to mice, swine show more similarities to humans in terms of skin and other organ structures, body size, and longevity, which can aid in toxicological, pharmaceutical, and longitudinal studies [35]. In particular, for NF1, the establishment of *NF1*$^{+/R1947X}$ minipigs shows complete penetrance of café au lait macules, which is the first time that cutaneous neurofibroma, myelin decompaction, OPG, and other manifestations have been demonstrated in animal models [41]. The extracted cells from this model also showed *NF1* LOH, loss of neurofibromin expression, and Ras pathway activation. The generation of $NF1^{+/ex42del}$ minipigs showed similar manifestations to the $NF1^{+/R1947X}$ minipig model and displayed obvious cognitive impairment and attention-deficit/hyperactivity disorder and a decreased pain threshold [42]. However, complex genetic engineering techniques and time-consuming reproductive methods limit the application of genetically modified pigs in the laboratory compared to the widespread use of mice.

5.2 Malignant PNST-MPNST

PNFs can transform into MPNSTs, especially in patients with a high tumor burden, and they may have relatively higher risks of becoming malignant. Throughout their lifespan, NF1 patients show an 8–13% risk of malignant transformation. Approximately 50% of MPNST masses develop from previous PNFs [7]. Once MPNSTs occur, they can significantly reduce life expectancy with no gender preference [43].

The model types of malignant PNST tumors, MPNSTs, are the same as benign PNFs, and they can be generally divided into cell culture models, including human MPNST cell lines [44] and primary MPNST cells, animal models, such as GEMMs, or other animal models, such as genetically engineered zebrafish, *Drosophila*, and PDX [45, 46]. These various types of models that are used to study the biology of MPNST are discussed in more detail below.

5.2.1 In Vitro MPNST Cell Culture Model

The method to obtain primary MPNST is similar to the protocol of PNF primary cells [47, 48]. Obtaining malignant tumor primary cells from MPNSTs can help to understand the characterization of various tumors excised from different patients and study the pathogenesis process of their malignancy and can be kept as a valuable resource for pharmacologic studies in vitro [47].

Similar to PNF, MPNST cell lines were obtained from nerves of patients with nonpathological or pathological NF1 with neurofibromas and immortalized using lentiviral transduction of human TERT and mouse Cdk4 transgenes [16, 22]. These immortalized MPNST cells can also be used to evaluate the effects of oncogenes, tumorigenesis and the development of MPNST [49, 50]. The main characteristics of cell lines isolated from either NF1-associated or sporadic MPNST are displayed in Table 5.2 [22]. Because of their malignant status and high replication ability, MPNST cell lines can also be used as xenograft models to evaluate new treatment strategies such as chemo/radiotherapy, targeted therapy, and immunotherapy [22, 52, 53]. Other applications include the role of tyrosine kinase receptors such as stem cell factor/KIT complex [54], EGFR [55], PDGFR [54] and HGFR [56] in MPNST, the role of oncogene mutations [57], oncogenes, modifier genes, or tumor suppressors, growth factors [58], steroid hormones, microRNAs [59], and in the pathogenesis of MPNST pathology.

However, 3D cell culture of MPNST is not widely used as 2D culture and is limited to the use of spheroids to analyze the growth or drug efficacy of MPNST cells [60]. 3D models, including spheroids, organoids, and iPSCs, also have important research value and may be the future directions of model establishment for a credible study.

MPNST cell lines are widely used in the study of this disease. However, MPNST cell lines have not been well characterized by whole exome/genome sequencing, gene expression profiling, etc. M MPNST patient-derived xenografts (PDXs) can also be used to establish an in vivo model in immunodeficient mice [17].

5.2.2 In Vivo MPNST-PDX Model

Minced tumor fragments from donor mice can be subcutaneously implanted close to the sciatic

Table 5.2 Human MPNST cell lines and origins [51]

		Recipient mouse species				Route of administration				
Cell line	Origin	Nude	SCID	NSG	NOD/SCID	NIH III	SC	IP	IN	TV
S462	NF1 MPNST	X					X		X	
S462.Ty	NF1 MPNST	X					X			
SNF02.2	Metastatic lung of NF1 MPNST									
SNF94.3	NF1 MPNST		X						X	
SNF96.2	NF1 MPNST		X				X			
ST88-14	NF1 MPNST	X	X	X		X	X	X		
T265	NF1 MPNST	X	X	X		X		X		
ST88-3	NF1 MPNST									
90-8	NF1 MPNST									
FMS-1	NF1 MPNST	X	X		X		X			
MPNST642	NF1 MPNST		X				X			
MPNST724	MPNST		X				X			
NMS-2	NF1 MPNST	X					X			
NMS-2PC	Metastatic retroperitoneal lesion of NF1 MPNST	X			X		X			
YST-1	Sporadic MPNST	X					X			
STS26T	Metastatic left scapula of sporadic MPNST	X	X	X		X	X	X	X	X
Hs-Sch-2	Sporadic MPNST	X			X		X			

IN intraneural, *IP* intraperitoneal, *SC* subcutaneous, *TV* tail vein

nerves of 6–8-week-old athymic nude (NF90.8) or NSG (PDX JH-2-002) mice to form a patient-derived xenograft (PDX) model [61]. For the MPNST-PDX model, biopsied tumor tissues are subcutaneously implanted directly into NSG mice. Tumors used for the study were grown in a donor mouse to a volume of approximately 2500 mm^3, which is larger than the volume of PNF-PDX due to their malignant nature. This process can be repeated, and tumor characteristics as well as genetic information can be transmitted to several passages. The tumor is harvested, cut into 1.5 mm^3 pieces, and inserted subcutaneously into the flank of each mouse. Models in which patient-derived tissues, including MPNST, are directly grafted into immunocompromised mice are increasingly being used to test therapeutics [62]. We describe a PDX model that contains common cancer-associated mutations found in patient MPNSTs—NF1, CDKN2A/B, and SUZ12 [63, 64]—and retains patient MPNST histological features [65].

5.2.3 In Vivo GEMMs of MPNST

For many other diseases, GEMMs of MPNST have been a workhorse for investigating the underlying biology of MPNST development and for preclinical drug testing. The main method of the establishment is the loss of the *Nf1* gene, while other methods, such as the oncogene activation of RAS along with the loss of PTEN [66] or the overexpression of growth factor ligands or their receptors in Schwann cells [67], can also induce the occurrence of MPNST.

Particularly, mice that have coinactivated mutations of *Nf1* and Trp53 placed in a cis configuration are the most commonly used $Nf1^{+/-}$ MPNST GEMMs; the $Trp53^{+/-}$ NPCis model is used for the evaluation of therapeutic targets/methods. As in the human genome, these genes in mice are closely linked on chromosome 11. The loss of heterozygous alleles of these two genes can eliminate each other and initiate various sarcomas, including MPNSTs [68, 69]. NPCis mice can develop into a variety of sarcomas, such as MPNSTs and gliomas [70]. However, this model does not recapitulate the actual tumor development process from benign PNF to atypical/ANNUBP and malignant MPNST [22]. This tendency of malignant transition was modeled in mice by the conditional deletion of the Arf and *Nf1* genes in neural crest-derived Schwann cells in mice. This model can recapitulate the process of transformation from ANNUBP to MPNST in humans [71] and can help understand the mechanisms underlying tumor progression and in the screening of drugs. Moreover, alterations in the p16INK4A/cyclin D/CDK4/Rb and p19ARF/Mdm/p53 pathways and chromosomes can lead to the activation of candidate oncogenes [72]. With the overexpression of neuregulin-1, a neuron growth factor of myelinating Schwann cells that can bind to the receptors ERBB3 and ERBB4, progression from neurofibroma to MPNST developed in a mouse model.

The Cre-loxP system is also widely used to conditionally inactivate *Nf1* in MPNST GEMMs. By knocking out Pten or Cdkn2a in *Nf1*-deficient Schwann cells, PNF can progress into MPNST [71, 73]. The overexpression of EGFR in *Nf1*-deficient Schwann cells can promote MPNST in a JAK2/STAT3-dependent manner [74]. Inactivation of the PTEN gene and overexpression of K-Ras or EGFR can promote plexiform neurofibroma transformation into *Nf1*-independent sporadic MPNST [67]. The overexpression of EGFR in Schwann cells in Trp53 mutant mice also showed increased neurofibroma and MPNST [75]. Adenovirus-Cre injections can also be used to establish mice with inactivation of the *Nf1* gene and other genes, such as p16INK4A and p19ARF [76], and generate $Nf1^{flox/flox}$ or $Nf1^{flox/-}$; Ink4a/Arf $^{flox/flox}$ mice. The CRISPR-Cas9 gene editing strategy and adenovirus delivery of *Nf1* and Trp53 to the location of sciatic nerves can also be used to generate MPNSTs [76].

Sleeping Beauty (SB) transposon-based mutagenesis can be used to find potential drivers of genes and pathways to tumorigenesis, additional phenotypes, metastatic potential, and treatment resistance. The results showed that Cnp-Cre-positive cells of loxP-STOP-loxP-Trp53R270H (LSL-Trp53R270H) plus Cnp-hEGFR background mice can increase MPNST tumor grade. The SB transposon screen also suggested potential drivers of MPNST, such as PTEN, Wnt/β--catenin signaling, RAS-MAPK, Hippo/Yap, and other pathways [77].

The major limitation of GEMMs is the lack of transformation from benign to malignant tumors in mice. The codeletions of Trp53 and Suz12 in *Nf1* germline mutation mice can also accelerate MPNST tumor formation and promote the coexistence of other malignant tumor types, including glioma and lymphoma. However, the stochastic tumor development of other types in this model limits the utilization of Suz12 germline mutations in the study of MPNST [17, 63].

5.2.4 *Nf1*-Mutant Fly Models

Due to the high (60%) similarity of the amino acid sequence in neurofibromin expression, *Drosophila* can also be used as an *Nf1*-mutant model of behavioral disabilities, such as growth problems [78–80], learning disability [51, 78, 79], or excessive behavior [81, 82]. The genetic closeness to humans of *Nf1*-mutant fly models can partially reflect signaling pathway changes, including the activated Ras-MAPK pathway [83] or the inactivated cAMP-protein kinase A pathway [78, 84].

5.2.5 Zebrafish Models of MPNST

Nf1-mutant zebrafish is another model that is used to evaluate MPNST-related biological activity, tumor suppression, and therapeutic strategies. Further p53-mutant zebrafish can be used to study the biology of MPNST in ribosomal protein (RP)-mutant zebrafish. The expression level of RPs affects the production of p53, and comutations can result in the tumorigenesis of MPNST [85]. RECK, a membrane-bound glycoprotein, was shown to be a tumor suppressor in this zebrafish model. Moreover, p53-knockout zebrafish on the $nf1a^{+/-}$; $nf1b^{-/-}$ ortholog background model exhibit earlier onset and increased penetrance, including pigmentation of MPNSTs [86], which is especially useful in studying NF1-related MPNSTs [46, 87]. Heterozygous loss of the RPrps6 gene can lead to a large deletion of chromosome 1, which can further cause MPNSTs [88].

Two vital problems need to be mentioned about MPNST models. Firstly, even the most used GEMM models cannot display all the genetic phenotypes of MPNSTs, particularly hard to recapitulate the malignant process from a preexisting PNF or ANNUBP. There are presently no corresponding models that mimic the benign, premalignant to malignant tumor transformation. Secondly, there are few established MPNST cell lines of human or mouse that are used for investigation in the field compared to other tumors. Better model systems should be expected to develop for future studies, including (1) better GEMMs to mimic malignant process, (2) genetically engineered human cells for MPNSTs with correct lineage(s), and (3) the application of clinic resources, including primary MPNSTs cell extracted from surgery and grown in vitro and masses grafted in vivo as PDX models. However, other limitations of these models should also be realized, such as few models to display metastasis, which is common in human MPNSTs, lack of PRC2 function to display, and unclear to recapitulate MPNSTs at the transcriptomic level or reflect human MPNST molecular subtypes.

References

1. Kraniak JM, et al. Development of 3D culture models of plexiform neurofibroma and initial application for phenotypic characterization and drug screening. Exp Neurol. 2018;299(Pt B):289–98.
2. Xu X, et al. Opportunities and challenges of glioma organoids. Cell Commun Signal. 2021;19(1):102.
3. Rebecca VW, Somasundaram R, Herlyn M. Preclinical modeling of cutaneous melanoma. Nat Commun. 2020;11(1):2858.
4. Lee J, et al. Tumor stem cells derived from glioblastomas cultured in bFGF and EGF more closely mirror the phenotype and genotype of primary tumors than do serum-cultured cell lines. Cancer Cell. 2006;9(5):391–403.
5. Bristow RG, Hill RP. Hypoxia and metabolism. Hypoxia, DNA repair and genetic instability. Nat Rev Cancer. 2008;8(3):180–92.
6. Torsvik A, et al. U-251 revisited: genetic drift and phenotypic consequences of long-term cultures of glioblastoma cells. Cancer Med. 2014;3(4):812–24.
7. Bryant HE, et al. Specific killing of BRCA2-deficient tumours with inhibitors of poly(ADP-ribose) polymerase. Nature. 2005;434(7035):913–7.
8. Olson B, et al. Mouse models for cancer immunotherapy research. Cancer Discov. 2018;8(11):1358–65.
9. Gunti S, et al. Organoid and spheroid tumor models: techniques and applications. Cancers (Basel). 2021;13(4):874.
10. Rosenbaum T, et al. Long-term culture and characterization of human neurofibroma-derived Schwann cells. J Neurosci Res. 2000;61(5):524–32.
11. Serra E, et al. Schwann cells harbor the somatic NF1 mutation in neurofibromas: evidence of two different Schwann cell subpopulations. Hum Mol Genet. 2000;9(20):3055–64.
12. Wallace MR, et al. Culture of cytogenetically abnormal Schwann cells from benign and malignant NF1 tumors. Genes Chromosomes Cancer. 2000;27(2):117–23.
13. Mazuelas H, Carrio M, Serra E. Modeling tumors of the peripheral nervous system associated with Neurofibromatosis type 1: reprogramming plexiform neurofibroma cells. Stem Cell Res. 2020;49:102068.
14. Li A, et al. Genomic changes and gene expression profiles reveal that established glioma cell lines are poorly representative of primary human gliomas. Mol Cancer Res. 2008;6(1):21–30.
15. Jiang W, et al. Preclinical assessment of the anticancer drug response of plexiform neurofibroma tissue using primary cultures. J Clin Neurol. 2015;11(2):172–7.
16. Li H, et al. Immortalization of human normal and NF1 neurofibroma Schwann cells. Lab Invest. 2016;96(10):1105–15.
17. Williams KB, Largaespada DA. New model systems and the development of targeted therapies for the treatment of neurofibromatosis type 1-associated malignant peripheral nerve sheath tumors. Genes (Basel). 2020;11(5):477.
18. Wang ZC, et al. Community-boosted neurofibromatosis research in China. Lancet Neurol. 2022;21(9):773–4.
19. Wang W, et al. Combined cyclin-dependent kinase inhibition overcomes MAPK/extracellular signal-regulated kinase kinase inhibitor resistance in plexiform neurofibroma of neurofibromatosis type I. J Invest Dermatol. 2022;142(3 Pt A):613–623 e7.
20. Jacob F, et al. A patient-derived glioblastoma organoid model and biobank recapitulates inter- and intra-tumoral heterogeneity. Cell. 2020;180(1):188–204.e22.
21. Moll C, et al. Tissue engineering of a human 3D in vitro tumor test system. J Vis Exp. 2013;78:50460.

22. Somatilaka BN, et al. Malignant peripheral nerve sheath tumor: models, biology, and translation. Oncogene. 2022;41(17):2405–21.
23. Rossi G, Manfrin A, Lutolf MP. Progress and potential in organoid research. Nat Rev Genet. 2018;19(11):671–87.
24. Anastasaki C, et al. Elucidating the impact of neurofibromatosis-1 germline mutations on neurofibromin function and dopamine-based learning. Hum Mol Genet. 2015;24(12):3518–28.
25. Wegscheid ML, Anastasaki C, Gutmann DH. Human stem cell modeling in neurofibromatosis type 1 (NF1). Exp Neurol. 2018;299(Pt B):270–80.
26. Carrio M, et al. Reprogramming captures the genetic and tumorigenic properties of neurofibromatosis type 1 plexiform neurofibromas. Stem Cell Rep. 2019;12(2):411–26.
27. Mazuelas H, et al. Modeling iPSC-derived human neurofibroma-like tumors in mice uncovers the heterogeneity of Schwann cells within plexiform neurofibromas. Cell Rep. 2022;38(7):110385.
28. Mo J, et al. Humanized neurofibroma model from induced pluripotent stem cells delineates tumor pathogenesis and developmental origins. J Clin Invest. 2021;131(1):e139807.
29. Lee AS, et al. Tumorigenicity as a clinical hurdle for pluripotent stem cell therapies. Nat Med. 2013;19(8):998–1004.
30. Lund RJ, Narva E, Lahesmaa R. Genetic and epigenetic stability of human pluripotent stem cells. Nat Rev Genet. 2012;13(10):732–44.
31. Shi Y, et al. Induced pluripotent stem cell technology: a decade of progress. Nat Rev Drug Discov. 2017;16(2):115–30.
32. Rowe RG, Daley GQ. Induced pluripotent stem cells in disease modelling and drug discovery. Nat Rev Genet. 2019;20(7):377–88.
33. Aldape K, et al. Challenges to curing primary brain tumours. Nat Rev Clin Oncol. 2019;16(8):509–20.
34. Bleijs M, et al. Xenograft and organoid model systems in cancer research. EMBO J. 2019;38(15):e101654.
35. Osum SH, Watson AL, Largaespada DA. Spontaneous and engineered large animal models of neurofibromatosis type 1. Int J Mol Sci. 2021;22(4):1954.
36. Brossier NM, Carroll SL. Genetically engineered mouse models shed new light on the pathogenesis of neurofibromatosis type I-related neoplasms of the peripheral nervous system. Brain Res Bull. 2012;88(1):58–71.
37. Li K, et al. Mice with missense and nonsense NF1 mutations display divergent phenotypes compared with human neurofibromatosis type I. Dis Model Mech. 2016;9(7):759–67.
38. Chen Z, et al. Spatiotemporal loss of NF1 in Schwann cell lineage leads to different types of cutaneous neurofibroma susceptible to modification by the hippo pathway. Cancer Discov. 2019;9(1):114–29.
39. Radomska KJ, et al. Cellular origin, tumor progression, and pathogenic mechanisms of cutaneous neurofibromas revealed by mice with Nf1 knockout in boundary cap cells. Cancer Discov. 2019;9(1):130–47.
40. Boyd KP, Korf BR, Theos A. Neurofibromatosis type 1. J Am Acad Dermatol. 2009;61(1):1–14. quiz 15–6.
41. Isakson SH, et al. Genetically engineered minipigs model the major clinical features of human neurofibromatosis type 1. Commun Biol. 2018;1:158.
42. White KA, et al. A porcine model of neurofibromatosis type 1 that mimics the human disease. JCI Insight. 2018;3(12):e120402.
43. Antoszczyk S, et al. Treatment of orthotopic malignant peripheral nerve sheath tumors with oncolytic herpes simplex virus. Neuro Oncol. 2014;16(8):1057–66.
44. Durbin AD, et al. Malignant peripheral nerve sheath tumors. Adv Exp Med Biol. 2016;916:495–530.
45. He S, et al. Synergy between loss of NF1 and overexpression of MYCN in neuroblastoma is mediated by the GAP-related domain. Elife. 2016;5:e14713.
46. Ki DH, et al. Mechanisms underlying synergy between DNA topoisomerase I-targeted drugs and mTOR kinase inhibitors in NF1-associated malignant peripheral nerve sheath tumors. Oncogene. 2019;38(39):6585–98.
47. Frahm S, et al. Genetic and phenotypic characterization of tumor cells derived from malignant peripheral nerve sheath tumors of neurofibromatosis type 1 patients. Neurobiol Dis. 2004;16(1):85–91.
48. Muir D, et al. Tumorigenic properties of neurofibromin-deficient neurofibroma Schwann cells. Am J Pathol. 2001;158(2):501–13.
49. Legius E, et al. TP53 mutations are frequent in malignant NF1 tumors. Genes Chromosomes Cancer. 1994;10(4):250–5.
50. Chau V, et al. Preclinical therapeutic efficacy of a novel pharmacologic inducer of apoptosis in malignant peripheral nerve sheath tumors. Cancer Res. 2014;74(2):586–97.
51. Buchanan ME, Davis RL. A distinct set of Drosophila brain neurons required for neurofibromatosis type 1-dependent learning and memory. J Neurosci. 2010;30(30):10135–43.
52. Ohishi J, et al. Imatinib mesylate inhibits cell growth of malignant peripheral nerve sheath tumors in vitro and in vivo through suppression of PDGFR-beta. BMC Cancer. 2013;13:224.
53. Wang W, et al. Effect of triptolide on malignant peripheral nerve sheath tumours in vitro and in vivo. J Int Med Res. 2012;40(6):2284–94.
54. Holtkamp N, et al. EGFR and erbB2 in malignant peripheral nerve sheath tumors and implications for targeted therapy. Neuro Oncol. 2008;10(6):946–57.
55. Hakozaki M, et al. Establishment and characterization of a novel human malignant peripheral nerve sheath tumor cell line, FMS-1, that overexpresses epidermal growth factor receptor and cyclooxygenase-2. Virchows Arch. 2009;455(6):517–26.
56. Torres KE, et al. Activated MET is a molecular prognosticator and potential therapeutic target for malignant peripheral nerve sheath tumors. Clin Cancer Res. 2011;17(12):3943–55.

57. Sun D, Tainsky MA, Haddad R. Oncogene mutation survey in MPNST cell lines enhances the dominant role of hyperactive Ras in NF1 associated pro-survival and malignancy. Transl Oncogenomics. 2012;5:1–7.
58. Demestre M, et al. Effects of pigment epithelium derived factor (PEDF) on malignant peripheral nerve sheath tumours (MPNSTs). J Neurooncol. 2013;115(3):391–9.
59. Byer SJ, et al. Tamoxifen inhibits malignant peripheral nerve sheath tumor growth in an estrogen receptor-independent manner. Neuro Oncol. 2011;13(1):28–41.
60. Thomas LE, et al. Evaluation of copy number variation and gene expression in neurofibromatosis type-1-associated malignant peripheral nerve sheath tumours. Hum Genomics. 2015;9(1):3.
61. Wang J, et al. Combined inhibition of SHP2 and MEK is effective in models of NF1-deficient malignant peripheral nerve sheath tumors. Cancer Res. 2020;80(23):5367–79.
62. Castellsague J, et al. Comprehensive establishment and characterization of orthoxenograft mouse models of malignant peripheral nerve sheath tumors for personalized medicine. EMBO Mol Med. 2015;7(5):608–27.
63. De Raedt T, et al. PRC2 loss amplifies Ras-driven transcription and confers sensitivity to BRD4-based therapies. Nature. 2014;514(7521):247–51.
64. Lee W, et al. PRC2 is recurrently inactivated through EED or SUZ12 loss in malignant peripheral nerve sheath tumors. Nat Genet. 2014;46(11):1227–32.
65. Ramkissoon A, et al. Targeted inhibition of the dual specificity phosphatases DUSP1 and DUSP6 suppress MPNST growth via JNK. Clin Cancer Res. 2019;25(13):4117–27.
66. Gregorian C, et al. PTEN dosage is essential for neurofibroma development and malignant transformation. Proc Natl Acad Sci U S A. 2009;106(46):19479–84.
67. Keng VW, et al. Conditional inactivation of Pten with EGFR overexpression in Schwann cells models sporadic MPNST. Sarcoma. 2012;2012:620834.
68. Vogel KS, et al. Mouse tumor model for neurofibromatosis type 1. Science. 1999;286(5447):2176–9.
69. Cichowski K, et al. Mouse models of tumor development in neurofibromatosis type 1. Science. 1999;286(5447):2172–6.
70. Mertens F, et al. Cytogenetic findings in malignant peripheral nerve sheath tumors. Int J Cancer. 1995;61(6):793–8.
71. Rhodes SD, et al. Cdkn2a (Arf) loss drives NF1-associated atypical neurofibroma and malignant transformation. Hum Mol Genet. 2019;28(16):2752–62.
72. Kazmi SJ, et al. Transgenic mice overexpressing neuregulin-1 model neurofibroma-malignant peripheral nerve sheath tumor progression and implicate specific chromosomal copy number variations in tumorigenesis. Am J Pathol. 2013;182(3):646–67.
73. Keng VW, et al. PTEN and NF1 inactivation in Schwann cells produces a severe phenotype in the peripheral nervous system that promotes the development and malignant progression of peripheral nerve sheath tumors. Cancer Res. 2012;72(13):3405–13.
74. Wu J, et al. EGFR-STAT3 signaling promotes formation of malignant peripheral nerve sheath tumors. Oncogene. 2014;33(2):173–80.
75. Rahrmann EP, et al. Trp53 haploinsufficiency modifies EGFR-driven peripheral nerve sheath tumorigenesis. Am J Pathol. 2014;184(7):2082–98.
76. Dodd RD, et al. NF1 deletion generates multiple subtypes of soft-tissue sarcoma that respond to MEK inhibition. Mol Cancer Ther. 2013;12(9):1906–17.
77. Rahrmann EP, et al. Forward genetic screen for malignant peripheral nerve sheath tumor formation identifies new genes and pathways driving tumorigenesis. Nat Genet. 2013;45(7):756–66.
78. The I, et al. Rescue of a Drosophila NF1 mutant phenotype by protein kinase A. Science. 1997;276(5313):791–4.
79. Gouzi JY, et al. The receptor tyrosine kinase Alk controls neurofibromin functions in Drosophila growth and learning. PLoS Genet. 2011;7(9):e1002281.
80. Walker JA, et al. Genetic and functional studies implicate synaptic overgrowth and ring gland cAMP/PKA signaling defects in the Drosophila melanogaster neurofibromatosis-1 growth deficiency. PLoS Genet. 2013;9(11):e1003958.
81. van der Voet M, et al. ADHD-associated dopamine transporter, latrophilin and neurofibromin share a dopamine-related locomotor signature in Drosophila. Mol Psychiatry. 2016;21(4):565–73.
82. King LB, et al. Developmental loss of neurofibromin across distributed neuronal circuits drives excessive grooming in Drosophila. PLoS Genet. 2020;16(7):e1008920.
83. Williams JA, et al. A circadian output in Drosophila mediated by neurofibromatosis-1 and Ras/MAPK. Science. 2001;293(5538):2251–6.
84. Tong J, et al. Neurofibromin regulates G protein-stimulated adenylyl cyclase activity. Nat Neurosci. 2002;5(2):95–6.
85. MacInnes AW, et al. Loss of p53 synthesis in zebrafish tumors with ribosomal protein gene mutations. Proc Natl Acad Sci U S A. 2008;105(30):10408–13.
86. Shin J, et al. Zebrafish neurofibromatosis type 1 genes have redundant functions in tumorigenesis and embryonic development. Dis Model Mech. 2012;5(6):881–94.
87. Oppel F, et al. Loss of atrx cooperates with p53-deficiency to promote the development of sarcomas and other malignancies. PLoS Genet. 2019;15(4):e1008039.
88. Astone M, et al. A GFP-tagged gross deletion on chromosome 1 causes malignant peripheral nerve sheath tumors and carcinomas in zebrafish. PLoS One. 2015;10(12):e0145178.

MRI in the Diagnosis of PNST: State-of-the-Art and Recent Advancements

6

Francesca Serpi, Domenico Albano, Salvatore Gitto, Carmelo Messina, and Luca Maria Sconfienza

6.1 Introduction

Peripheral nerve sheath tumors (PNSTs) constitute an array of different lesions [1], ranging from benign to malignant PNSTs (MPNSTs). PNSTs may arise sporadically, with an incidence of around 1 in 100,000 individuals per year, or associated with neurocutaneous syndromes such as neurofibromatosis type 1 (NF1), neurofibromatosis type 2 (NF2), and schwannomatosis [1, 2]. Malignant transformation is particularly seen in NF1 tumors.

F. Serpi
Unit of Diagnostic and Interventional Radiology, IRCCS Istituto Ortopedico Galeazzi, Milan, Italy

D. Albano
Unit of Diagnostic and Interventional Radiology, IRCCS Istituto Ortopedico Galeazzi, Milan, Italy

Department of Biomedical, Surgical and Dental Sciences, Università degli Studi di Milano, Milan, Italy

S. Gitto · L. M. Sconfienza (✉)
Dipartimento di Scienze Biomediche per la Salute, Università degli Studi di Milano, Milan, Italy

Unit of Diagnostic and Interventional Radiology, IRCCS Istituto Ortopedico Galeazzi, Milan, Italy
e-mail: io@lucasconfienza.it

C. Messina
Dipartimento di Scienze Biomediche per la Salute, Università degli Studi di Milano, Milan, Italy

Unit of Radiology, ASST Gaetano Pini - CTO, Milan, Italy

Ultrasound is a useful, cost-effective technique for nerve evaluation, with good performance and high spatial resolution, especially for peripheral and superficial nerves.

Computed tomography (CT) has a marginal role in PNSTs, usually represented as mass-like, with low attenuation reflecting the contest of lipid of Schwann cells myelin, with variable post-contrast enhancement.

Magnetic resonance imaging (MRI) is the modality of choice for the noninvasive evaluation of PNSTs and has been used for presurgical planning and imaging-guided biopsy. In fact, MRI can establish a nerve-tumor relationship and therefore exclude other diagnoses, as well as define the size of the tumor and its association with adjacent structures. Moreover, MRI has been suggested as a screening tool in individuals with NF1 for the early diagnosis of malignant degeneration in preexisting PNSTs to avoid unnecessary biopsy [3].

Despite advances in MRI imaging, there are still no clear imaging criteria or a single feature which enable distinguishing between different PNSTs, because characteristics overlap, especially between neurofibromas and schwannomas. With respect to differentiating benign PNSTs and malignant PNSTs, it is suggested that if the characteristic findings described for benign PSNT are not present, then an MPNST or other soft tissue neoplasm cannot be excluded, and a carefully planned biopsy should be obtained [2] (Table 6.1).

I. G. Vetrano, V. Nazzi (eds.), *Peripheral Nerve Sheath Tumors*,
https://doi.org/10.1007/978-3-031-76382-3_6

Table 6.1 Major imaging features at the imaging of benign and malignant PNSTs

Images	BPNSTs	MPNSTs
T1-weighted images	Isointense to muscle Homogeneous	Isointense to muscle Homogeneous or heterogeneous
T2-weighted images	Hyperintense Target sign Fat-split sign Fascicular sign (more common in neurofibroma than schwannoma)	Hyperintense, heterogeneous No target sign No fat-split sign or rim incomplete No fascicular sign
Contrast enhancement	Variable (targetoid, peripheral or heterogenous) Dynamic delayed contrast enhancement	Variable (heterogenous) Dynamic early contrast enhancement
Size/morphology	<5 cm Neurofibroma: encapsulated and centrally located Schwannoma: capsulated and eccentrically located	>5 cm
Margin	Well-defined	Well or ill-defined
Invasion	Absent	Can be present
Necrosis/cystic changes	Uncommon	Common
Hemorrhage	Uncommon (usually ancient schwannoma)	Common
Calcification	Uncommon	Associated to necrosis
Perilesional edema	Absent	Common
DWI/ADC and DTI	Minimum ADC values usually >1.0–1.1 × 10^{-3}	Minimum ADC values usually <1.0–1.2 × 10^{-3}
Tractography	Near normal or partial tract disruption	Partial or complete tract disruption
Spectroscopy	Trimethylamine fraction usually <50%	Trimethylamine fraction usually >50%

6.2 Benign Peripheral Nerve Sheath Tumors

Among benign PNSTs, the majority are represented by neurofibromas, schwannomas, and perineuriomas.

6.2.1 Neurofibroma

Neurofibromas represent around 5% of all benign soft tissue tumors. They mostly affect people 20–30 years old, with no sex predilection [4]. They occur both sporadically (95%) or related to neurofibromatosis type 1 (NF1) [5]. On pathologic examination, these lesions are intimately imbricated with their nerve of origin and are not capsulated, unlike schwannomas. Therefore, complete surgical excision includes at least some parts of the involved nerves.

There are three types of neurofibromas: solitary, diffuse, or plexiform.

Solitary neurofibromas (also known as localized, sporadic, or nodular) represent the most common type (90%), mainly encountered in patients without NF1 [2]. They are usually slow-growing and superficial, affecting cutaneous or subcutaneous small nerves; however, deep-seated locations with the involvement of larger nerves can also occur. Symptoms are variable, mostly connected with deep-seated lesions rather than superficial neurofibromas. In concomitance with NF1, they tend to be larger, multiple, and deeper in location, with an increased risk of malignancy transformation [2, 4].

Diffuse neurofibromas are an uncommon but distinct entity that primarily affects children and young adults and most frequently involves the head and neck subcutaneous fat [2]. They present with a plaque-like elevation of the skin with thickening of the subcutaneous fat.

Plexiform neurofibromas are typically related to NF1. They manifest as a diffuse tumor with tortuous expansion along the branches of multiple fascicles, nerves, or even plexuses, spreading the axons, with a gross appearance defined as a "bag of worms." In NF1, the most common location is the trunk, including the paraspinal region (41%), followed by the neck/upper trunk (24%), and the extremities (17%) [2]. They may be present at birth or become apparent later in life, usually preceding cutaneous neurofibromas [2, 6]. While the growth of plexiform lesions is normal in childhood, rapid growth in adults should raise concern for possible malignant transformation [1, 7]. Plexiform neurofibroma may be associated with massive and disfiguring widening of an extremity (*elephantiasis neuromatosa*), a condition associated with enlargement of the affected extremity, hypertrophy of bone related to chronic hyperemia, and redundant skin [8, 9].

On MRI, **solitary neurofibromas** appear as unencapsulated fusiform or round masses, typically <5 cm in size, with intermediate intensity on T1w images and heterogeneous high signal intensity on T2w images. After contrast injection, small lesions have intense and homogeneous or targetoid contrast enhancement, whereas larger ones can show predominantly peripheral, central, or heterogeneous nodular enhancement [2].

Other characteristics suggestive but not pathognomonic of neurogenic origin are:

- *Tail sign*: fusiform shape related to the growth along the course of the affected nerve. It's easier to be seen in lesions affecting larger nerves than small or superficial lesions (Fig. 6.1).
- *Split-fat sign*: as neurovascular bundles are generally surrounded by fat, slow-growing lesions are separated from other structures by a complete thin rim of fat, which is better depicted on T1w images. This is not specific for PNSTs but is suggestive that the tumor originates in the intermuscular space, in which neurogenic lesions are one of the most frequent causes. This sign is more commonly associated with neurofibromas than schwannomas or other slow-growing lesions. In malignant PNSTs, the rim is typically incom-

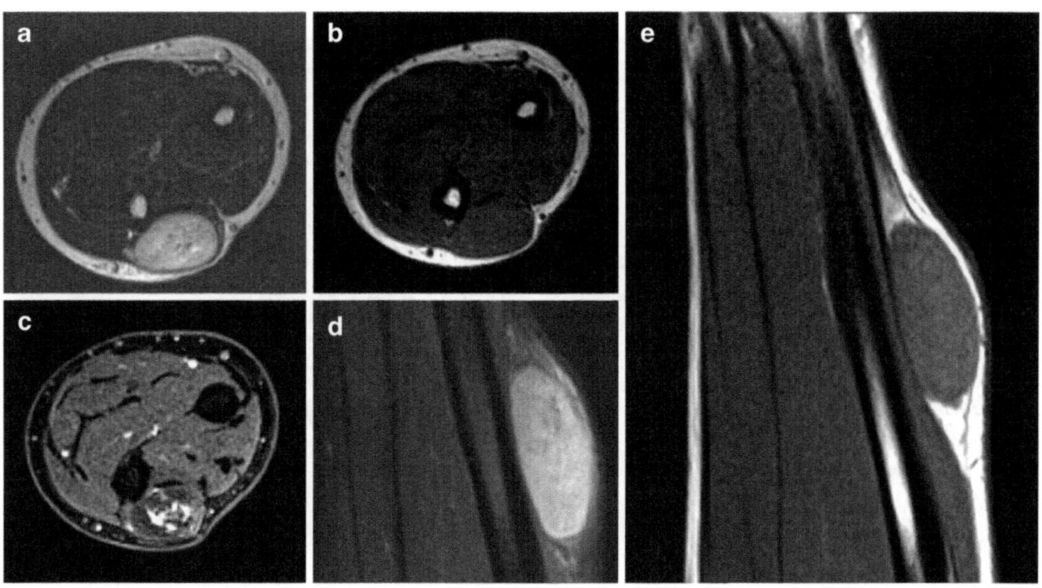

Fig. 6.1 Dorsal forearm schwannoma with hyperintense heterogeneous signal on axial T2w (**a**) and isointense on T1w (**b**), with inhomogeneous contrast uptake on axial T1w fs (**c**). (**d**, **e**) Sagittal PDw fs and T1w. It has a typical fusiform ovoid shape related to the growth along the course of the affected nerve, resembling a tail coming out of the tumor (tail sign). On sagittal T1w (**e**) it is surrounded by a small amount of fat (split-fat sign)

plete, reflecting the growing infiltrative pattern (Fig. 6.2).
- *Target sign* refers to a central low-intensity area surrounded by a high-intensity rim on T2w images. This pattern reflects the histological composition of myxomatous tissue surrounding the fibro-collagenous core. Rarely, on T1w images, it may show a reverse intensity pattern. After contrast administration, enhancement may show a reverse or a similar T2 pattern. Usually, the target sign is typical of deep-seated lesions; superficial neurofibromas show a more homogeneous pattern instead. Target signs can also be found in schwannomas and MNPSTs (Figs. 6.3 and 6.4).

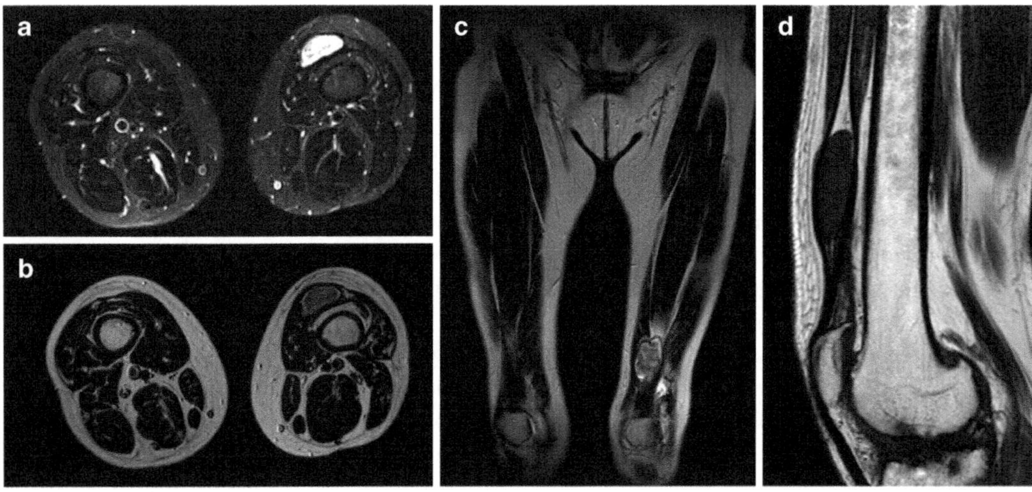

Fig. 6.2 Schwannoma at the level of the left quadriceps tendon, between the fibers of the rectus femoris and vastus intermedius, represented by a capsulated lesion with hyperintense signal on axial STIR (**a**) and axial and coronal (**b**, **c**) T2w image. On sagittal T1w image (**d**), it is isointense to muscle, surrounded by a small amount of fat (split-fat sign)

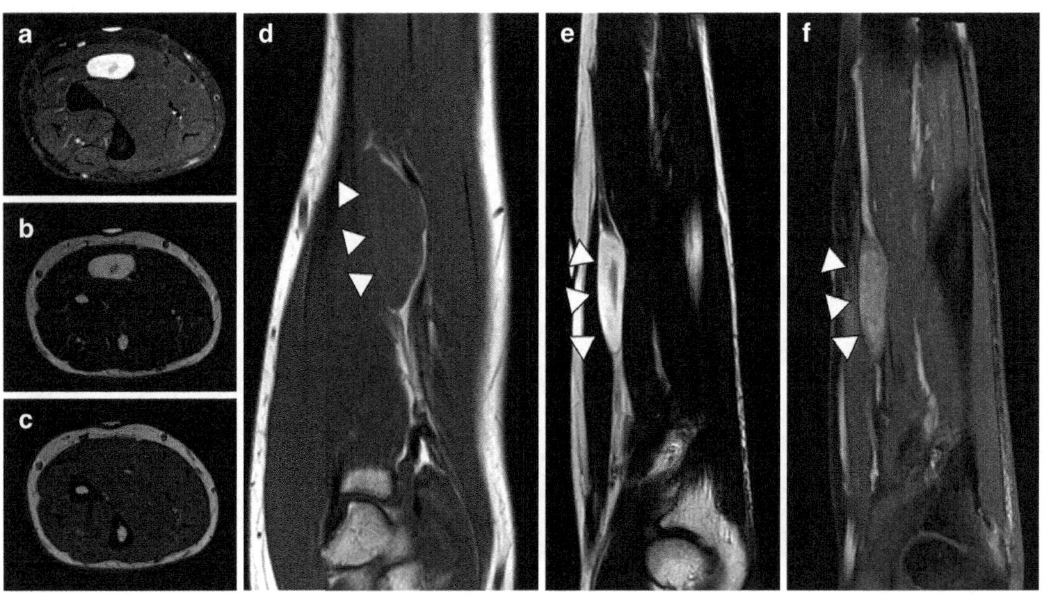

Fig. 6.3 Forearm neurofibroma. Ovoid lesion with high signal intensity on axial PDw fs (**a**) and T2w (**b**), with a central hypointense area (also known as target sign). (**c**) Axial T1w. The lesion is isointense to the muscle. On coronal T1w image (**d**), it has a fusiform ovoid shape, following the course of the neurovascular bundle (tail sign) separated from muscle by a thin rim of fat (split-fat sign). Sagittal T2w (**e**) and sagittal T1w fs post-contrast (**f**) with slight heterogeneous enhancement

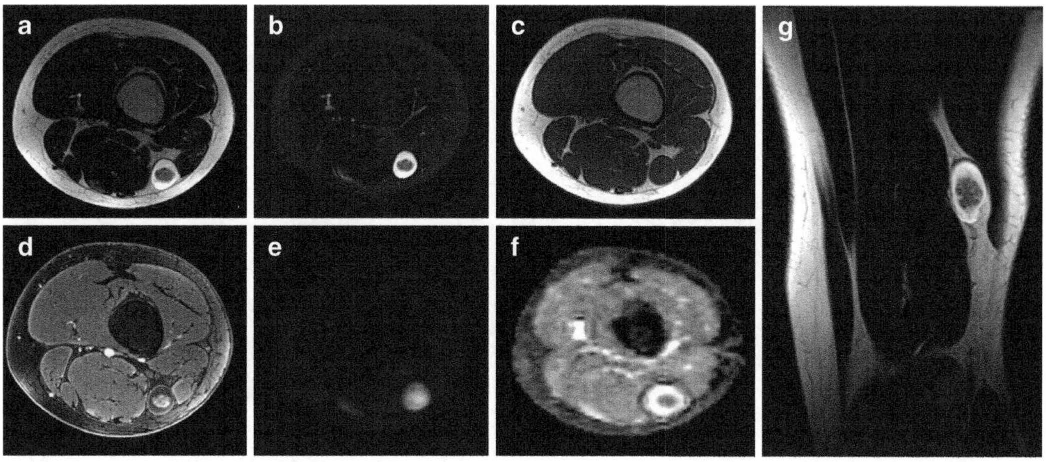

Fig. 6.4 Poster leg schwannoma with a target sign on axial T2w (**a**) and STIR (**b**) images, isointense to muscle on axial T1w (**c**) with a complete split-fat sign. On DWI (**e**) and ADC map (**f**) it shows a targetoid restriction with similar peripheral enhancement after contrast media on axial T1w fs (**d**). The lesion is fusiform on coronal T2w (**g**), following the neurovascular bundle (tail sign)

- *Fascicular sign*: this sign is the reflection of the physiological nerve structure of fascicular bundles surrounded by different layers of connective tissue, respectively, endoneurium, perineurium, and epineurium. This configuration on axial T2w images appears as multiple ringlike hypointense structures within the lesion.
- *Muscle denervation*: MRI features of denervation of the muscles innervated by the involved nerve are represented as an edema-like signal (high signal on fluid-sensitive sequences), fatty infiltration (high T1 and T2 signal), or/ and atrophy (loss of muscle belly volume). Sometimes these findings are subtle, especially at the beginning, and require comparison with the normal contralateral side.

Diffuse neurofibromas demonstrate similar signal and enhancement characteristics as solitary lesions, but they are often ill-defined and spread extensively along connective tissue septa and between adipose tissue, involving the subcutaneous fat.

Plexiform neurofibromas mainly appear as a tortuous multinodular lesion that involves multiple nerve branches, giving the appearance of a serpentine "bag of worms." Moreover, plexiform neurofibromas present three different types of growing patterns: superficial, displacing, and invasive. The last two forms are related to MPNSTs.

- Superficial lesions have cutaneous or subcutaneous localization, with asymmetric extension and no clear demarcated borders. They are isointense to the skin on T1w images, homogeneously hyperintense on fluid-sensitive images, with intense and homogeneous enhancement after contrast injection. They can mimic venous malformation on conventional MRI; sometimes, MR arthrography or Doppler ultrasound can be necessary for differentiation (Fig. 6.5).
- Displacing lesions develop along primary nerves, feature multinodular, smoothly defined borders, and compress adjacent structures.
- Invasive lesions have more aggressive behavior. They can penetrate muscle, fasciae, joints, and surrounding tissues.

6.2.2 Schwannoma

Schwannomas (also known as neurilemmoma, neurinoma, or perineural fibroblastoma) are encapsulated benign sheath tumors composed of Schwann cells. They can develop at any age, with a peak incidence in the fourth to sixth decade, but they are rare in the pediatric age group [10]. They

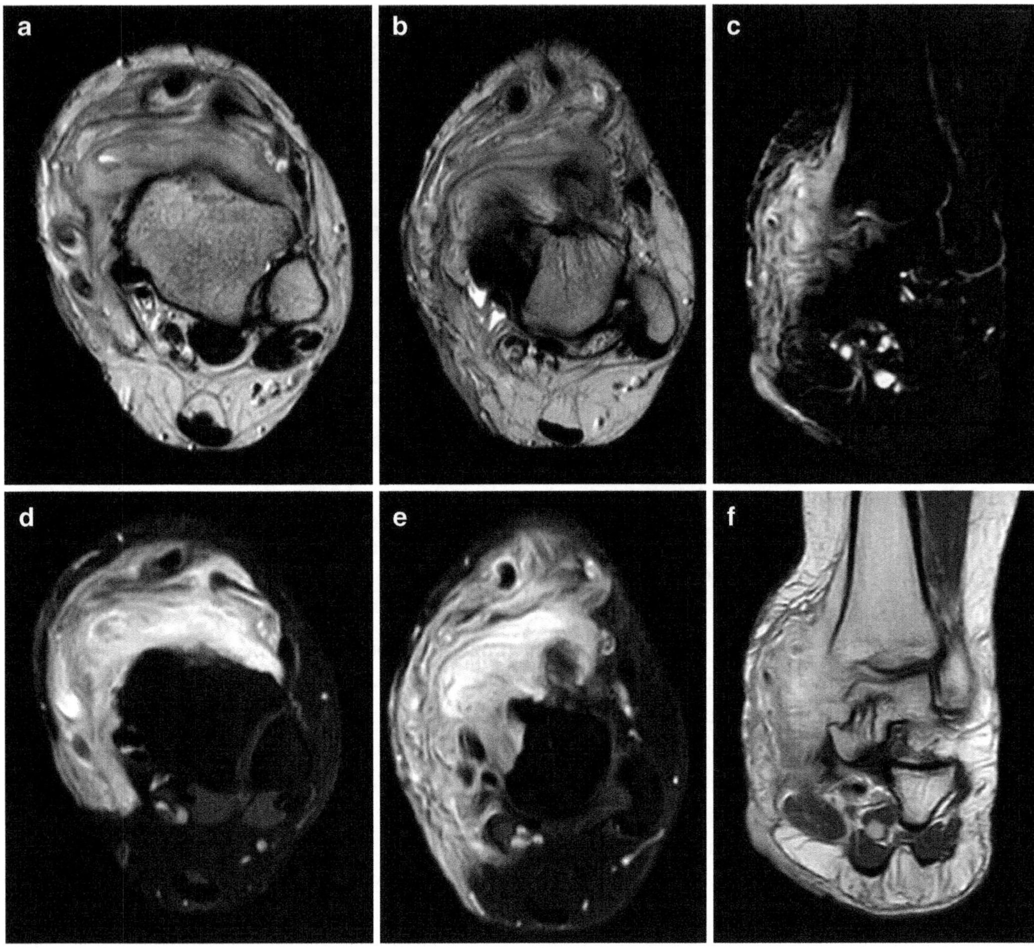

Fig. 6.5 Peripheral superficial neurofibroma of the antero-medial side of the ankle. Opposite to central neurofibroma, this superficial neurofibroma has an inhomogeneous appearance and is scarcely capsulated with infiltrating appearance but without bone erosion or muscle and tendons infiltrations. It is hyperintense on axial T2w (**a**, **b**) and hyperintense on coronal STIR (**c**) with diffuse contrast uptake on axial T1w fs (**d**, **e**) and coronal T1w (**f**)

usually present as a solitary slow-growing mass, multiple in schwannomatosis and occasionally in NF2 [2]. In NF2, they can be plexiform, usually subcutaneous, mostly in the trunk or the upper extremities. They frequently involved the head and neck areas and the major nerve trunks. They usually appear on the flexor aspects of the limbs, especially near the elbow, wrist, and knee [11]. The spinal roots, the cervical plexus, the vagus, the peroneal, and the ulnar nerves are the most commonly affected [11]. Relatively few Schwannomas appear on the trunk, in contrast with neurofibroma in NF1. Pain and symptoms are more common in larger lesions than small ones, which are usually asymptomatic. They can rarely undergo malignant degeneration but may be locally aggressive and tend to recur.

At MRI, they share common findings with solitary neurofibromas, such as an isointense signal on T1w and hyperintense signal on T2w images, split-fat sign, fascicular sign, and sometimes target sign (Figs. 6.6, 6.7, and 6.8). After contrast administration, small lesions have intense homogeneous enhancement, while larger lesions may have heterogeneous enhancement (Figs. 6.9 and 6.10). However,

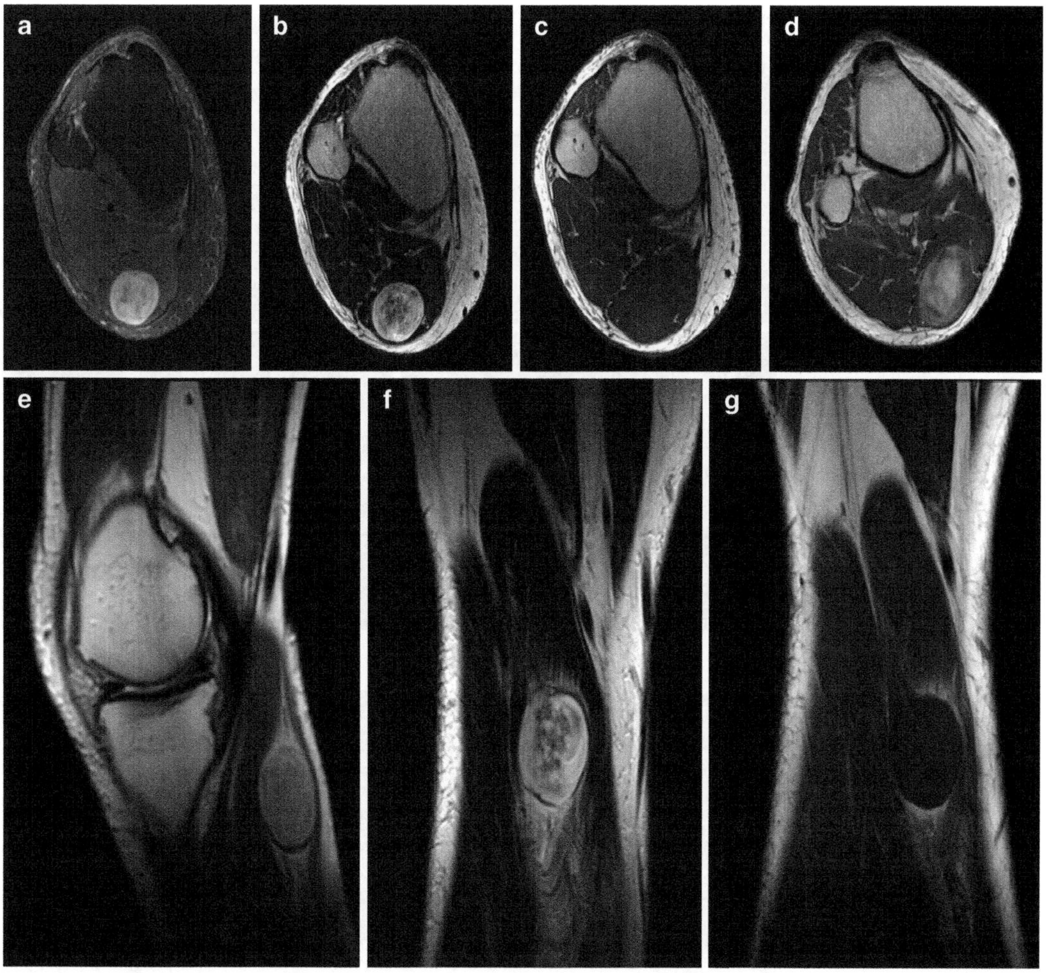

Fig. 6.6 Posterior knee schwannoma. It has an inhomogeneous hyperintense signal on axial PDw fs (**a**) and T2w (**b**) and isointense to muscle on T1w (**c**). Contrast enhancement after gadolinium on axial T1w is heterogeneous (**d**). Sagittal PDw (**e**) and coronal T2w (**f**) show a capsulated fusiform lesion with ovoid shape, in continuity with the nerve fascicles (tail-sign) and heterogeneous high signal. Coronal T1w (**g**) shows an isointense lesion with a split-fat sign

unlike neurofibromas, they are capsulated. Therefore, a thin hypointense rim (the epineurium) can be identified on T2w images, especially if the lesion is surrounded by fat (Figs. 6.11 and 6.12) [11]. According to the nerve of origin, schwannomas are eccentrically positioned, whereas a centrally located mass suggests a neurofibroma [11]. Moreover, heterogeneous appearance due to hemorrhagic or cystic degeneration or calcification is more common in schwannomas than in neurofibromas (Figs. 6.13, 6.14, and 6.15).

Because of their eccentric location to the original nerve, the nerve is usually spared during surgical excision, compared to neurofibromas.

6.2.3 Perineurioma

Perineuriomas are underrecognized benign tumors exclusively derived from perineurial cells. They are arranged in concentric layers around the central axon and Schwann cells, creating an onion-bulb appearance on the axial histo-

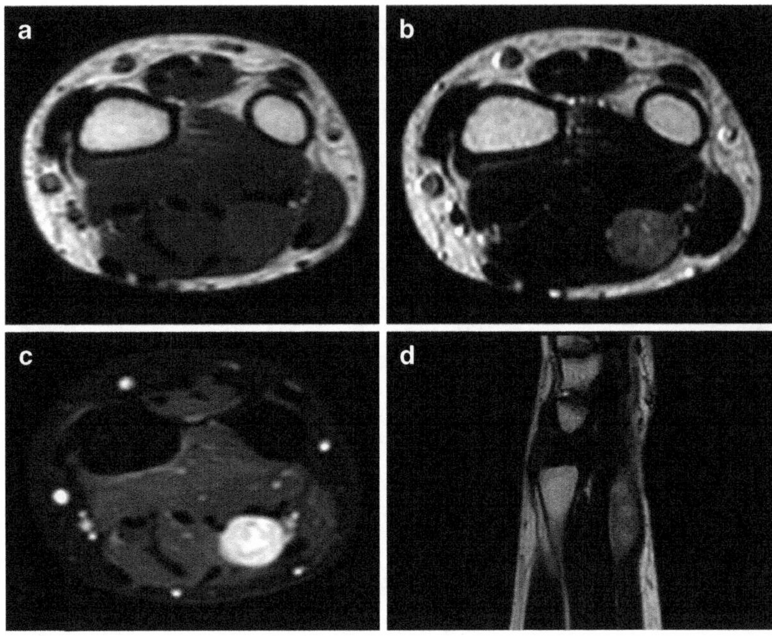

Fig. 6.7 Volar wrist schwannoma is represented by as a well-circumscribed lesion between flexor muscles with an isointense signal to them on axial T1w (**a**) and slightly hyperintense on axial T2w (**b**), with homogeneous contrast enhancement on axial T1w fs. On sagittal T2w (**d**) the lesion has a fusiform shape

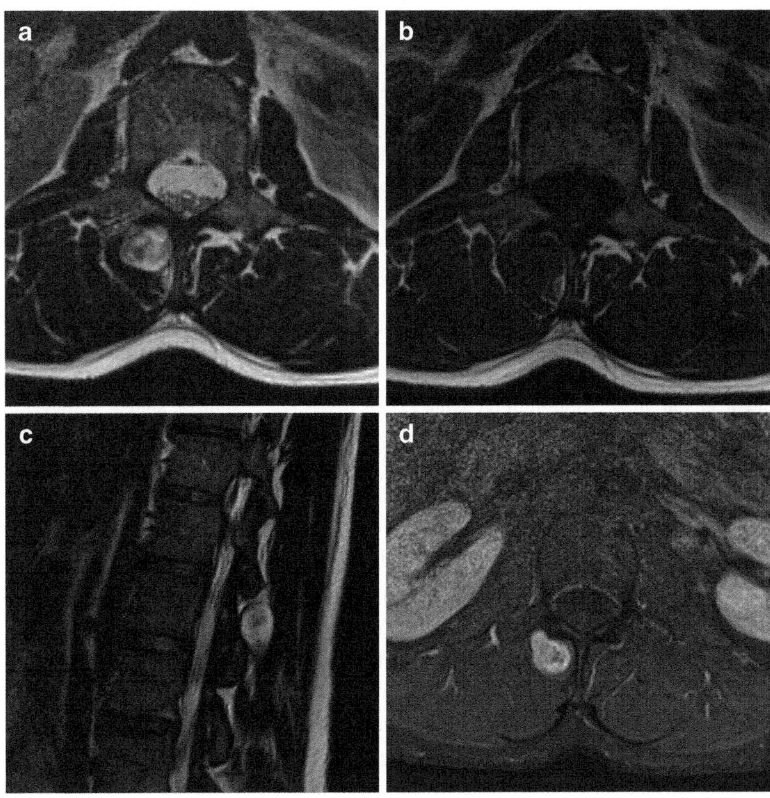

Fig. 6.8 Posterior paraspinal schwannoma with a heterogeneous hyperintense signal on axial T2w (**a**) and isointense signal on axial T1w (**b**), with fusiform shape on coronal T2w (**c**). Axial T1w fs post-contrast (**d**) shows vivid enhancement

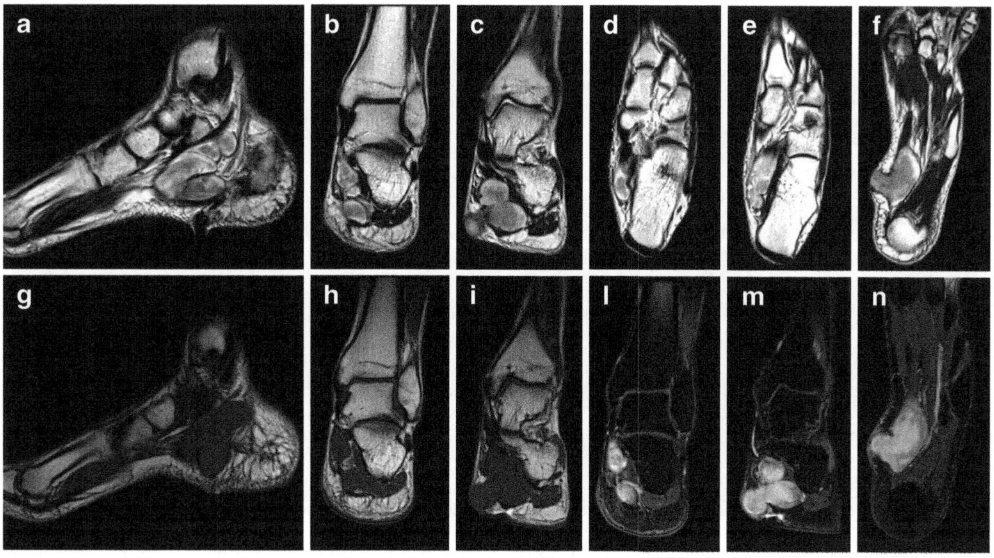

Fig. 6.9 Schwannoma of the tibial nerve. The lesion has a polylobate capsulated appearance, following the anatomical course of the ankle's tibial nerve branches at the posteromedial side, with extension on the planter aspect of the foot. It has a hyperintense heterogeneous signal on sagittal (**a**) coronal (**b, c**) and axial (**d–f**) T2w images. It is isointense to muscle on sagittal (**g**) and coronal (**h, i**) T1w images. After contrast injection, it has diffuse enhancement on T1w fs coronal (**j–m**) and axial plane (**n**)

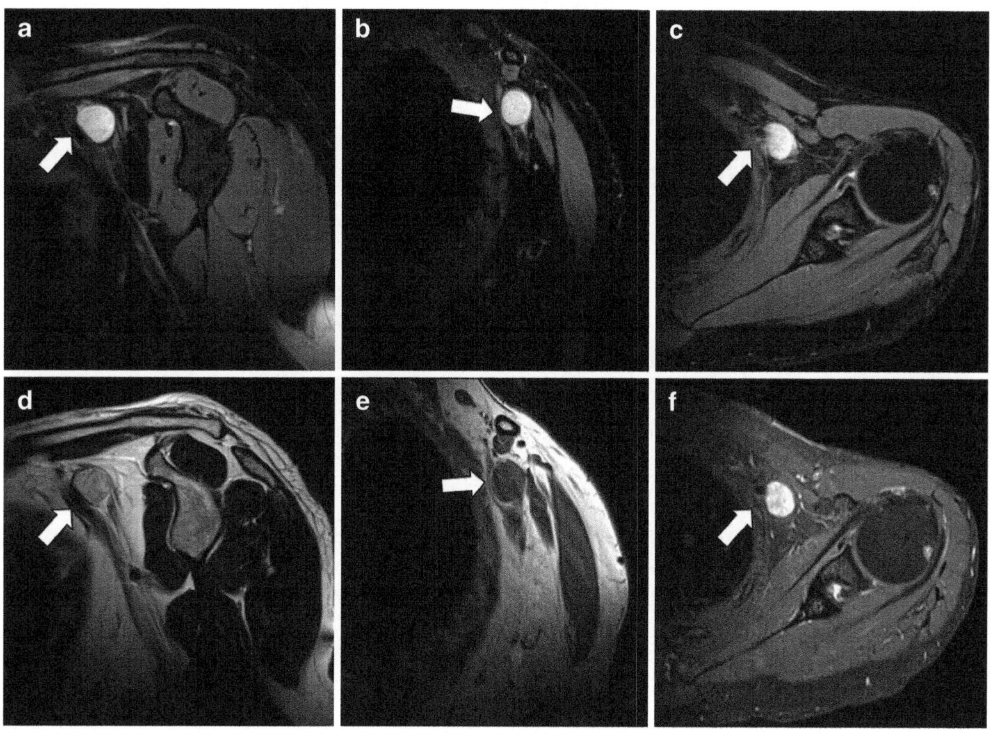

Fig. 6.10 Shoulder schwannoma (arrows) is derived from branches of the brachial plexus. A rounded well-capsulated lesion with high signal on PD fs sagittal (**a**) coronal (**b**) and axial (**c**) plane. Heterogeneous high signal on sagittal T2w (**d**) and isointense signal on coronal T1w (**e**). After contrast administration, it has vivid homogeneous enhancement on axial T1 fs (**f**)

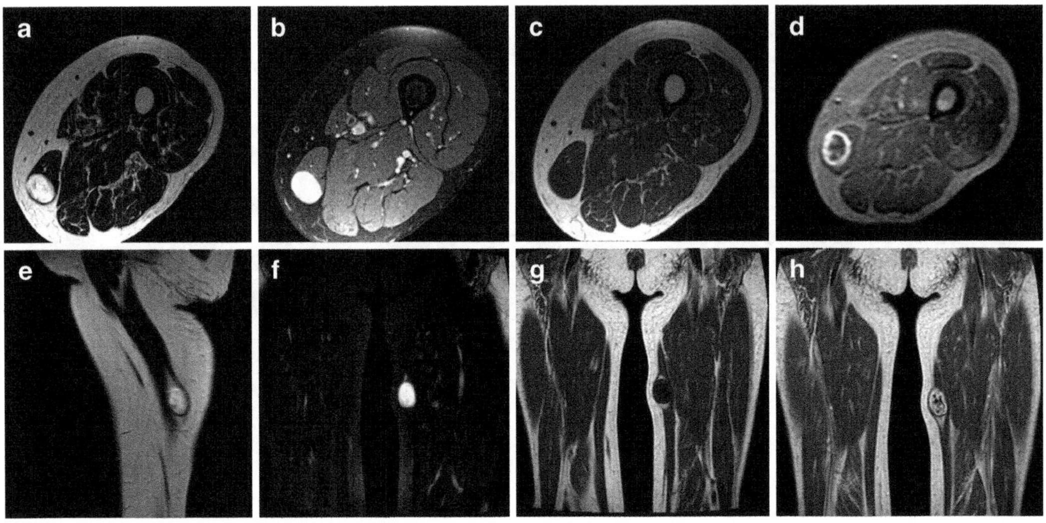

Fig. 6.11 Intramuscular schwannoma of the gracile muscle represented as a well-circumscribed and capsulated lesion with a heterogeneous hyperintense signal on axial (**a**) and sagittal (**e**) T2 and high signal on axial PDw fs (**b**), isointense to muscle on axial T1w (**c**). After contrast administration, it shows peripheric contrast enhancement on axial T1w fs (**d**). Coronal STIR (**f**) and T1w (**g**) have fusiform shapes with fat-split signs on T1w. (**h**) Coronal T1w post-contrast

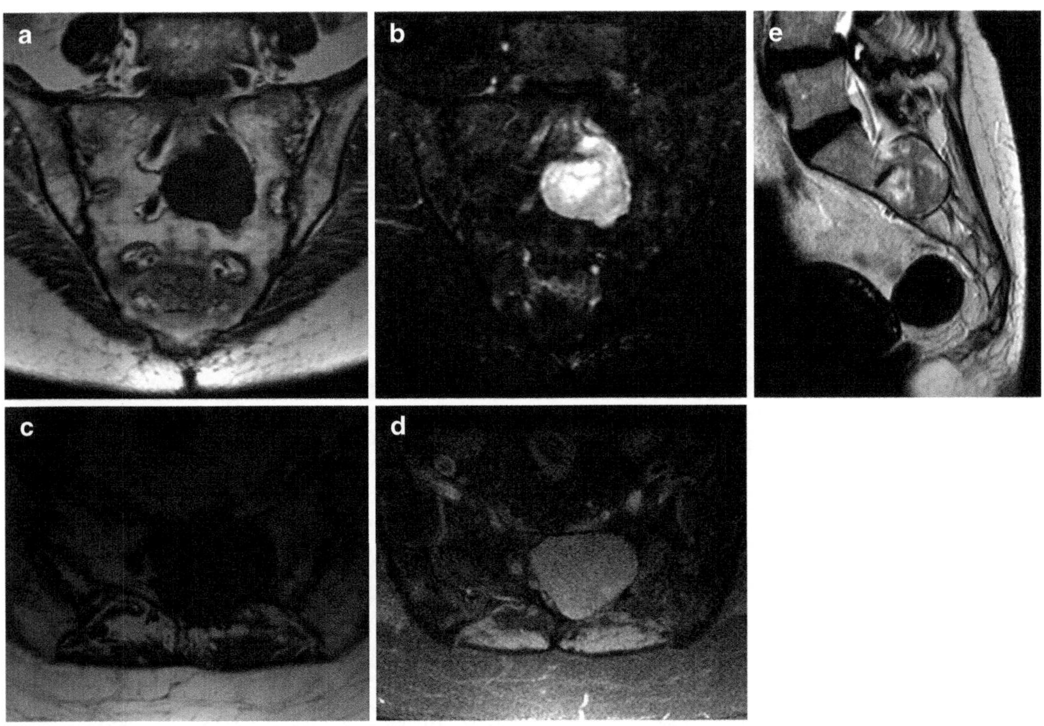

Fig. 6.12 Sacral schwannoma. Coronal T1w (**a**) and axial T1w (**c**) with low homogeneous intensity. Coronal STIR (**b**) with a heterogeneous high signal. Sagittal T2w (**e**) shows a capsulated lesion with a thin hypointense rim and heterogeneous high signal. Axial T1w fs post-contrast (**d**) shows homogeneous vivid enhancement

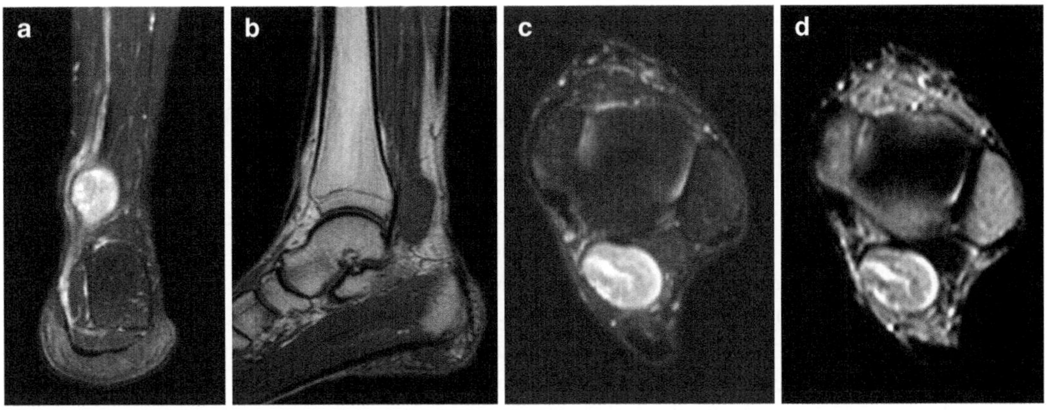

Fig. 6.13 Ankle schwannoma. A focal lesion with ovoid structure, following the course of the vasculonervous bundle at the posteromedial side of the ankle with hyperintense signal on coronal PDw fs (**a**), isointense on sagittal T1w (**b**). On axial PDw fs (**c**) and axial T2w (**d**), it has a heterogenous high-intensity signal with internal cystic components, suggestive of an old schwannoma

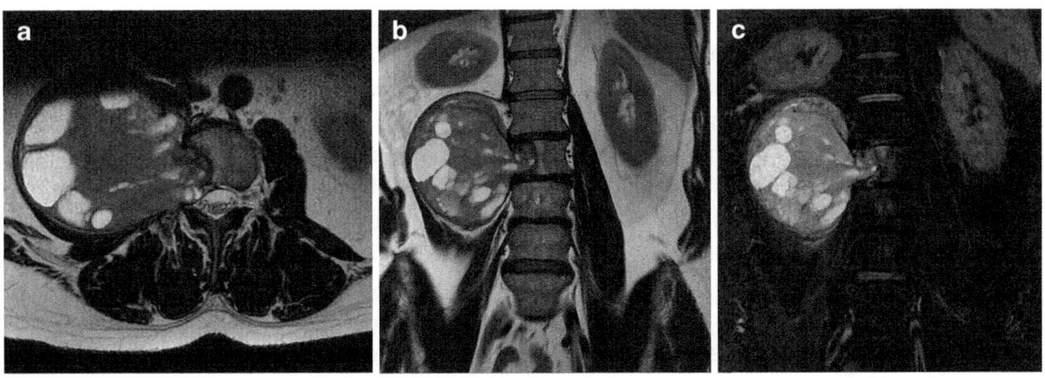

Fig. 6.14 Ancient schwannoma. Right paraspinal capsulated mass at the level of L2, with multiple cystic internal areas, with contiguous bone erosion. (**a**) Axial T2w, (**b**) coronal T2w, and (**c**) coronal STIR

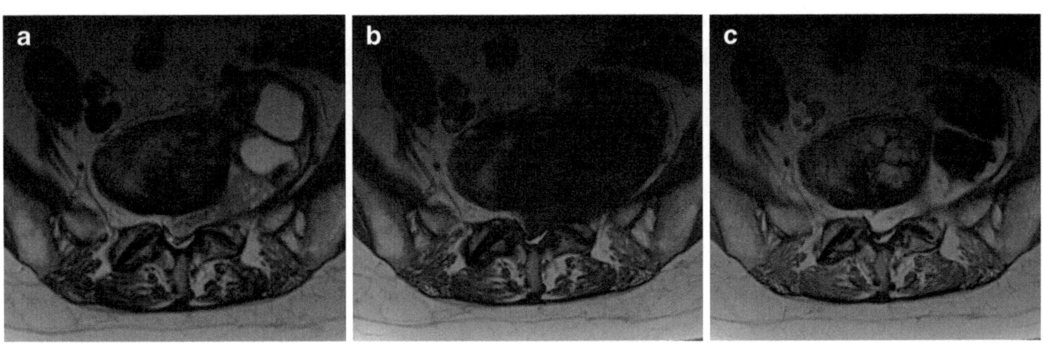

Fig. 6.15 Ancient schwannoma with inner cystic areas with hyperintense signal on axial T2w (**a**) and hypointense on axial T1w (**b**). After contrast injection, central areas don't enhance (**c**, axial T1w)

logic section. Based on the location, they are divided into intraneural and extraneural perineuriomas. *Intraneural perineuriomas* (also known as localized hypertrophic neuropathy) are restricted to the boundaries of the nerve and determine localized cylindrical or fusiform enlargement of the nerve, with variable extension from a few centimeters to more than 30 cm. *Extraneural perineuriomas* present as painless nodules in soft tissues e and skin [2].

At MRI, the nerve is typically increased in size, with a normal fascicular structure on axial images. Perineuriomas typically have low to intermediate signals on T1w images, heterogeneous signal on T2w images, and avid contrast enhancement.

The diagnosis is suggested by slowly progressive mononeuropathy at a young age, with a lack of tumor syndromes. Extraneural perineuriomas typically have a benign prognosis, and surgical resection with a margin free of neoplasm is curative. In contrast, for intraneural perineuriomas, the treatment of choice is resection of neoplasm with end-to-end grafting. However, despite extensive surgery, patients usually do not recover from symptoms.

6.3 Malignant Peripheral Nerve Sheath Tumors

Malignant peripheral nerve sheath tumors (MPNSTs) are spindle cell sarcomas arising from nerve or neurofibroma or demonstrating nerve tissue differentiation. They have also been referred to as malignant schwannomas, neurogenic sarcomas, malignant neurilemmomas, and neurofibrosarcomas [11]. MPNSTs account for 3–10% of all soft tissue sarcomas, with 15–70% occurring in patients with NF1 [2, 12, 13]. Among patients with NF1, 3–13% develop MPNSTs after a long latent period of 10–20 years, with a mean age of 30 years old [2, 14]. In patients with NF1, a sudden increase in the size of a stable neurofibroma should be viewed with great suspicion of malignant transformation and lead to immediate biopsy [11]. Around 11% of MPNSTs are secondary to radiation therapy, with a latent period of 10–20 years. MPNSTs usually involve large nerves, including the sciatic nerve, the brachial, and the lumbosacral plexus [5].

Patients present with pain and neurological symptoms of motor weakness and sensory deficits more frequently than patients with benign PNSTs [11]. They usually have a poor prognosis, with 5-year survival ranging from 34 to 60% [15, 16]. Local recurrence (22–42%) and distal metastasis (20–30%) are common, usually in the lungs, bone, pleura, and retroperitoneum. Lymph node metastasis is rarely seen [16].

On conventional MRI, they are usually heterogeneous with indistinct margins. The signal is heterogeneous on T2w images with hyperintense features and isointense to muscle on T1w images. T1 signal might also be high, suggesting the presence of hemorrhagic content (Fig. 6.16). Other features suggesting MPNSTs are large size (>5 cm), peripheral contrast enhancement, perilesional edema, intratumoral cystic areas, and lack of the target and fascicular sign (Figs. 6.17, 6.18, and 6.19) [2, 5]. Heterogeneous appearance with cystic-hemorrhagic changes can also be present in ancient larger schwannoma. On dynamic contrast-enhanced MRI, MPNSTs display early arterial enhancement, a rare feature in benign PNSTs [2, 5, 17, 18].

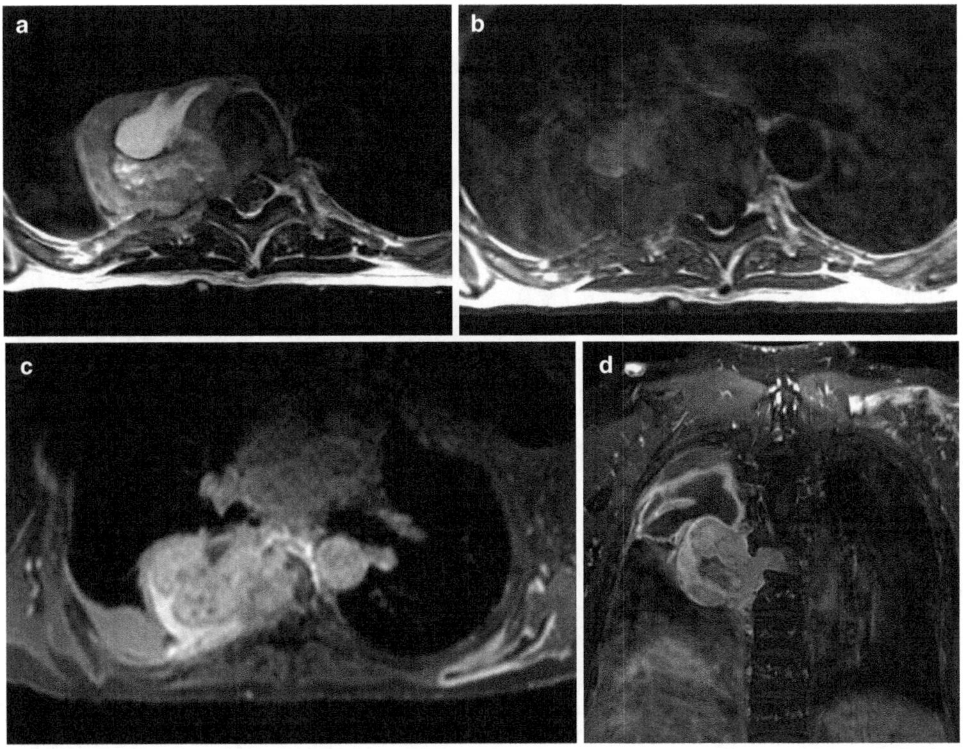

Fig. 6.16 Dorsal MPNST in a patient with NF1. Voluminous mass lesion at the right paraspinal side with a central colliquative-necrotic area with high signal intensity on axial T2w image (**a**) and bone erosion. Axial T1w (**b**), the central area, has a slight hyperintense signal, suggestive of some hemorrhagic components. After contrast administration (**c**, **d**), it shows enhancement of the peripheral part with a lack in the central necrotic area. On coronal T1w fs post-contrast (**d**), there is also right pleural effusion near the lesion

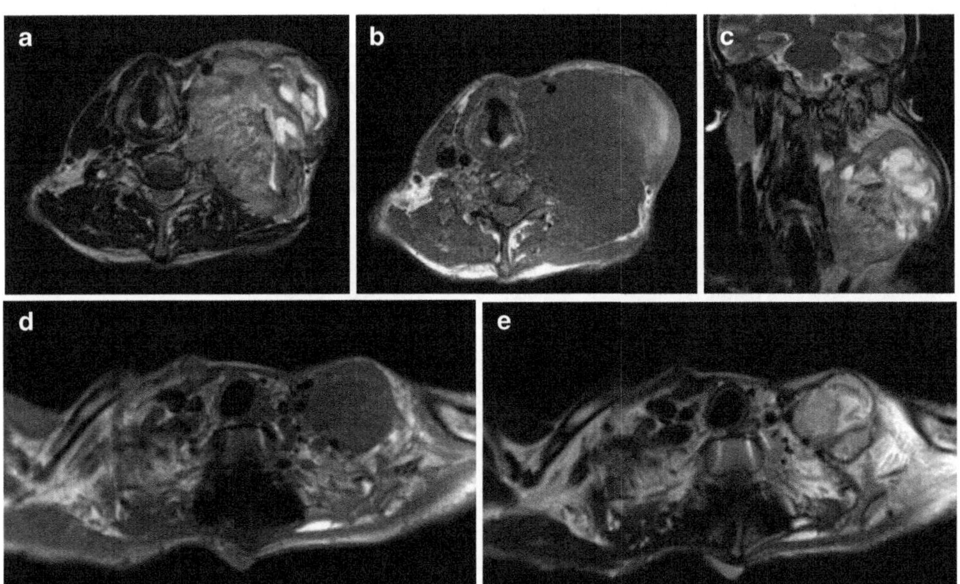

Fig. 6.17 Malignant transformation of a left cervical Schwannoma on axial (**a**) and coronal (**c**) T2w shows a large mass with a markedly heterogeneous signal. (**b**) Axial T1w image, with a heterogeneous isointense signal. After surgical resection, there was locoregional recurrence on axial T1w (**d**) and axial T2w (**e**)

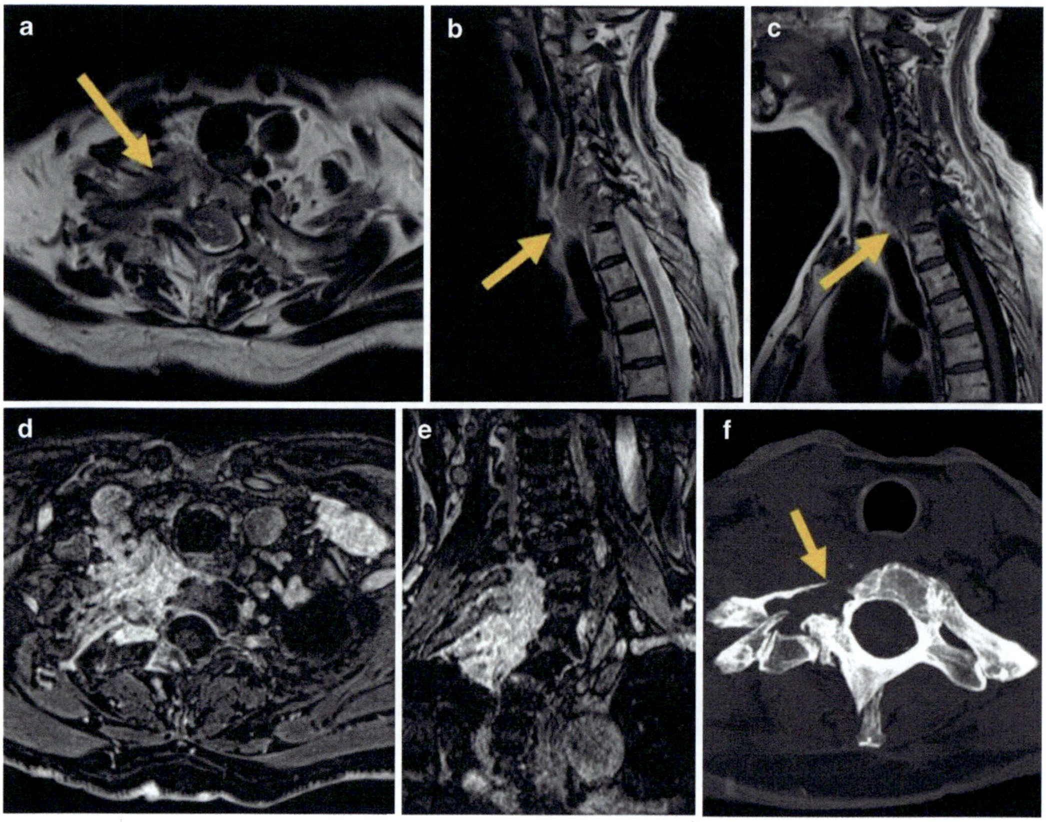

Fig. 6.18 Malignant PNST of C7 and D1 right nerves, with bone erosion. Axial T2w image (**a**), sagittal T2w (**b**), and sagittal T1w (**c**) show an inhomogeneous lesion with ill-defined margins and bone erosion (arrows). After contrast injection, it has vivid enhancement on axial (**d**) and coronal (**e**) T1w fs. Axial MIP reconstruction CT (**f**) showing erosion of the first and second right ribs

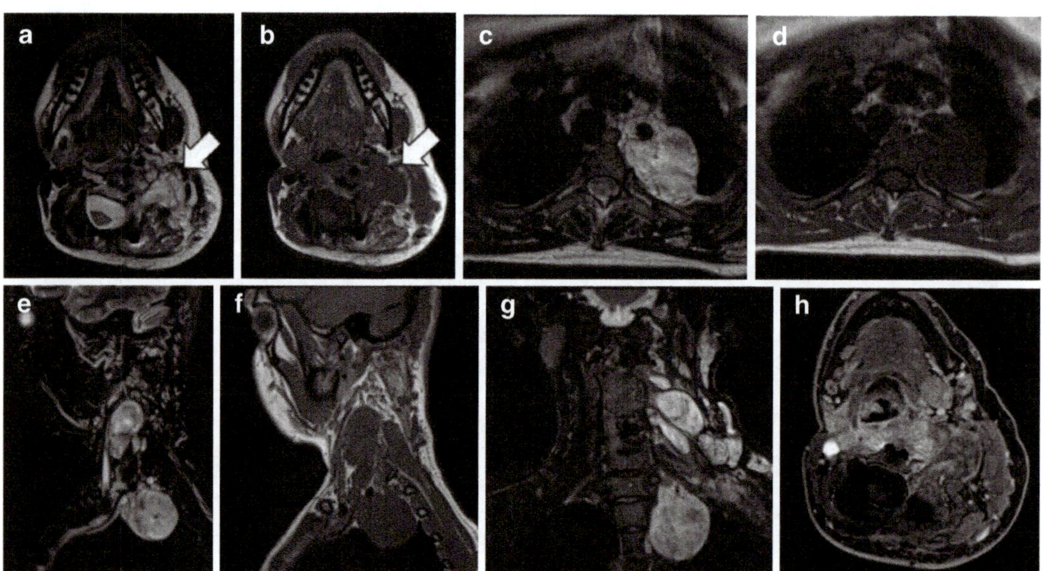

Fig. 6.19 Malignant PNST in a patient with NF1 with motile masses that develop in the paraspinal left space with heterogeneous high signal on axial T2w axial (**a–c**) and isointense on T1w (**b–d**). (**e**) Sagittal STIR. (**f**) Sagittal T1w. (**g**) Coronal T2w fs. (**h**) Axial T1w fs post-contrast

6.4 Peripheral Nerve Sheath Tumor Mimics

There are some entities that can mimic PNSTs:

- *Post-traumatic neuroma.* It usually develops after injury or amputation. It shows perineural scarring, with an intermediate signal on T1w images and a heterogeneous intermediate to high signal on T2w images [1]. Contrast enhancement is a variable [1]. Traumatic neuromas have a bulbous end, lie peripherally, and appear to be in continuity with a normal nerve proximally [1, 19]. These lesions develop within 1–12 months after injury, and their size depends on the original nerve's caliber [1]. History of trauma is fundamental to make the proper diagnosis.
- *Nerve sheath ganglion.* Intraneural ganglion is considered a benign entity that mostly develops at the external peroneal nerve at the head of the fibula. The etiology is controversial. An intraneural theory considers it a degenerative process rather than an actual benign tumor due to chronic irritation or trauma with consequent myxoid degeneration of the nerve with fluid cyst formation [11]. Other theories proposed that extraneural ganglia, usually from a joint capsule, could invade epineurium, leading to intraneural ganglia formation [1]. It typically presents with a palpable local mass with neurological symptoms from nerve compression. At imaging, it represents a cystic mass and may have internal septations. The MRI T1w signal, typically hypointense for fluid cystic formations, may be slightly hyperintense due to myxoid and protein content. The rim and septations may show enhancement after contrast media administration.
- *Morton's neuroma.* It is considered a perineural fibrosis of the digital plantar nerve due to chronic compression against the intermetatarsal ligament rather than a true neoplasm. It typically occurs between the third and fourth metatarsal head. At MRI, the signal reflects the fibrotic content with an intensity similar to post-traumatic neuroma (Fig. 6.20). Ultrasound with dynamic maneuvers while squeezing the metatarsal heads is also a useful diagnostic tool.
- *Peripheral nerve lipomatosis.* It is an intraneural fibrofatty proliferation that creates a "spaghetti-like" configuration [20].
- *Charcot-Marie-Tooth disease.* It is a hereditary disease that appears as a mass-like symmetrical enlargement of peripheral nerves and/or cauda equina.
- *Amyloidosis.* It presents as unilateral or bilateral, focal (amyloidoma), or diffuse enlargement of lumbosacral plexus segments or sciatic nerves, with the involved nerve branches featuring prominent or disrupted fascicles and occasionally T2 hypointense foci. Imaging is not specific, and biopsy is often required.
- *Intraneural metastatic disease.* They are typically located in the brachial plexus from breast or lung cancer. Neural lymphoma usually occurs in patients with known primary or secondary lymphoma, mostly B-cell lymphoma. These lesions show markedly restricted diffu-

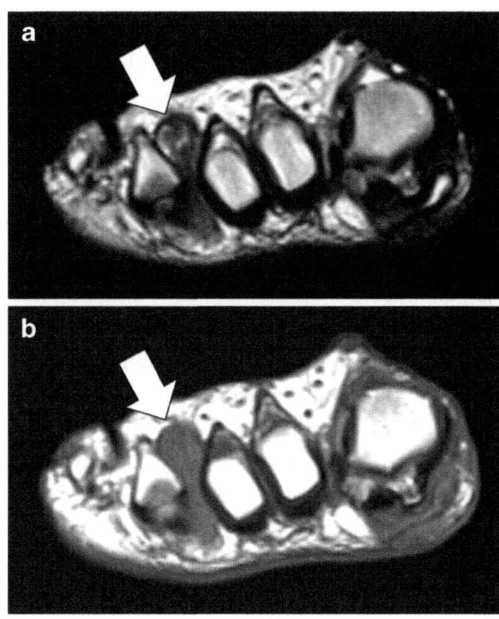

Fig. 6.20 Morton's neuroma. Fibrotic perineural proliferation of the interdigital plantar nerve between the third and fourth metatarsal heads (Arrows). Axial T2-w (**a**) and axial T1-w (**b**)

sion and early arterial avid contrast enhancement.

- *Vascular malformations.* They are commonly found in the head, neck, and distal extremities. These lesions tend to be poorly circumscribed and are hypointense or isointense to skeletal muscles on T1w imaging, hyperintense on T2w imaging, and enhanced markedly with contrast [1, 21]. They can also mimic plexiform PNSTs [22]. MR angiography or ultrasound are sometimes warranted to discriminate between the two entities.

6.5 Advanced Magnetic Resonance Neurography Techniques

6.5.1 DWI/ADC

The common theoretical framework for diffusion-weighted imaging (DWI) with apparent diffusion coefficient (ADC) is built upon the physical principles of free molecular diffusion, described as the Brownian motion. Highly cellular tissues, such as tumors, have greater free diffusion restriction along the diffusion gradient axis and, consequently, high signal intensity. DWI with ADC mapping has been shown to be a useful qualitative and quantitative tool for distinguishing benign from malignant PNSTs [1]. The qualitative presence of a "target sign" on DWI and ADC maps are more indicative of benign PNSTs and uncharacteristic of MPNSTs [1, 23]. In addition, the quantitative measurement of minimum ADC values is also useful and offers high diagnostic accuracy. Benign lesions, such as neurofibromas, schwannomas, and perineuriomas, usually have high ADC values ($>1.1 \times 10^{-3}$ mm^2/s) or lack substantial restricted diffusion [2]. Conversely, MPNSTs have low ADC values ($<1.0–1.1 \times 10^{-3}$ mm^2/s) or significantly restricted diffusion [2]. In ancient schwannomas, internal hemorrhagic areas have low ADC values and can mimic malignant lesions. Thus, these areas should be avoided when calculating the diffusion values carefully watching other MR sequences to detect the presence of blood [2].

In individuals with NF1 who are at a higher risk for the development of MPNST, whole-body MR imaging (WBMRI), when performed in conjunction with DWI/ADC mapping, enables the simultaneous determination of internal PNST burden, characterization of PNSTs as benign versus malignant, and surveillance strategy over time [1, 24].

6.5.2 Magnetic Resonance Spectroscopy (MRS)

Trimethylamine concentrations and the trimethylamine fraction were found to be relatively lower in benign neoplasms. A trimethylamine fraction threshold of 50% resulted in 100% sensitivity and 72.2% specificity in distinguishing BPNSTs from MPNSTs [2, 25]. The presence of trimethylamine (total choline content) within a PNST could be considered a sensitive marker of malignant disease, especially for differentiating neurofibromas from MPNST [1, 25]. Unfortunately, MRS requires significant additional scan and post-processing time and, thus far, has not been integrated successfully into routine clinical practice.

6.5.3 Diffusion Tensor Imaging

Diffusion tensor imaging (DTI) aids in determining the directional information of predominant diffusion and has been shown to be particularly useful for preoperative planning of these tumors by delineating healthy and pathologic fascicles [26]. Opposite of DWI, DTI can provide direct information of anisotropic diffusion by sampling diffusion across six or more gradient directions, thereby allowing visualization of anisotropic diffusion across multiple axes simultaneously [26–28]. The degree of anisotropy of tissue is obtained through fractional anisotropy (FA) calculation [26]. FA is indexed from 0 to 1, with FA = 0 representing free isotropic diffusion and FA = 1 representing complete anisotropy. The FA of pathologic nerves is usually lower than that seen in healthy contralateral nerves, possibly reflecting myelin loss and/or axonal degeneration of involved nerve segments [18, 26]. Moreover, fiber tractography can be used to delineate healthy and pathological fascicles accurately. Tract disruption is often more severe in MPNSTs than benign PNSTs. Benign PNSTs usually exhibit a near-normal appearance and partial rather than complete tract disruption [18, 26, 29]. Preoperative visualization of the fascicular course and integrity using fiber tractography is likely to be of great value for surgical planning and subsequent preservation of existing nerve function following tumor resection [26]. DTI and tractography are also useful for discriminating between neurofibromas and schwannomas, as the latter have an eccentric distribution in relation to the nerve fibers. Neurofibromas, instead, are embedded with fascicles of the nerve of origin [30]. In addition, radiomics could potentially allow a detailed characterization and analysis of structures in radiological images by using quantitative image features (e.g., textures or homo-/heterogeneity of signal intensities), which are frequently not visible to even radiologically trained eyes. In combination with machine-learning (ML) techniques, radiomics allows automatic extraction and differentiation of typical radiomics signatures or feature expressions. ML MRI-based radiomics is a promising tool for creating classification models, especially to differentiate benign and malignant PNSTs [31]. However, further studies are needed to implement this sector in clinical practice.

References

1. Debs P, Fayad LM, Ahlawat S. MR neurography of peripheral nerve tumors and tumor-mimics. Semin Roentgenol. 2022;57:232–40.
2. Soldatos T, Fisher S, Karri S, Ramzi A, Sharma R, Chhabra A. Advanced MR imaging of peripheral nerve sheath tumors including diffusion imaging. Semin Musculoskelet Radiol. 2015;19:179–90.
3. Chhabra A, Soldatos T, Durand DJ, Carrino JA, McCarthy EF, Belzberg AJ. The role of magnetic resonance imaging in the diagnostic evaluation of malignant peripheral nerve sheath tumors. Indian J Cancer. 2011;48:328–34.
4. Chee DWY, Peh WCG, Shek TWH. Pictorial essay: imaging of peripheral nerve sheath tumours. Can Assoc Radiol J. 2011;62:176–82.
5. Ahlawat S, Chhabra A, Blakely J. Magnetic resonance neurography of peripheral nerve tumors and tumorlike conditions. Neuroimaging Clin N Am. 2014;24:171–92.
6. Mautner VF, Hartmann M, Kluwe L, Friedrich RE, Fünsterer C. MRI growth patterns of plexiform neurofibromas in patients with neurofibromatosis type 1. Neuroradiology. 2006;48:160–5.
7. Dombi E, Solomon J, Gillespie AJ, et al. NF1 plexiform neurofibroma growth rate by volumetric MRI: relationship to age and body weight. Neurology. 2007;68:643–7.
8. Hertzanu Y, Hirsch M, Peiser J, Avinoach I. Computed tomography of elephantiasis neuromatosa. J Comput Assist Tomogr. 1989;13:156–8.
9. Stevens KJ, Ludman CN, Sully L, Preston BJ. Magnetic resonance imaging of elephantiasis neuromatosa. Skelet Radiol. 1998;27:696–701.
10. Abreu E, Aubert S, Wavreille G, Gheno R, Canella C, Cotten A. Peripheral tumor and tumor-like neurogenic lesions. Eur J Radiol. 2013;82:38–50.
11. Pilavaki M, Chourmouzi D, Kiziridou A, Skordalaki A, Zarampoukas T, Drevelengas A. Imaging of peripheral nerve sheath tumors with pathologic correlation. Eur J Radiol. 2004;52:229–39.
12. Murphey MD, Smith WS, Smith SE, Kransdorf MJ, Temple HT. From the archives of the AFIP. Imaging of musculoskeletal neurogenic tumors: radiologic–pathologic correlation. Radiogr Rev Publ Radiol Soc N Am Inc. 1999;19:1253–80.
13. Ducatman BS, Scheithauer BW, Piepgras DG, Reiman HM, Ilstrup DM. Malignant peripheral nerve sheath tumors. A clinicopathologic study of 120 cases. Cancer. 1986;57:2006–21.
14. Woertler K. Tumors and tumor-like lesions of peripheral nerves. Semin Musculoskelet Radiol. 2010;14:547–58.
15. Stucky C-CH, Johnson KN, Gray RJ, Pockaj BA, Ocal IT, Rose PS, Wasif N. Malignant peripheral nerve sheath tumors (MPNST): the Mayo Clinic experience. Ann Surg Oncol. 2012;19:878–85.
16. Kamran SC, Shinagare AB, Howard SA, Hornick JL, Ramaiya NH. A–Z of malignant peripheral nerve sheath tumors. Cancer Imaging. 2012;12:475–83.
17. Demehri S, Belzberg A, Blakeley J, Fayad LM. Conventional and functional MR imaging of peripheral nerve sheath tumors: initial experience. Am J Neuroradiol. 2014;35:1615–20.
18. Chhabra A, Thakkar RS, Andreisek G, Chalian M, Belzberg AJ, Blakeley J, Hoke A, Thawait GK, Eng J, Carrino JA. Anatomic MR imaging and functional diffusion tensor imaging of peripheral nerve tumors and tumorlike conditions. AJNR Am J Neuroradiol. 2013;34:802–7.
19. Tagliafico A, Isaac A, Bignotti B, Rossi F, Zaottini F, Martinoli C. Nerve tumors: what the MSK radi-

ologist should know. Semin Musculoskelet Radiol. 2019;23:076–84.
20. Vetrano IG, Sconfienza LM, Albano D, Chianca V, Nazzi V. Recurrence of carpal tunnel syndrome in isolated non-syndromic macrodactyly: DTI examination of a giant median nerve. Skelet Radiol. 2019;48:989–93.
21. Parmar V, Haldeman C, Amaefuna S, Hanna AS. A vascular malformation presenting as a peripheral nerve sheath tumor. J Brachial Plex Peripher Nerve Inj. 2016;11:e38–41.
22. Jalali-Farahani S, Blakeley JO, Belzberg AJ, Carrino JA, Chhabra A. Plexiform nerve sheath tumor or vascular malformation—role of advanced MR neurography and diffusion tensor imaging. Skelet Radiol. 2013;42:1007–10.
23. Ahlawat S, Fayad LM. Imaging cellularity in benign and malignant peripheral nerve sheath tumors: utility of the "target sign" by diffusion weighted imaging. Eur J Radiol. 2018;102:195–201.
24. Ahlawat S, Baig A, Blakeley JO, Jacobs MA, Fayad LM. Multiparametric whole-body anatomic, functional, and metabolic imaging characteristics of peripheral lesions in patients with schwannomatosis. J Magn Reson Imaging. 2016;44:794–803.
25. Fayad LM, Wang X, Blakeley JO, Durand DJ, Jacobs MA, Demehri S, Subhawong TK, Soldatos T, Barker PB. Characterization of peripheral nerve sheath tumors with 3T proton MR spectroscopy. AJNR Am J Neuroradiol. 2014;35:1035–41.
26. Mazal AT, Ashikyan O, Cheng J, Le LQ, Chhabra A. Diffusion-weighted imaging and diffusion tensor imaging as adjuncts to conventional MRI for the diagnosis and management of peripheral nerve sheath tumors: current perspectives and future directions. Eur Radiol. 2019;29:4123–32.
27. Basser PJ, Mattiello J, LeBihan D. MR diffusion tensor spectroscopy and imaging. Biophys J. 1994;66:259–67.
28. Mori S, Zhang J. Principles of diffusion tensor imaging and its applications to basic neuroscience research. Neuron. 2006;51:527–39.
29. Schmidt M, Kasprian G, Amann G, Duscher D, Aszmann OC. Diffusion tensor tractography for the surgical management of peripheral nerve sheath tumors. Neurosurg Focus. 2015;39:E17.
30. Gersing AS, Cervantes B, Knebel C, et al. Diffusion tensor imaging and tractography for preoperative assessment of benign peripheral nerve sheath tumors. Eur J Radiol. 2020;129:109110.
31. Ristow I, Madesta F, Well L, et al. Evaluation of magnetic resonance imaging-based radiomics characteristics for differentiation of benign and malignant peripheral nerve sheath tumors in neurofibromatosis type 1. Neuro-Oncol. 2022;24:1790–8.

The Developing Role of Molecular Imaging PET in Peripheral Nerve Sheath Tumors

7

Annachiara Arnone, Ignazio Gaspare Vetrano, and Pierpaolo Alongi

7.1 Background

The current management of patients harboring peripheral nerve sheath tumors (PNST), in particular in the case of asymptomatic patients affected by neurofibromatosis type 1 (NF1), is based on physical and neurological examination and traditional imaging, with the aim of early detection of malignant tumors [1].

Medical imaging tools such as computed tomography (CT), magnetic resonance imaging (MRI), and positron emission tomography/computed tomography (PET/CT) play essential roles in many oncological settings. Currently, structural radiologic methodologies (CT or MRI) are the standard techniques for diagnosis in the suspicion of PNST [2]. In particular, MRI has higher soft-tissue resolution and provides better anatomical information than CT images in these and other diseases [2]. For example, benign neurofibroma is a well-delimited lesion characterized by high intensity on T2-weighted MRI images and sometimes by a central area of low intensity (the "target sign") due to the presence of myxoid material [3]. Conversely, MPNSTs usually present a more heterogeneous signal on the T1-WI and edema reaction surrounding the tumor, which generally appears more extensive with an infiltrative margin compared with the benign forms [4].

Nevertheless, although several key features can be adapted to distinguish MPNST from benign lesions, the differential diagnosis based on MRI features remains insufficiently accurate, with reported sensitivity and specificity of 61% and 90%, respectively [5, 6]. CT imaging represents a helpful tool for observing bone, joint, and soft tissue lesions, but no definitive diagnostic features have been reported until now [7]. Consequently, based on morphological imaging findings alone, it is challenging to discriminate among these lesions [8]. A correct pretreatment classification is also crucial to optimize diagnosis and set the best care for patients with PNSTs. Moreover, because of the limitations related to clinical presentation, anatomical imaging, and invasive techniques, malignant degeneration is often identified late during the disease, compromising the potential for treatment. Therefore, the need for alternative techniques for detecting malignancy has led to evaluate the role of PET as

A. Arnone
Nuclear Medicine Unit, Department of Experimental and Clinical Biomedical Sciences "Mario Serio", University of Florence, Florence, Italy

I. G. Vetrano
Department of Neurosurgery, Oncological Neurosurgery Unit, Fondazione IRCCS Istituto Neurologico Carlo Besta, Milano, Italy
e-mail: ignazio.vetrano@istituto-besta.it

P. Alongi (✉)
Nuclear Medicine Unit, A.R.N.A.S. Ospedale Civico Di Cristina Benfratelli, Palermo, Italy
e-mail: pierpaolo.alongi@arnascivico.it

I. G. Vetrano, V. Nazzi (eds.), *Peripheral Nerve Sheath Tumors*,
https://doi.org/10.1007/978-3-031-76382-3_7

a noninvasive and suitable in-vivo imaging tool in different phases of the clinical management of PNSTs, including the differential diagnosis between benign and malignant tissues, treatment planning, and follow-up. However, as PNSTs are relatively rare tumors, the use of this imaging has not yet been entirely determined.

7.2 PET Imaging with 18F-Fluorodeoxiglucose (18F-FDG)

The PET scan is an imaging methodology using radiolabeled compounds and, in the oncological practice, can simultaneously inform about the biochemical changes and the morphological structure of the lesion, demonstrating to have wide application in many clinical settings. To date, many studies show the high accuracy of ^{18}F-fluorodeoxyglucose (FDG) PET/CT in differentiating benign and malignant PNST. Most of them are based on significant differences in uptake levels, also assessed by semiquantitative parameters such as the maximal standardized uptake value (SUVmax), even if the lack of guidelines results in a heterogeneous adoption of these technologies [8–16]. Noninvasive techniques for detecting malignancy are crucial, particularly in patients with NF1, as they typically show multiple simultaneous PNSTs and usually develop synchronous malignant and benign forms [17]. Cook et al. reported a higher FDG uptake in MPNSTs than benign symptomatic neurofibromas, as defined by SUV parameters, and also more heterogeneous, as defined by first- and high-order heterogeneity parameters [9].

7.2.1 18F-FDG PET in the Diagnosis of PNSTs: Differentiation of Malignant Lesions

One of the limitations of FDG PET/CT is that there is a significant overlap in SUV values between benign and malignant lesions. On this scope, in their review, Tovmassian et al. proposed the mean SUVmax values 1.93 vs. 7.48 to discriminate benign and malignant lesions, respectively [8]. Additionally, the authors reported that, although the existence of an overlapping, all malignant lesions showed SUVmax values greater than 3.15 [8]. Warbey et al. proposed an SUVmax threshold value of 3.5 to distinguish neurofibroma and MPNST [15]. However, most of the reviewed studies presented a small cohort and suffered from not providing histological data on all lesions, which underwent FDG PET [8]. Ferneret and coworkers showed a statistically significant SUVmax increase in malignant lesions compared to benign lesions (mean 5.4 ± 2.4 vs. 1.54 ± 0.7, respectively, $p = 0.002$) [18]. Similar findings have been reported by Bredella et al. [12]. Benz and coauthors proposed an optimal SUVmax threshold of 6.1 g/mL to differentiate malignant from benign PNST, with a sensitivity and specificity of 94% and 91%, respectively [11]. Lowering this threshold to 4.5 ng/mL increased the sensitivity to 100%, but the specificity reduced to 83%. Conversely, raising the threshold to 8.5 ng/mL, the specificity increased to 100%, but the sensitivity reduced to 65%. Nevertheless, some benign lesions, such as schwannomas, have a wide range of FDG uptake, so many authors recommended that schwannoma should be suspected in the case of a long history of symptoms or atypical imaging findings [19]. Adel et al. observed high FDG uptake in 8/22 patients of their population with schwannomas of the extremities [20]. The high FDG uptake in some schwannomas is supposed to be related to the presence of peritumoral lymphoid cuffs and glucose transporters (GLUTs) [21]. Wang et al. reported higher SUVmax values of schwannomas with peritumoral lymphoid cuffs ($r = 0.456$, $p = 0.000$) than that of the schwannomas without peritumoral lymphoid cuffs [22].

The possibility of false-positive results, probably related to the extent of heterogeneity in the histological findings of PNSTs, had suggested carefully interpreting the images and globally evaluating the clinical presentation. The range of overlapping SUVmax values between benign and malignant lesions represents the most significant issue regarding using FDG PET/CT in this context. Efforts to improve the diagnostic performance of the test have also included trials based on serial acquisitions [14].

Ferner et al. tried to evaluate late delayed images 4 h after FDG administration, reporting 1.5 and 5.7 as mean SUVmax values of benign and malignant lesions, respectively, and no malignant tumors presented SUVmax <1.5 [14]. The implementation of delayed imaging was demonstrated to reduce the number of false positives, whereas it would require further analyses to validate the observed findings. In this way, other authors demonstrated a statistically significant difference between early and delayed imaging for malignant lesions, with a sensitivity and specificity of 97% and 87%, respectively, also adopting the SUV normalized by lean body mass (SUL) and not the more common SUVmax derived from patient weight, as a more reliable measure of tissue activity to account for the variation of individual body habitus [15, 16] (Fig. 7.1). Several factors can affect SUVmax measurement, including blood glucose level, respiratory effort, the acquisition time after the administration, and technical factors such as scanner variability, reconstruction parameters, and inter-patient differences [23]. Novel semiquantitative parameters such as metabolic tumor volume (MTV) and total lesion glycolysis (TLG) have already been tested as useful prognostic markers compared to SUVmax, showing promising but not still adequate evidence to justify their routine use in these tumors [24, 25]. Combemale et al. reported a 98.8% negative predictive value (PPN) for a SUVmax tumor/liver <1.5, but 25/65 lesions

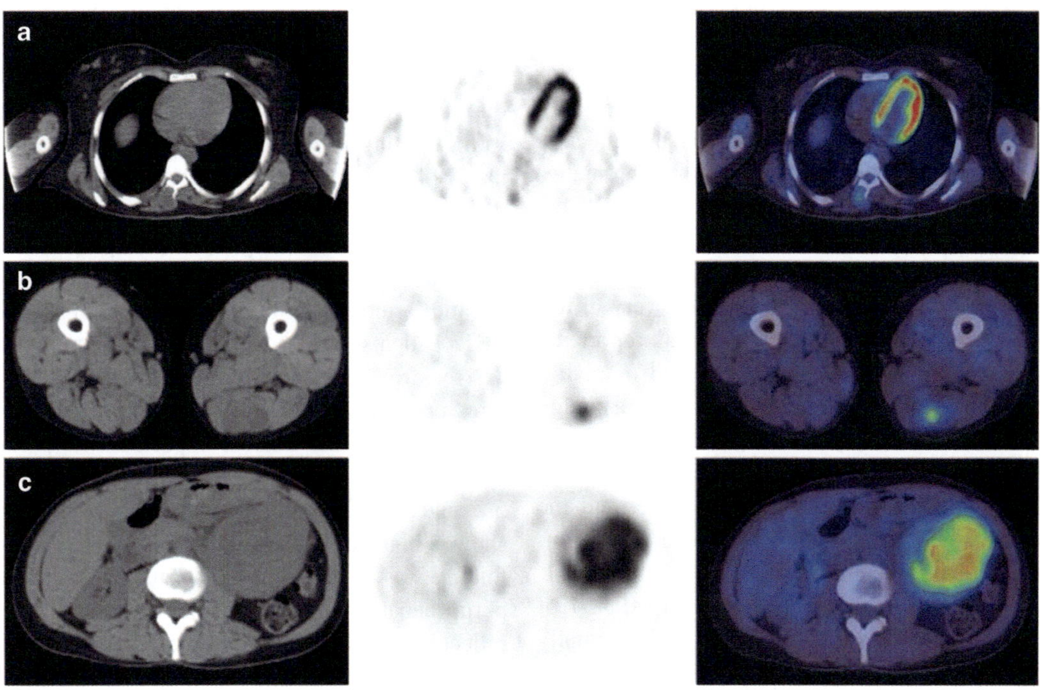

Fig. 7.1 Axial CT, PET, and PET/CT images at 4 h. (**a**) Biopsy-proven atypical neurofibroma in the right spinal erector muscle (SUVmax 4.2) shows the focal area of increased [18F]FDG uptake. (**b**) Biopsy-proven low-grade MPNST within the posterior musculature of the left thigh (SUVmax 6.2) shows the focal area of increased [18F]FDG. (**c**) Biopsy-proven high-grade MPNST occupying the left flank (SUVmax 10.0) shows heterogeneously increased [18F]FDG uptake. [*From* [15]*: Warbey VS, Ferner RE, Dunn JT, Calonje E, O'Doherty MJ. [18F] FDG PET/CT in the diagnosis of malignant peripheral nerve sheath tumours in neurofibromatosis type-1. Eur J Nucl Med Mol Imaging. 2009 May;36(5):751-7. doi: 10.1007/s00259-008-1038-0. Epub 2009 Jan 14. PMID: 19142634*]

characterized by a ratio >1.5 were revealed to be benign tumors [26]. In another study, SUVmax lesion/SUVmean liver cutoff >2.6 was suggested, reaching a sensitivity of 100% and specificity of 87% [24]. In line with these two cutoff points, Simsek et al. described SUVmax lesion/SUVmax liver = 1.0 and SUVmax lesion/SUVmean liver = 1.5, in correspondence of solid lesions extended from mediastinum to axillary region and neck, in a 6-year-old male patient with NF1, revealed as benign neurofibroma [27]. Therefore, despite its limitations, SUVmax is currently the most adopted and widely used FDG PET/CT parameter, according to available data in the literature [28].

FDG PET has been helpful in the differential diagnosis between malignant solitary pulmonary nodules and benign ones. In this context, intrapulmonary schwannomas are rare entities, accounting for 0.2% of all pulmonary neoplasms [29, 30]. However, they are clinically meaningful and require a careful differential diagnosis compared to malignant lung tumors [29, 30]. Bronchial schwannomas primarily present as a well-defined mass, potentially mimicking malignant lesions. Some studies have identified low and moderate SUVmax values, although the degrees of cellularity, microvascular density, or vascular permeability can determine very different uptake levels [31, 32]. Zhou et al. reported a case of a bronchial lesion suspected of malignancy (SUVmax value 8.5) diagnosed as a benign schwannoma only after surgical resection and histological characterization [33]. Therefore, SUV value alone has a minimal role, and the diagnostic gold standard for bronchial schwannoma remains the histological examination [34].

Saidi et al. reported a rare case of MPNST of pleura, an extremely rare neoplasm, incidentally discovered performing an FDG PET/CT scan for restaging colon cancer [35]. The images showed a hypermetabolic mass involving the entire right pleura, with no evidence of distant metastasis, emphasizing the role of FDG PET/CT in evaluating the extent of the disease, potentially playing a role also for postsurgical or post-chemotherapy assessment [35].

7.2.2 18F-FDG PET in Rare PNSTs

FDG PET/CT has proven to determine the lesion's metabolic activity, detecting distant sites, possibly changing the initial staging, and properly guiding the management [36]. Brandão et al. reported a case of a patient with a highly infrequent cardiac MPNST diagnosed by FDG PET/CT [37]. This type of neoplasm may cause heart failure, arrhythmias, cardiac tamponade, dyspnea, and peripheral embolism [38]. In this case, the exam showed 13.4 as SUVmax value in correspondence with the cardiac mass and also with other sites, so surgical resection was ruled out in favor of chemotherapy [37].

Schwannomas also account for 2–7% of gastrointestinal mesenchymal tumors, and the rate of misdiagnosis and mistreatment is high [39, 40]. Among them, esophageal schwannoma may present both benign and malignant features originating from the intermuscular and submucosal plexus [41]. In a case report, the authors described the utility of FDG PET in diagnosing this rare form, limiting its role in the definition of malignancy [42].

Another infrequent entity, the porta hepatic schwannoma, originating from the sympathetic and parasympathetic nerve fibers along the hepatic artery, has been evaluated by FDG PET. Patients affected may present symptoms derived from the compression of nearby structures or may be asymptomatic. In the study by Nakaya et al. the authors reported two porta hepatic schwannomas with SUVmax values of 5.2 and 9.4, respectively, and the pathology showed high cell density in both cases [43]. This finding was in line with other studies in which porta hepatic schwannomas are supposed to reflect high cell density and/or inflammation [44, 45]. In their study, Xiao et al. presented a case of a man suffering from recurrent chest and back pain who underwent imaging examinations [46]. Chest CT revealed bone destruction of the first thoracic vertebra, giving the suspicion of a primary malignant bone tumor or metastasis from an unknown primary. FDG PET/CT examination was performed for staging, showing elevated uptake levels (SUVmax 9.15) in correspondence with the CT reported site but also in distant organs. The vertebral tumor resection biopsy supported a diagnosis of MPNST, and the lymph node aspiration biopsy indicated MPNST with necrosis. This case demonstrated that, although intraosseous MPNST is relatively rare, a solitary lytic vertebral lesion with intense FDG uptake could suggest MPNST in the differential diagnosis [47, 48].

7.3 Other Radiopharmaceuticals

Other PET radiopharmaceuticals have revealed their utility in the management of patients affected by different PNST subtypes.

Dondi et al. described a 73-year-old patient with a previous history of prostate adenocarcinoma who underwent an ^{18}F-choline PET/CT for restaging after radiotherapy and during hormonotherapy [49]. The scan showed two incidental and focal uptakes, one into an intervertebral foramen and another close to the temporal bone. After an MRI scan of the spine and brain, the images confirmed the two expansive lesions, both highly suspicious of neurinomas [49] (Fig. 7.2).

Arico et al. described the first case of an ^{18}F-choline PET/CT demonstrating a relapse of a vestibular schwannoma after stereotactic radiotherapy in a patient evaluated for metachronous prostate cancer a previously known meningioma [50].

Calabria et al. also highlighted the usefulness of radiolabeled choline in the management of patients with suspected brain tumor relapse, especially in association with MRI [51].

Other authors emphasized the role of ^{68}Ga-PSMA PET/CT, a transmembrane glycoprotein located in the prostate epithelial cells and highly overexpressed in prostate adenocarcinoma and other benign and malignant tumors [52–55]. Alçin and coauthors incidentally identified an intense focal PSMA uptake in the left cerebellopontine angle in a patient with a history of left-sided sensorineural hearing loss and a newly diagnosed prostate adenocarcinoma [56]. Following examinations, the mass was compatible with a vestibular schwannoma, also known as acoustic neuromas, originating from the vestibulocochlear nerve (CN VIII) [55, 56].

Benign neurofibromatosis could also show PSMA uptake, as reported during a ^{68}Ga-PSMA PET/CT acquisition in a patient with known neurofibromatosis and suspicion of prostate cancer. This finding has been possibly explained as the leakage of the radiotracer due to the increased vessel permeability, considering that ^{67}Ga-citrate, ^{99m}Tc-DTPA, ^{201}Tl, and ^{131}I uptakes have been reported in neurofibromatosis [57–59]. Moreover, it is well-known that PSMA uptake could be related to cross-reaction mechanisms or that other benign lesions showing PSMA uptake (fibrous dysplasia, angiolipoma, sarcoidosis, bone fracture, osteoid osteoma, accessory submandibular salivary gland) were widely reported [60–64].

^{68}Ga-PSMA PET/CT has also been proposed in the surveillance of NF1, improving its specificity when complementary with ^{11}C-methionine PET/CT [12, 65].

Some authors have used amino acid PET imaging instead of FDG due to the high proliferative rate of malignant cells that requires an increase in protein synthesis [66]. Watanabe et al. compared an amino acid tracer, fluorine-18 labeled alpha-methyltyrosine (FMT), with FDG, showing a significant correlation in the uptake for all lesions but a higher accuracy for FMT (81.3% vs. 68%) [67].

^{68}Ga-DOTA-compounds PET/CT scan has been reported as a valuable imaging tool in the noninvasive diagnosis and management of meningiomas. Al Feghali et al. described the case of a patient with optic nerve sheath meningioma, an orbital space-occupying lesion representing 2% of orbital tumors and 1–2% of all intracranial meningiomas [68, 69]. The authors demonstrated the role of this examination in a multidisciplinary tumor board, mainly in a case of unusual clinical presentation and of lack of typical MRI findings [68]. Vay et al. also underlined the value of this

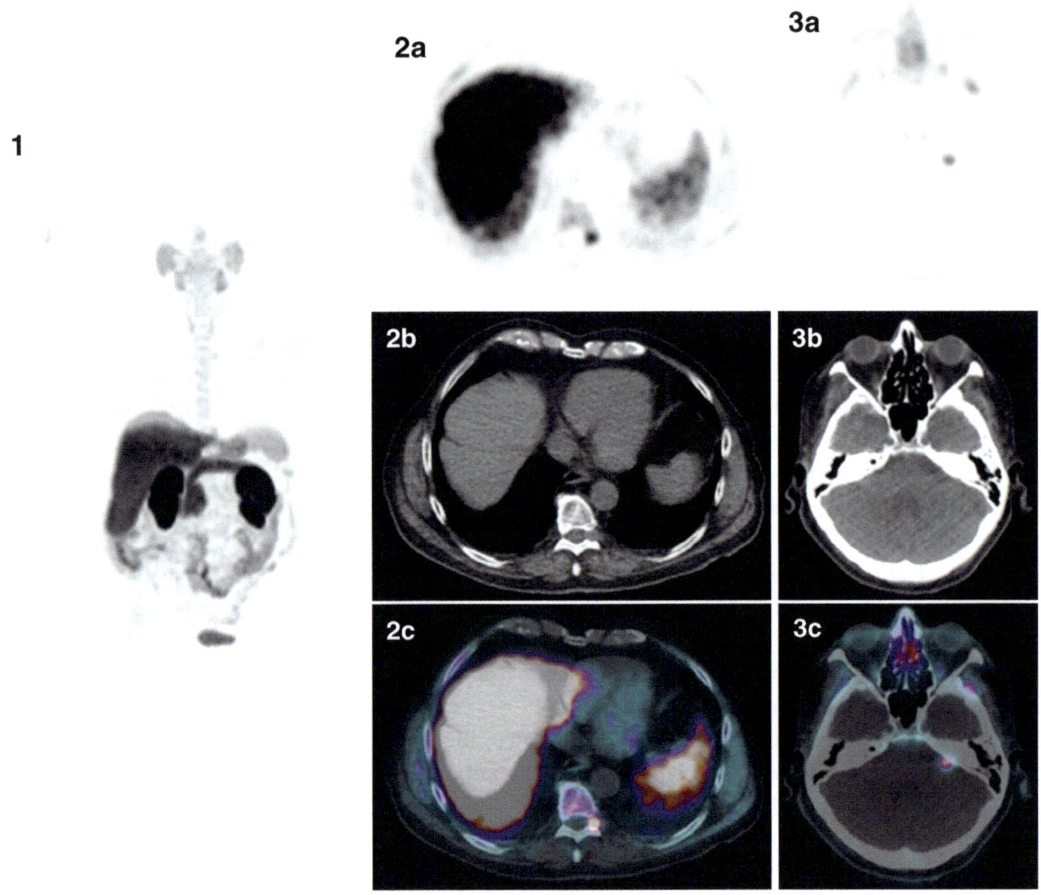

Fig. 7.2 Maximum intensity projection (MIP, **1**) of PET/CT scan. Axial PET (**2a**), axial CT (**2b**), and axial fused PET/CT (**2c**) images show 18F-choline uptake in the left intervertebral foramen between the 9th and 10th dorsal vertebra, in correspondence with new growing tissue. Axial PET (**3a**), axial CT (**3b**), and axial fused PET/CT (**3c**) depict focal uptake between the left temporal bone and the left cerebellar lobe. [*From Article Open Access* [49]*: Dondi F, Albano D, Prandini F, Bertagna F, Giubbini R. Incidental double neurinoma detected by 18F-choline PET/CT scan in a prostate cancer patient. Nucl Med Rev Cent East Eur. 2020;23(1):40-41. doi: 10.5603/NMR.2020.0006*]

examination in a case of suspected optic neuritis, confirming the differential diagnosis of an optic sheath meningioma, thus influencing the treatment choice. After radiotherapy, the ocular movement discomfort was rapidly relieved [70].

In another study, the intense ^{68}Ga-DOTATOC uptake, indicating high expression of somatostatin receptors, helped define a mass lesion with an atypical gadolinium enhancement of a meningioma [71]. Meningiomas represent one-third of brain tumors, and the neoplastic cells have been shown to express somatostatin receptor subtype 2 (SSTR2), supporting the differential diagnosis with astrocytomas and other entities [72]. Previously, other somatostatin receptor radioligands were adopted in the management of patients with meningiomas and even optic nerve sheath meningioma; ^{111}Indium(In)-octreotide scintigraphy was used in the differential diagnosis between meningiomas and other dural-originated tumors, increasing the diagnostic accuracy of MRI imaging but lacking in specificity [73, 74]. More recently, two analogous ^{68}Ga-radiolabeled compounds, ^{68}Ga-DOTATOC and ^{68}Ga-DOTATATE, have been found to show higher lesion detectability and spatial resolution, also significantly reducing radiation exposure [75–77]. In a study including 21 patients with primary or recurrent meningiomas, Rachinger et al. evaluated the diagnostic accuracy of ^{68}Ga-DOTATATE PET/CT compared to MRI imaging, reporting sensitivity of 90.1% vs. 79%, respectively, and similar specificity [78]. Klingenstein et al. reported, even though in a small sample size of 13 patients, high ^{68}Ga-DOTATATE uptake in patients with meningiomas (mean SUVmax = 14.3 ± 15.4) compared to other lesions of the anterior optic pathway (mean SUVmax = 2.1 ± 1.0) [79]. In addition to its diagnostic role, further studies are needed to better evaluate the complementary applications of ^{68}Ga-DOTATATE PET/CT and ^{68}Ga-DOTATOC PET/CT to MRI in the management and follow-up of these tumors [80].

Arif et al. reported a case of pancreatic gastrinoma associated with NF1, assessed by ^{68}Ga-DOTATOC PET/CT during the restaging [81]. Head and neck paragangliomas, which majorly express somatostatin receptors SSTR 2 and 3, determine a basis for radionucleotide imaging and treatment with somatostatin analogs [82].

The interest of researchers is also focused on the emerging extraosseous applications of ^{18}F-NaF PET/CT, such as in the case of tumors with calcifications. Liu and coworkers reported two contemporary incidentalomas (a retroperitoneal neurofibroma and a uterine leiomyoma) in a patient with intermittent lower back pain [83]. Other authors have described some interesting cases of ^{18}F-NaF uptake from uterine leiomyoma [84, 85]. In contrast, calcifications within neurofibromas are rare and of unclear origin [86].

The recent introduction of whole-body combined PET and MRI examinations might allow better accuracy in tumor management than PET/CT alone, adding the high resolution and functional information of MRI to the metabolic activity evaluation of PET. Clinical studies of PET/MRI in the differential diagnosis between benign and malignant neurofibromas are still rare. In a study, Reinert et al. evaluated FDG PET/MRI images in patients affected by NF1 and reported significantly higher values in the malignant group than in the benign neurofibromas [87]. The study of Raad et al. assessed similar sensitivity and strong correlation between PET/CT and PET/MRI, suggesting that PET/MRI is a promising alternative imaging tool for the evaluation of MPNSTs, with all the advantages compared to TC images, such as diffusion-weighted imaging (DWI) sequences, spectroscopy, and perfusion analysis on dynamic contrast-enhanced sequences (Fig. 7.3) [88–90].

Combining these two imaging modalities also simplifies the clinical workflow and limits the number of sedations in younger children and radiation exposure [87].

Sérézal et al. proposed three interesting examples: a schwannoma detected by both MRI and

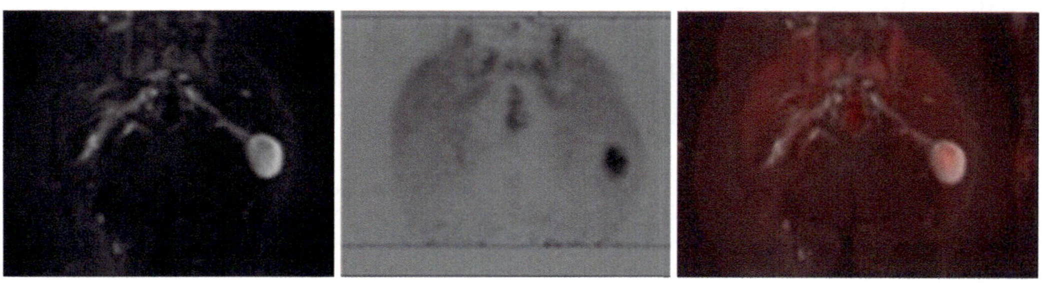

Fig. 7.3 Coronal short T1 inversion recovery (left), positron emission tomography (center), and fused post hoc (right) images of a left gluteus muscle mass demonstrating high T2 signal intensity and mild fludeoxyglucose uptake (maximum standardized uptake value = 3.7). Imaging findings were consistent with a benign or atypical plexiform neurofibroma. [*From Article Open Access* [88]*: Raad RA, Lala S, Allen JC, Babb J, Mitchell CW, Franceschi AM, Yohay K, Friedman KP. Comparison of hybrid 18F-fluorodeoxyglucose positron emission tomography/magnetic resonance imaging and positron emission tomography/computed tomography for evaluation of peripheral nerve sheath tumors in patients with neurofibromatosis type 1. World J Nucl Med. 2018 Oct-Dec;17(4):241-248. doi: 10.4103/wjnm.WJNM_71_17 © Thieme*]

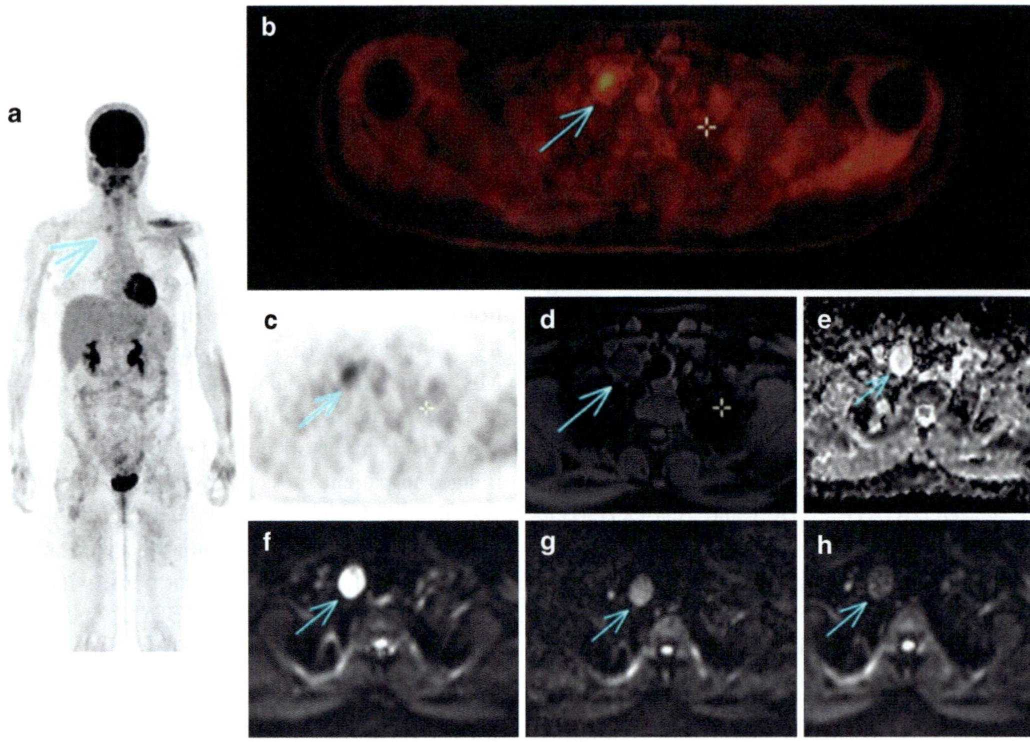

Fig. 7.4 Example of a schwannoma detected by MRI and $^{[18F]}$FDG-PET. From top left to down right: (**a**) Coronal maximum intensity projection (MIP) $^{[18F]}$FDG-PET image, (**b**) axial fused TEP/T1w image, (**c**) axial $^{[18F]}$FDG-PET image, (**d**) axial T1-weighted (T1w) fat-suppressed image, (**e**) axial apparent diffusion coefficient (ADC) image, and (**f**–**h**) axial diffusion-weighted images with 3 *b* values (50, 400 and 800 s/mm^2). The arrows indicate the same schwannoma in the different images. [*From* [91] *Creative Commons CC BY: Gallais Sérézal I, Ferkal S, Lerman L, Mulé S, Funalot B, Wolkenstein P.* $^{[18F]}$*FDG Positron emission tomography with whole body magnetic resonance imaging (*$^{[18F]}$*FDG-PET/MRI) as a diagnosis tool in Schwannomatosis. Orphanet J Rare Dis. 2021 Jan 28;16(1):49. doi: 10.1186/s13023-021-01680-0*]

FDG PET, a schwannoma detected by FDG but not MRI, and a schwannoma identified by MRI but no FDG [91] (Fig. 7.4) (Figure caption from 91, licensed under CC-BY 4.0). Even though most studies in this field lack large cohorts, include heterogeneous populations, and are often in the early stages, current data have proposed the promising role of medical radiological imaging in the differential diagnosis of MPNST from benign lesions, improving the knowledge of these diseases. Prospectively, further insights can give relevance to these imaging techniques in the clinical diagnosis, follow-up, and treatment of these rare but deserving attention tumors.

References

1. Ahlawat S, Blakeley JO, Rodriguez FJ, Fayad LM. Imaging biomarkers for malignant peripheral nerve sheath tumors in neurofibromatosis type 1. Neurology. 2019;93(11):e1076–84.
2. Abul-Kasim K, Thurnher MM, McKeever P, Sundgren PC. Intradural spinal tumors: current classification and MRI features. Neuroradiology. 2008;50(4):301–14.
3. Well L, Salamon J, Kaul MG, Farschtschi S, Herrmann J, Geier KI, et al. Differentiation of peripheral nerve sheath tumors in patients with neurofibromatosis type 1 using diffusion-weighted magnetic resonance imaging. Neuro Oncol. 2019;21(4):508–16.
4. Wasa J, Nishida Y, Tsukushi S, Shido Y, Sugiura H, Nakashima H, et al. MRI features in the differentiation

of malignant peripheral nerve sheath tumors and neurofibromas. Am J Roentgenol. 2010;194(6):1568–74.
5. Ogose A, Hotta T, Morita T, Yamamura S, Hosaka N, Kobayashi H, et al. Tumors of peripheral nerves: correlation of symptoms, clinical signs, imaging features, and histologic diagnosis. Skelet Radiol. 1999;28(4):183–8.
6. Derlin T, Tornquist K, Münster S, Apostolova I, Hagel C, Friedrich RE, et al. Comparative effectiveness of 18F-FDG PET/CT versus whole-body MRI for detection of malignant peripheral nerve sheath tumors in neurofibromatosis type 1. Clin Nucl Med. 2013;38(1):e19–25.
7. Wei CJ, Yan C, Tang Y, Wang W, Gu YH, Ren JY, et al. Computed tomography-based differentiation of benign and malignant craniofacial lesions in neurofibromatosis type I patients: a machine learning approach. Front Oncol. 2020;10:1192.
8. Tovmassian D, Abdul Razak M, London K. The role of [18F]FDG-PET/CT in predicting malignant transformation of plexiform neurofibromas in neurofibromatosis-1. Int J Surg Oncol. 2016;2016:6162182.
9. Cook GJR, Lovat E, Siddique M, Goh V, Ferner R, Warbey VS. Characterisation of malignant peripheral nerve sheath tumours in neurofibromatosis-1 using heterogeneity analysis of 18F-FDG PET. Eur J Nucl Med Mol Imaging. 2017;44(11):1845–52.
10. Azizi AA, Slavc I, Theisen BE, Rausch I, Weber M, Happak W, et al. Monitoring of plexiform neurofibroma in children and adolescents with neurofibromatosis type 1 by [18F]FDG-PET imaging. Is it of value in asymptomatic patients? Pediatr Blood Cancer. 2018;65(1):26733.
11. Benz MR, Czernin J, Dry SM, Tap WD, Allen-Auerbach MS, Elashoff D, et al. Quantitative F18-fluorodeoxyglucose positron emission tomography accurately characterizes peripheral nerve sheath tumors as malignant or benign. Cancer. 2010;16(2):451–8.
12. Bredella MA, Torriani M, Hornicek F, Ouellette HA, Palmer WE, Williams Z, et al. Value of PET in the assessment of patients with neurofibromatosis type I. Am J Roentgenol. 2007;189(4):928–35.
13. Cardona S, Schwarzbach M, Hinz U, Dimitrakopoulou-Strauss A, Attigah A, Mechtersheimer G, et al. Evaluation of F18-deoxyglucose positron emission tomography (FDG-PET) to assess the nature of neurogenic tumours. Eur J Surg Oncol. 2003;29(6):536–41.
14. Ferner RE, Golding JF, Smith M, Calonje E, Jan W, Sanjayanathan V, et al. [18F]2-Fluoro-2-deoxy-D--glucose positron emission tomography (FDG PET) as a diagnostic tool for neurofibromatosis 1 (NF1) associated malignant peripheral nerve sheath tumours (MPNSTs): a long-term clinical study. Ann Oncol. 2008;19(2):390–4.
15. Warbey VS, Ferner RE, Dunn JT, Calonje E, O'Doherty MJ. FDG PET/CT in the diagnosis of malignant peripheral nerve sheath tumours in neurofibromatosis type-1. Eur J Nucl Med Mol Imaging. 2009;36:751–7.
16. Chirindel A, Chaudhry M, Blakeley JO, Wahl R. 18F-FDG PET/CT qualitative and quantitative evaluation in neurofibromatosis type 1 patients for detection of malignant transformation: comparison of early to delayed imaging with and without liver activity normalization. J Nucl Med. 2015;56(3):379–85.
17. Ferner RE, Gutmann DH. Neurofibromatosis type 1 (NF1): diagnosis and management. In: Handbook of clinical neurology; 2013.
18. Ferner RE, Lucas JD, O'Doherty MJ, Hughes RAC, Smith MA, Cronin BF, et al. Evaluation of 18-fluorodeoxyglucose positron emission tomography (18FDG PET) in the detection of malignant peripheral nerve sheath tumours arising from within plexiform neurofibromas in neurofibromatosis 1. J Neurol Neurosurg Psychiatry. 2000;68(3):353–7.
19. Kang S. Benign Schwannoma mimicking metastatic lesion on F-18 FDG PET/CT in differentiated thyroid cancer. Nucl Med Mol Imaging. 2013;47:138–40.
20. Goldman RC, Baizman ER, Longley CB, Branstrom AA. Schwannoma of the extremities: the role of PET in preoperative planning. Eur J Nucl Med. 2001;28:1541–51.
21. Miyake KK, Nakamoto Y, Kataoka TR, Ueshima C, Higashi T, Terashima T, et al. Clinical, morphologic, and pathologic features associated with increased FDG uptake in schwannoma. Am J Roentgenol. 2016;207(6):1288–96.
22. Wang SY, Liu JH, Yao S, Wang SX, Shao D. PET/CT and contrast-enhanced CT imaging findings in benign solitary schwannomas. Eur J Radiol. 2021;141:109820.
23. Velasquez LM, Boellaard R, Kollia G, Hayes W, Hoekstra OS, Lammertsma AA, et al. Repeatability of 18F-FDG PET in a multicenter phase I study of patients with advanced gastrointestinal malignancies. J Nucl Med. 2009;50(10):1646–54.
24. Salamon J, Papp L, Tóth Z, Laqmani A, Apostolova I, Adam G, et al. Nerve sheath tumors in neurofibromatosis type 1: assessment of whole-body metabolic tumor burden using F-18-FDG PET/CT. PLoS One. 2015;10(12):e0143305.
25. van der Gucht A, Zehou O, Djelbani-Ahmed S, Valeyrie-Allanore L, Ortonne N, Brugières P, et al. Metabolic tumour burden measured by 18F-FDG PET/CT predicts malignant transformation in patients with neurofibromatosis type-1. PLoS One. 2016;11(3):e0151809.
26. Combemale P, Valeyrie-Allanore L, Giammarile F, Pinson S, Guillot B, Goulart DM, et al. Utility of 18F-FDG PET with a semi-quantitative index in the detection of sarcomatous transformation in patients with neurofibromatosis type 1. PLoS One. 2014;9(2):e85954.
27. Simsek FS, Akarsu S, Narin Y. Can we differentiate malignant peripheral nerve sheath tumor from benign neurofibroma without invasive sampling. World J Nucl Med. 2019;18(01):66–8.
28. Wong CS, Gong N, Chu YC, Anthony MP, Chan Q, Lee HF, et al. Correlation of measurements from

diffusion weighted MR imaging and FDG PET/CT in GIST patients: ADC versus SUV. Eur J Radiol. 2012;81(9):2122–6.
29. Ohtsuka T, Nomori H, Naruke T, Orikasa H, Yamazaki K, Suemasu K. Intrapulmonary schwannoma. J J Thorac Cardiovasc Surg. 2005;53:154–6.
30. Roviaro G, Montorsi M, Varoli F, Binda R, Cecchetto A. Primary pulmonary tumours of neurogenic origin. Thorax. 1983;38(12):942–5.
31. Nakamura R, Ishikawa S, Sakai M, Goto Y, Minami Y. Increased fluorodeoxyglucose-uptake in positron emission tomography with an endobronchial schwannoma occluding the left main stem bronchus. J Thorac Oncol. 2009;4(9):1183–4.
32. Ushiyama T, Katsuda E, Tanaka M, Numanami H, Takahashi E, Baba K, et al. 18F-Fluorodeoxyglucose uptake and apparent diffusion coefficient in lung schwannoma. Clin Imaging. 2012;36(1):65–7.
33. Zhou D, Xing X, Fan J, Zhang Y, Liu J, Gong Y. PD-1/PD-L1 negative schwannoma mimicking obstructive bronchial malignancy: a case report. Thorac Cancer. 2020;11(8):2335–8.
34. Boré P, Descourt R, Ollivier L, Le Roux PY, Abgral R. False positive 18F-FDG positron emission tomography findings in Schwannoma—a caution for reporting physicians. Front Med. 2018;5:275.
35. Saidi B, Fallahi B, Abdi A, Rakhshani N, Eftekhari M. FDG PET/CT of an advanced case of malignant nerve sheath tumor of pleura. Clin Nucl Med. 2021;46(3):246–7.
36. Assadi M, Velez E, Najafi MH, Matcuk G, Gholamrezanezhad A. PET imaging of peripheral nerve tumors. PET Clin. 2019;14(1):81–9.
37. Soares Brandão SC, Dompieri LT, Tonini RC, Grativvol PS, Gama JD, Calado EB, et al. Cardiac malignant peripheral nerve sheath tumor accessed by 18F-FDG PET/CT. Can J Cardiol. 2020;36(6):967.e17.
38. Arava S, Bagmar S, Jain P, Kumaran M, Kumar S, Ray R. Primary intracardiac malignant peripheral nerve sheath tumor: a rare case report. Indian J Pathol Microbiol. 2015;58(4):531–3.
39. de Araújo Souza LC, Pinto TD, de Figueiredo Cavalcanti HO, Rezende AR, Nicoletti AL, Leão CM, et al. Esophageal schwannoma: case report and epidemiological, clinical, surgical and immunopathological analysis. Int J Surg Case Rep. 2019;55:69–75.
40. Ahn D, Sohn JH, Kim H, Yeo CK. Large esophageal schwannoma mimicking thyroid tumor with egg-shell calcification on preoperative ultrasonography. Asian J Surg. 2017;40(3):236–9.
41. Powley TL, Mittal RK, Baronowsky EA, Hudson CN, Martin FN, McAdams JL, et al. Architecture of vagal motor units controlling striated muscle of esophagus: peripheral elements patterning peristalsis? Auton Neurosci Basic Clin. 2013;179(1–2):90–8.
42. Gao ZY, Liu XB, Pandey S, Gao B, Liu P, Zhang QH, et al. Clinicopathological features of esophageal schwannomas in mainland China: systematic review of the literature. Int J Clin Oncol. 2021;26:284–95.
43. Nakaya M, Kurokawa R, Watadani T, Morisaka H, Higuchi M, Toda Y, et al. Schwannoma in and around the porta hepatis: radiological features in eight cases and literature review. Abdom Radiol (New York). 2022;47(6):1917–28.
44. Kim YC, Park MS. Primary hepatic schwannoma mimicking malignancy on fluorine-18 2-fluoro-2-deoxy-D-glucose positron emission tomography–computed tomography. Hepatology. 2010;51(3):1080–1.
45. Yamamoto M, Hasegawa K, Arita J, Maeno R, Akamatsu N, Kaneko J, et al. Primary hepatic schwannoma: a case report. Int J Surg Case Rep. 2016;29:146–50.
46. Xiao L, Li Y, Li L. Solitary lymph node metastasis from a malignant peripheral nerve sheath tumor of the thoracic vertebra revealed on FDG PET/CT. Clin Nucl Med. 2022;47(5):e419–20.
47. Liu W, Zhang S, Liu J, Shao Z. Intraosseous malignant peripheral nerve sheath tumor of 2 consecutive lumbar vertebrae: a case report and literature review. World Neurosurg. 2019;130:459–66.
48. Muthusamy S, Conway SA, Pitcher JD, Temple HT. Primary intraosseous malignant peripheral nerve sheath tumor of the medial cuneiform: a case report and review of the literature. J Foot Ankle Surg. 2017;56(1):129–34.
49. Dondi F, Albano D, Prandini F, Bertagna F, Giubbini R. Incidental double neurinoma detected by 18F-choline PET/CT scan in a prostate cancer patient. Nucl Med Rev. 2020;23(1):40–1.
50. Arico D, Giardina C, Tuscano C, Picone A, Militano V. First case of 18F-choline uptake in acoustic schwannoma after stereotactic radiotherapy. World J Nucl Med. 2019;18(02):187–8.
51. Calabria FF, Barbarisi M, Gangemi V, Grillea G, Cascini GL. Molecular imaging of brain tumors with radiolabeled choline PET. Neurosurg Rev. 2018;41(1):67–76.
52. Chang SS, Reuter VE, Heston WDW, Bander NH, Grauer LS, Gaudin PB. Five different anti-prostate-specific membrane antigen (PSMA) antibodies confirm PSMA expression in tumor-associated neovasculature. Cancer Res. 1999;59(13):3192–8.
53. Bilgin R, Ergül N, Çermik TF. Incidental meningioma mimicking metastasis of prostate adenocarcinoma in 68Ga-labeled PSMA ligand PET/CT. Clin Nucl Med. 2016;41(12):956–8.
54. Spatz S, Tolkach Y, Jung K, Stephan C, Busch J, Ralla B, et al. Comprehensive evaluation of prostate specific membrane antigen expression in the vasculature of renal tumors: implications for imaging studies and prognostic role. J Urol. 2018;199(2):370–7.
55. Sathekge M, Lengana T, Modiselle M, Vorster M, Zeevaart JR, Maes A, et al. 68Ga-PSMA-HBED-CC PET imaging in breast carcinoma patients. Eur J Nucl Med Mol Imaging. 2017;44:689–94.
56. Alçin G, Gündoğan C, Mutlu İN, Çermik TF. 68Ga-prostate-specific membrane antigen-11 PET/CT: incidental finding of a vestibular Schwannoma. Clin Nucl Med. 2019;44:883–5.

57. Holbert BL, Lamki LM, Holbert JM. Uptake of bone scanning agent in neurofibromatosis. Clin Nucl Med. 1987;12(1):66–7.
58. Behrad S, Sohanian S, Ghanbarzadegan A. Solitary intraosseous neurofibroma of the mandible: report of an extremely rare histopathologic feature. Indian J Pathol Microbiol. 2020;63(2):276–8.
59. Lee JH, Bae JH, Kim KS. A case of solitary neurofibroma of the nasal dorsum: resection using an external rhinoplasty approach. Eur Arch Oto-Rhino-Laryngol. 2005;262:813–5.
60. Zucker RJ, Bradley YC. Indium-111 capromab pendetide (ProstaScint) uptake in a meningioma. Clin Nucl Med. 2001;26(6):568–9.
61. Sasikumar A, Joy A, Pillai MRA, Alex TM, Narayanan G. 68Ga-PSMA PET/CT in osteosarcoma in fibrous dysplasia. Clin Nucl Med. 2017;42(6):446–7.
62. Castello A, Lopci E. Incidental identification of osteoid osteoma by 68Ga-PSMA PET/CT. Eur J Nucl Med Mol Imaging. 2018;45(3):509–10.
63. Zhang W, Zhang L, Zhao Y, Chen Y. Avid 68Ga-PSMA uptake in accessory submandibular salivary gland. Clin Nucl Med. 2019;44(7):591–3.
64. Kobe C, Maintz D, Fischer T, Drzezga A, Chang DH. Prostate-specific membrane antigen PET/CT in splenic sarcoidosis. Clin Nucl Med. 2015;40(11):897–8.
65. Gulhane B, Ramsay S, Fong W. 68Ga-PSMA uptake in neurofibromas demonstrated on PET/CT in a patient with neurofibromatosis type 1. Clin Nucl Med. 2017;42(10):776–8.
66. Jager PL, Vaalburg W, Pruim J, De Vries EGE, Langen KJ, Piers DA. Radiolabeled amino acids: basic aspects and clinical applications in oncology. J Nucl Med. 2001;42(3):432–45.
67. Endo K, Takagishi K, Watanabe H, Inoue T, Shinozaki T, Yanagawa T, et al. PET imaging of musculoskeletal tumours with fluorine-18 alpha-methyltyrosine: comparison with fluorine-18 fluorodeoxyglucose PET. Eur J Nucl Med. 2000;27(10):1509–17.
68. Al Feghali KA, Yeboa DN, Chasen B, Gule MK, Johnson JM, Chung C. The use of 68Ga-DOTATATE PET/CT in the noninvasive diagnosis of optic nerve sheath meningioma: a case report. Front Oncol. 2018;8:454.
69. Sepahdari AR, Aakalu VK, Setabutr P, Shiehmorteza M, Naheedy JH, Mafee MF. Indeterminate orbital masses: restricted diffusion at MR imaging with echo-planar diffusion-weighted imaging predicts malignancy. Radiology. 2010;256(2):554–64.
70. Vay SU, Werner JM, Kabbasch C, Schmidt M, Drzezga A, Fink GR, et al. Uncovering an optic nerve sheath meningioma using 68Ga-DOTATATE PET/CT. Clin Nucl Med. 2021;46(9):e464–5.
71. Mairot K, Sahakian N, Salgues B, Levy N, Gascon P, Denis D. Somatostatin receptor PET/CT scan as a helpful diagnostic tool for optic nerve sheath meningioma. J Fr Ophtalmol. 2021;44(10):e619–21.
72. Kiviniemi A, Gardberg M, Kivinen K, Posti JP, Vuorinen V, Sipilä J, et al. Somatostatin receptor 2A in gliomas: association with oligodendrogliomas and favourable outcome. Oncotarget. 2017;8(30):49123–32.
73. Nathoo N, Ugokwe K, Chang AS, Li L, Ross J, Suh JH, et al. The role of 111indium-octreotide brain scintigraphy in the diagnosis of cranial, dural-based meningiomas. J Neurooncol. 2007;81(2):167–74.
74. Andrews DW, Faroozan R, Yang BP, Hudes RS, Werner-Wasik M, Kim SM, et al. Fractionated stereotactic radiotherapy for the treatment of optic nerve sheath meningiomas: preliminary observations of 33 optic nerves in 30 patients with historical comparison to observation with or without prior surgery. Neurosurgery. 2002;51(4):890–902.
75. Antunes P, Ginj M, Zhang H, Waser B, Baum RP, Reubi JC, et al. Are radiogallium-labelled DOTA-conjugated somatostatin analogues superior to those labelled with other radiometals? Eur J Nucl Med Mol Imaging. 2007;34(7):982–93.
76. Hofmann M, Maecke H, Börner AR, Weckesser E, Schöffski P, Oei ML, et al. Biokinetics and imaging with the somatostatin receptor PET radioligand 68Ga-DOTATOC: preliminary data. Eur J Nucl Med. 2001;28(12):1751–7.
77. Walker RC, Smith GT, Liu E, Moore B, Clanton J, Stabin M. Measured human dosimetry of 68Ga-DOTATATE. J Nucl Med. 2013;54(6):855–60.
78. Rachinger W, Stoecklein VM, Terpolilli NA, Haug AR, Ertl L, Pöschl J, et al. Increased 68Ga-DOTATATE uptake in PET imaging discriminates meningioma and tumor-free tissue. J Nucl Med. 2015;56(3):347–53.
79. Klingenstein A, Haug AR, Miller C, Hintschich C. Ga-68-DOTA-TATE PET/CT for discrimination of tumors of the optic pathway. Orbit (London). 2015;34(1):16–22.
80. Galldiks N, Albert NL, Sommerauer M, Grosu AL, Ganswindt U, Law I, et al. PET imaging in patients with meningioma—report of the RANO/PET Group. Neuro Oncol. 2017;19(12):1576–87.
81. Arif AA, Kim PTW, Melck A, Churg A, Schwartz Z, Stuart HC. Pancreatic gastrinoma, gastrointestinal stromal tumor (GIST), pheochromocytoma, and hürthle cell neoplasm in a patient with neurofibromatosis type 1: a case report and literature review. Am J Case Rep. 2021;22:e927761.
82. Leijon H, Remes S, Hagström J, Louhimo J, Mäenpää H, Schalin-Jäntti C, et al. Variable somatostatin receptor subtype expression in 151 primary pheochromocytomas and paragangliomas. Hum Pathol. 2019;86:66–75.
83. Liu H, Chen Y, Yang L, Fu W. 18F-NaF uptake in retroperitoneal neurofibroma and uterine leiomyoma calcifications. Clin Nucl Med. 2019;44(12):991–2.
84. Kothekar E, Raynor WY, Werner TJ, Alavi A, Baker JF. 18F-NaF uptake in calcified uterine leiomyoma. Clin Nucl Med. 2019;44(11):e620–1.
85. Al-Zaghal A, Werner TJ, Høilund-Carlsen PF, Alavi A. The detection of uterine leiomyoma (fibroid) calcifications on 18F-NaF PET/CT. Clin Nucl Med. 2018;43(8):e287–8.

86. Manning MA, Paal EE, Srivastava A, Mortele KJ. Nonepithelial neoplasms of the pancreas, part 2: malignant tumors and tumors of uncertain malignant potential. Radiographics. 2018;38(4):1047–72.
87. Reinert CP, Schuhmann MU, Bender B, Gugel I, la Fougère C, Schäfer J, et al. Comprehensive anatomical and functional imaging in patients with type I neurofibromatosis using simultaneous FDG-PET/MRI. Eur J Nucl Med Mol Imaging. 2019;46(3):776–87.
88. Raad R, Lala S, Allen J, Babb J, Mitchell C, Franceschi A, et al. Comparison of hybrid 18F-fluorodeoxyglucose positron emission tomography/magnetic resonance imaging and positron emission tomography/computed tomography for evaluation of peripheral nerve sheath tumors in patients with neurofibromatosis type 1. World J Nucl Med. 2018;7(4):241–8.
89. Heacock L, Weissbrot J, Raad R, Campbel N, Friedman KP, Ponzo F, et al. PET/MRI for the evaluation of patients with lymphoma: initial observations. Am J Roentgenol. 2015;204(4):842–8.
90. Heusch P, Buchbender C, Köhler J, Nensa F, Gauler T, Gomez B, et al. Thoracic staging in lung cancer: prospective comparison of 18F-FDG PET/MR imaging and 18F-FDG PET/CT. J Nucl Med. 2014;55(3):373–8.
91. Gallais Sérézal I, Ferkal S, Lerman L, Mulé S, Funalot B, Wolkenstein P. [18F]FDG Positron emission tomography with whole body magnetic resonance imaging ([18F]FDG-PET/MRI) as a diagnosis tool in Schwannomatosis. Orphanet J Rare Dis. 2021;16(1):49.

A Clinical Summary of PNST in Adults

8

Marica Eoli

8.1 Clinical Presentation

The majority of primary PNS originates from Schwann cells and more rarely perineural cells; however, paragangliomas arises from specialized neural crest in the cauda equina region. Primary PNS can occur sporadically or as part of a familiar tumor predisposition syndrome such as neurofibromatosis type 1, schwannomatosis, Carney complex familial, a melanoma-astrocytoma syndrome. Based on the histologic type of the tumor, it is possible to hypothesize a distinct tumor syndrome and to set up a specific comprehensive genetic screening.

Patients may present with coincidental findings of an indolent mass, pain, paresthesia, muscle atrophy, and focal weakness. The presentation and clinical or imaging findings will depend on the nerve involved and a careful neurological examination is needed to localize the lesion. The following aspects should be considered:

- Strength and trophy of muscle: if the nerve transmission is affected, the muscle supplied by the nerve may become weak and wasting. Electromyography recording (EMG) may be useful; if the clinical examination is uncertain, the findings of fibrillation or high amplitude and polyphasic motor unit potentials may suggest whether the lesion is radicular or located more distally.
- Presence of fasciculations or depression of tendon reflexes may suggest nerve lesion.
- Through sensory examination hyper- or hypoesthesia can be detected, suggesting the nerve is affected. Neurophysiological examination may show decreased amplitude of sensory action potential and slowing of sensory nerve conduction.
- Localization and features of pain: pain may have a radicular or a truncal distribution, but also the percussion of the mass or the nerve can elicit a shouting pain (Tinel's sign). The neuropathic pain is typically burning and shock-like.
- Palpation of the mass. Tender mass and shouting pain at palpation may suggest malignant transformation.

The location of the mass may indicate the type of lesion; if the median nerve is involved, a neural lipoma may be the cause; schwannomas and neurofibroma tend to involve large nerve trunks, when a mass is present beneath the fingernail, a glomus tumor may be suspected.

Multiple masses are suggestive of a predisposition tumor syndrome. Benign PNS are similar in imaging. The X-ray is usually normal or shows a nonspecific soft tissue mass; calcifications are rare

M. Eoli (✉)
Experimental Neuro-Oncology Unit, Fondazione IRCCS Istituto Neurologico Carlo Besta, Milan, Italy
e-mail: marica.eoli@istituto-besta.it

I. G. Vetrano, V. Nazzi (eds.), *Peripheral Nerve Sheath Tumors*,
https://doi.org/10.1007/978-3-031-76382-3_8

and may suggest an ancient schwannoma. At sonography, they appear as solid noncompressible mass well-defined hypoechoic mass [1]. MRI may better define the fusiform shape and visualizes the exiting nerve; the mass signal is T1 isointense to muscle and T2 variable increased. The enhancement is variable for both benign and malignant tumors [2]. However, MRI is useful in distinguishing benign from malignant PNS tumors. Benign neoplasms appear as a well-circumscribed homogenous low T1 signal and variable high T2 signal and may show the "target sign" a central zone of low signal with surrounding increased signal on T2 images [2]. Malignant PNS tumors appear as an heterogenous mass with area of hyperintensity due to hemorrhage and necrosis with infiltrative margins. The target sign on T2 W imaging and post-gadolinium is less often observed [3].

Primary peripheral nerve tumors are rare; peripheral schwannoma is mostly located in head and neck nerves or anterior aspect upper limbs. Small peripheral nerve tumors are often asymptomatic. Paresthesia, pain hypoesthesia, muscle atrophy, and weakness may appear when they grow and compress neural structure; in benign tumors, pain is often elicited by percussion; in malignant tumors, pain tends to be spontaneous.

Benign tumors are similar from a clinical point of view but include different histologic types that may be associated with different frequencies with various predisposition tumor syndrome, in fact primary peripheral nervous.

When a Genetic Tumor Risk Syndrome (TRS) Should Be Suspected in Adult Patients?
A genetic tumor risk syndrome, also called tumor predisposition syndrome, is an inherited disorder that increases the risk of several benign or malignant cancers compared to the risk for the general population. In this condition, germline pathogenic gene variants or somatic gene variants are passed off from parents to children. The pathogenic variants can function as dominant or recessive. The genetic tumor risk syndrome should be suspected when some features are present such as:

- Unusual early-onset of tumor disease
- Multiple primary tumors in an individual
  - No relapse residual tumors, but independent primary tumors
  - Cave: combination of frequent tumors and/or tumors primarily caused by exogenous factors skin (frequent; UV radiation), prostate (frequent), cervical (HPV), lung/bladder (nicotine abuse)
  - Secondary malignancies followed years after antineoplastic and antihormonal therapy
- Familial clustering
- Typical tumor spectrum
- Rare tumors

Some benign and malignant neoplasms of the peripheral nerve are the hallmark of a wide range of tumor predisposition syndromes that can be diagnosed by a specific comprehensive genetic testing.

8.1.1 Main Etiologies

8.1.1.1 Schwannoma

Schwannoma is a benign nerve sheath tumor composed by differentiated neoplastic Schwann cells (WHO grade 1). They represent about 5% of all benign soft tissue tumor [4]. After total resection, they do not recur; malignant transformation is exceptionally rare, there are some cases reported, and they often become epithelioid MPNST [5, 6].

More than 90% are sporadic and solitary, they affected people of all age, and their peak is incidence is in the fourth to sixth decade. The probability of an association with schwannomatosis is higher, when the lesion appears at young age, often before age 30 in NF2-related schwannomatosis and later in LZTR1- and SMARCB1-related

Schwannomatosis. A prior irradiation is a well-established risk factor to development of a schwannoma [7, 8]. The most common locations are peripheral nerves in the skin and subcutaneous tissues of the head and neck and along flexor surfaces of the limbs. Another frequent location are spinal roots, forming an intradural extramedullary lesion, when they are multiple and paraspinal NF2-related schwannomatosis should be suspected.

Schwannomas are common in NF2-related schwannomatosis and *LZTR1*- or *SMARCB1*-related schwannomatosis, three genetic tumor predisposition syndromes with distinct genetic etiologies and overlapping phenotypes [9]. In both syndromes, they can also occur also in mosaic or segmental form [10].

In addition to bilateral vestibular schwannomas, fifth cranial nerve schwannoma, and spinal dorsal roots, patients affected by NF2-related schwannomatosis show numerous small schwannoma on peripheral nerves and spinal nerve [11].

Peripheral nerve schwannomas are more frequent in LZTR1- and SMARCB1-related schwannomatosis than in NF2-related schwannomatosis [12].

Plexiform schwannoma, a subtype with a macroscopic appearance characterized by multinodular/multifascicular nerve expansion, can be sporadic or can occur in patients with NF2-related Schwannomatosis; in those cases, it is associated with a pigmented plaque-like lesion [13]. When a scattered atypical to bizarre appearing nuclei is present, the Schwannoma is called ancient schwannoma, because this feature is considered degenerative.

8.1.1.2 Neurofibroma

Neurofibroma is a benign peripheral nerve sheath tumor composes of mature neoplastic Schwannoma cells intermixed by nonneoplastic cells, such as fibroblast, mast cells, perineural-like cells, and residual cell. Several subtypes have been described: cellular neurofibroma, atypical neurofibroma, atypical neurofibromatous neoplasm of uncertain potential, plexiform neurofibroma, diffuse neurofibroma, nodular neurofibromas, and massive soft tissue neurofibroma. All subtypes are classified as WHO grade 1, except atypical neurofibromatous neoplasm of uncertain potential (ANNUB), which is not assigned a grade.

Neurofibroma is the most frequent peripheral nerve sheath tumors and in the majority of cases is solitary. Skin is the most frequent location, usually in the dermal, with different appearances such as flat, sessile, globular, pedunculated, or diffuse like a large plaque [14]. However, neurofibromas can affect nerves, nerve plexus, and spinal roots and dorsal root ganglia.

Multiple neurofibromas are a key feature of neurofibromatosis type 1 and are present in more than 90% of affected individuals in adult age.

The presence of bilateral neurofibromas involving all spinal roots and few, if any, cutaneous manifestations characterize a distinct subtype of NF1 entailing greater morbidity than the classical form of the disease and associated with missense pathogenic variants of the gene [15].

Neurofibromas, at variance with previous descriptions, are not present in *NF2*-related Schwannomatosis; most of the reported cases on histological review were hybrid sheath tumors [16].

Plexiform neurofibroma is characterized by the involvement of multiple nerve fascicles, each surrounded by perineurium. It is often localized in a large nerve or plexus. It can be sporadic [17] but is always associated with NF1. In fact, it is present in 30–50% of NF1 cases, and it is probably congenital [18]. At present, there is not an unambiguous PNF definition, but a new classification, MRI-based, was recently proposed [19]. The article also provides a comprehensive review on diagnostic evaluation, surveillance strategy, and treatment indications in this tumor [19]. Cutaneous and subcutaneous neurofibromas are benign lesions that don't transform into malignancies. However, NF1 patients have an 8–16% lifetime risk of developing a malignant peripheral nerve sheath tumor (MPNST), a highly aggressive soft-tissue sarcoma, often arising from preexisting benign plexiform neurofibromas, atypical neurofibromas, and ANNUBP. ANF and ANNUBP are considered premalignant lesions [20], even if the exact risk of malignant transformation is not clear and the clinical presentation and the natural history of both neoplasms remain understudied.

ANFs, different from both PNF and MPNST, exhibit one or more histological atypical features but do not fulfill the criteria to be classified as an MPNST, lacking brisk mitotic activity or necrosis [21, 22].

ANNUBPs are characterized by at least two of the following features: cytological atypia, hypercellularity, loss of neurofibroma architecture, and an increased mitotic index [19, 22]. Biallelic NF1 inactivation and a recurrent loss of the CDKN2A/B locus at 9p21.3 are present both in ANFs and ANNUBPs [21], while ANF, ANNUBP, and MPNST have distinct methylation profiles [23]. ANFs may be multiple, and the pain is the most common symptom [21]. It is not clear if total resection can prevent MPNST appearance. A phase 1/2 clinical trial on abemaciclib, a cyclin-dependent kinase 4 and 6 inhibitor, is ongoing to investigate tolerability and efficacy in patients with NF1-associated ANFs (NCT04750928).

The approval of the MEK inhibitor selumetinib by FDA and EMA for the treatment of symptomatic, inoperable plexiform neurofibromas in children ≥3 years has changed the clinical management of those tumors in childhood. A randomized, double-blind, placebo controlled, phase III study is ongoing to assess the efficacy and safety of selumetinib compared with placebo in adults with symptomatic, inoperable PNFs. Recent studies showed an objective partial response and a reduction of pain both in adults and children with PNFs using other MEK inhibitor, such as mirdametinib [24] and binimetinib (NCT03231306) [19]. Furthermore, also cabozantinib, a small tyrosine kinase inhibitor of c-Kit, VEGFR2, MET, RET, FLT3, and the TAM family receptors, proved efficacy in a recent phase 2 trial for adolescents and adults (NCT02101736).

8.1.1.3 Hybrid Nerve Sheath Tumors

Hybrid nerve sheath tumors (HNST) are benign PNS neoplasms that rarely recur locally [25]. A diagnostic consensus criterion for those tumors has not yet been established; the diagnosis is usually proposed when, within the same lesion, there are present intermingled areas with features of more than one conventional type neurofibroma, schwannoma, and perineurioma. HNST can be subdivided in schwannoma/perineurioma, neurofibroma/schwannoma, and neurofibroma/perineurioma. They occur over a wide age range but are more common in young adult, and the sex distribution is equal. The most common type is schwannoma/perineurioma that occurs sporadically in the extremities [25]. The second type by frequency is hybrid neurofibroma/schwannoma, and it is strongly associated both with neurofibromatosis type 1 and Schwannomatosis [26]. Hybrid neurofibroma/perineurioma, the rarest, has been observed in association with neurofibromatosis 1 [27, 28].

8.1.1.4 Perineurioma

Perineurioma are benign PNS tumors composed of neoplastic perineural cells that don't recur or metastasize; they can be subdivided into intraneural perineurioma when they are located within the nerve and soft tissue perineurioma when they are located in the deep soft tissue. Both are rare, accounting for 1% of nerve sheath and soft tissue tumors, respectively; intraneural perineurioma occurs in young adult and rarely in children [29, 30].

Both perineurioma have been observed in NF1- [31, 32] and in *NF2*-related Schwannomatosis patients [33, 34]. However, it is not clear if this a true association, because only case reports have been reported, and those tumors are not mentioned in large population study [35].

In patients with intraneural perineurioma, peripheral nerves of the extremities are often affected, and a reduced muscular strength is more often observed than sensory disturbances [30]. A solitary mass is observed in patients with soft tissue perineurioma.

8.1.1.5 MMNST

Malignant melanotic nerve sheath tumor (MMNST) is composed uniformly of tumor cells with features both of Schwann cells and cells with melanocytic differentiation. About 50% of cases are affected by Carney complex [36]; however, sporadic cases have also been described [37] Data on MMNST behavior are conflicting, in the past it was considered a benign lesion with a low metastatic risk [36], more recently a high recurrence and metastatic rate have been reported and the tumor was reconsidered as malignant [38]. It is rare and occurs in young adults at an earlier age in patients with Carney complex. The presence of multiple MMNST should raise suspicion of Caney complex.

MMNST arises from spinal or autonomic nerves, but it has been reported also in the gastrointestinal tract. Pain sensory abnormalities and mass effect are the most common symptoms. In most cases, histopathological features are not predictive of tumor behavior. A long follow-up is needed, because about 50% of cases are free of disease progression at 5 years [38].

8.1.1.6 MPNST

Malignant peripheral nerve sheath tumor is a malignant spindle cell sarcoma. It can occur sporadically, after radiation exposure or in association with NF1.

Patients with NF1 have a lifetime incidence risk of 8–15.8% of developing MPNST compared with an incidence of 0.001% in the general population [35]. It can invade surrounding soft tissue and metastasize, accounting for about 35% of the cancer death in NF1 patients [22]. Commonly it occurs between 20 and 50 years, but earlier in NF1 patients, and arises from a preexisting PNF, ANF, or ANNUB [39]. Common locations are the extremities, the trunk, the head, and the neck. MPNSTs often appear as painful rapidly enlarging masses, but it is difficult to differentiate PNF, ANF, and MPNST because the majority of clinical symptoms are the same. MRI and FDG-PET can be useful in the diagnosis, but the accuracy is not so high [40, 41]. Because an early detection is crucial in the management of those tumors, most recent studies focused on the possible use of the liquid biopsy in this setting. Molecular hallmarks are biallelic inactivation of NF1 gene, followed by pathogenic somatic variants in the polycomb repressive complex with or without loss of H3K27 trimethylation [42, 43]. Analysis of circulating tumor (ctDNA) or cell-free (cfDNA) DNA on plasma may reveal the same molecular alterations observed in the tumor, making possible the diagnosis of lesions at different stages of malignant transformation and sparing the patient from repeated biopsies [43–45].

To date, complete surgical resection with wide negative margins is the only curative treatment, and prognosis remains poor, although radiotherapy and chemotherapy [46] with 5 years survival rate of 35% and of 65% has been reported in of MPNST associated with NF1 and in sporadic tumors respectively [47]. The prognosis of radiation-induced is even worse at 23.5% at 5 years [48].

8.1.1.7 Intraneural and Extraneural Lipomas and Lipomatosis of Nerve

Adipose lesions of the nerve include intraneural and extraneural lipomas and lipomatosis of nerve.

Intraneural lipomas are relatively rare soft tissue tumors that often occur in the forearm and the wrist and particularly within the median nerve [49]. Patients refer neuropathic symptoms; if the lesion is large enough, tit may cause progressive compression neuropathy.

Intraneural lipoma appears as a distinct mass using MRI, CT, and ultrasonography. In particular at MRI, it is bright on T1-weighted sequences and dark on fat suppressed images, without enhancement on post-gadolinium sequences [50].

Extraneural lipomas are fatty lesion located outside the epineurium. They can arise throughout the body and can cause compressive neuropathy. Their imaging characteristics are similar with those observed in intraneural lipomas, except for the relationship to the nerve.

Lipomatosis of nerve is characterized by the presence of abundant fibroadipose tissue within the epineurium, resulting in a massive nerve enlargement [50]. The hallmark MRI feature is a cable-like appearance on axial images and a spaghetti-like appearance on longitudinal planes [50, 51].

It is diagnosed in young adults commonly before the age of 30. It is often located in the median nerve at the wrist in the ulnar nerves. No malignant transformation has been observed.

LN is included among the PIK3CA-related overgrowth spectrum (PROS) of disorders [51].

8.1.1.8 Paraganglioma

Paraganglioma is a benign grade 1 WHO neuroendocrinal tumor, commonly located in the adrenal medulla. In the nervous system, it is very rare, arising from specialized neural crest cells and located exclusively in the cauda equina region, comprising 3–4% of all tumors affecting the region. It is entirely intradural attached either to the filum terminale or to a caudal nerve root [52]. The most common symptoms are low back pain and sciatica, less often numbness, paraparesis, and sphincter disturbances. The imaging feature at MRI is not distinguishable from that or schwannoma or ependymoma [53]. The majority of those lesions are curable by complete resection; only 4% recur after gross total removal [54].

8.1.1.9 Glomus

Glomus tumor is a benign tumor composed of cells from the glomus body, which is a contractile neuromyoarterial structure that affects blood pressure and thermoregulation by modifying cutaneous blood flow. It represents 2% of all soft neoplasms and is more frequent in women [55]. It commonly occurs beneath the nail but, more rarely, may be located on the arm or the leg or the trunk. Single or multiple glomus tumors are often observed in patients with NF1, affecting about 5% of patients [56]. Sonography and magnetic resonance imaging can be performed in the diagnosis; both have a similarly high sensitivity [57].

Excision is the definitive treatment for a glomus tumor. Persistent or recurrent pain after glomus tumor excision may be an indicator of a residual neoplasm or a recurrent tumor.

Genetic Tumor Risk Syndromes in the Peripheral Nervous System Tumors

Peripheral nervous system can occur in the context of several genetic tumor predisposition syndromes. TRS are highly complex conditions for which specific diagnostic criteria and clinical guidelines have been established [58, 59]. To be aware of when to raise the possibility of a genetic tumor syndromes, we summarize those that should be suspected when a PNS tumors is present.

Neurofibromatosis Type 1

Neurofibromatosis type 1 (NF1) is an autosomal dominant disorder caused by a pathological sequence variant of the *NF1* tumor suppressor gene. It is the most common rare disease with a birth incidence of one in 3000 live births [60]. The main clinical features are café-au-lait spots, iris hamartomas (Lisch nodules), axillary, and inguinal freckling, gliomas in particular affecting optic pathway multiple neurofibromas, including plexiform neurofibromas and specific osseous lesion [61]. It is characterized by a high variable expressivity [11]. Specific diagnostic criteria have established and recently updated [58]. Other neoplasms observed at increase frequency are juvenile myelomonocytic leukaemia, glomus tumours of the digits, gastrointestinal stromal tumours, pheochromocytoma, female breast cancer, and duodenal neuroendocrine tumours [62].

NF2-Related Schwannomatosis

NF2-related schwannomatosis is an autosomal dominant disorder caused by a pathological sequence variant of the *NF2* tumor suppressor gene, with an incidence of one in 33000 live births [63]. In 95% of patients, this disorder is associated with the development of bilateral vestibular schwannomas. *NF2*-related schwannomatosis patients can also develop schwannomas of the cranial, spinal, peripheral, and cutaneous nerves (NF2 plaques), meningiomas , ependymomas, ocular abnormalities such as retinal hamartoma, and cataracts; café-au-lait spots may be present [9]. About 50% of affected individuals have genetic mosaicism, in which a somatic mutation in NF2 gene occurs during embryogenesis, showing variable phenotypes overlapping with those observed in schwannomatosis *LZTR1*- and *SMARCB1*-related [9].

LZTR1- and SMARCB1-Related Schwannomatosis

LZTR1- and *SMARCB1*-related schwannomatosis are two genetically distinct but clinically similar TRS caused by pathogenic variants of *LZTR1* and S*MARCB1* (both on chromosome22q119) genes, respectively. Their birth incidence is approximately one in 69,000 [63]. The hallmark is the presence of multiple schwannomas (spinal cutaneous and cranial) and less commonly cranial or spinal meningiomas. More than 80% of cases are sporadic; in familial form, an autosomal dominant inheritance with completed penetrance has been observed [64].

In about 30% of cases, the tumors have a segmental distribution suggesting the presence of a genetic mosaicism [64]. Because vestibular schwannomas often unilateral may be present, their phenotypic features overlap those observed in *NF2*-related schwannomatosis [9]. Unlike *NF2*-related schwannomatosis, most patients refer chronic pain associated with the tumors, while neurological deficits and polyneuropathy are un common.

Carney Complex

Carney complex (CNC) is an autosomal dominant syndrome affecting less than 700 individuals [65], in which cardiac, mammary, and cutaneous myxomas, myxoma, endocrinopathy, pigmented skin lesions, and malignant melanotic nerve sheath tumors are present. In about 70% of cases, a heterozygous inactivating pathogenic variant of the PRkAR1A gene is detected [37].

Melanoma-Astrocytoma Syndrome

Melanoma-astrocytoma syndrome is an autosomal dominant TRS caused by a germline pathogenic variant of the *CDKN2A* tumor suppressor gene. It is very rare, less than 100 genetically conformed cases are reported at the present, and the exact incidence is unknown. It confers an increased risk for melanoma, pancreatic carcinoma astrocytoma, nerve sheath tumors, and squamous cell carcinoma of the oropharynx [66, 67].

References

1. Walker FO, Cartwright MS, Alter KE, Visser LH, Hobson-Webb LD, Padua L, et al. Indications for neuromuscular ultrasound: expert opinion and review of the literature. Clin Neurophysiol. 2018;129(12):2658–79.
2. Soldatos T, Fisher S, Karri S, Ramzi A, Sharma R, Chhabra A. Advanced MR imaging of peripheral nerve sheath tumors including diffusion imaging. Semin Musculoskelet Radiol. 2015;19(2):179–90.
3. Koeller KK, Shih RY. Intradural extramedullary spinal neoplasms: radiologic–pathologic correlation. Radiographics. 2019;39(2):468–90.
4. Kransdorf MJ. Benign soft-tissue tumors in a large referral population: distribution of specific diagnoses by age, sex, and location. AJR Am J Roentgenol. 1995;164(2):395–402.
5. Woodruff JM, Selig AM, Crowley K, Allen PW. Schwannoma (neurilemoma) with malignant transformation. A rare, distinctive peripheral nerve tumor. Am J Surg Pathol. 1994;18(9):882–95.
6. McMenamin ME, Fletcher CD. Expanding the spectrum of malignant change in schwannomas: epithelioid malignant change, epithelioid malignant peripheral nerve sheath tumor, and epithelioid angiosarcoma: a study of 17 cases. Am J Surg Pathol. 2001;25(1):13–25.

7. Ron E, Modan B, Boice JD, Alfandary E, Stovall M, Chetrit A, et al. Tumors of the brain and nervous system after radiotherapy in childhood. N Engl J Med. 1988;319(16):1033–9.
8. Preston DL, Ron E, Yonehara S, Kobuke T, Fujii H, Kishikawa M, et al. Tumors of the nervous system and pituitary gland associated with atomic bomb radiation exposure. J Natl Cancer Inst. 2002;94(20):1555–63.
9. Halliday D, Parry A, Evans DG. Neurofibromatosis type 2 and related disorders. Curr Opin Oncol. 2019;31(6):562–7.
10. Goetsch Weisman A, Weiss McQuaid S, Radtke HB, Stoll J, Brown B, Gomes A. Neurofibromatosis- and schwannomatosis- associated tumors: approaches to genetic testing and counseling considerations. Am J Med Genet A. 2023;191(10):2467–81.
11. Bettegowda C, Upadhayaya M, Evans DG, Kim A, Mathios D, Hanemann CO, et al. Genotype–phenotype correlations in neurofibromatosis and their potential clinical use. Neurology. 2021;97(7 Suppl 1):S91–8.
12. Kehrer-Sawatzki H, Farschtschi S, Mautner V-F, Cooper DN. The molecular pathogenesis of schwannomatosis, a paradigm for the co-involvement of multiple tumour suppressor genes in tumorigenesis. Hum Genet. 2017;136(2):129–48.
13. Castellanos E, Plana A, Carrato C, Carrió M, Rosas I, Amilibia E, et al. Early genetic diagnosis of neurofibromatosis type 2 from skin plaque plexiform schwannomas in childhood. JAMA Dermatol. 2018;154(3):341–6.
14. Ortonne N, Wolkenstein P, Blakeley JO, Korf B, Plotkin SR, Riccardi VM, et al. Cutaneous neurofibromas: current clinical and pathologic issues. Neurology. 2018;91(2 Suppl 1):S5–13.
15. Paterra R, Bettinaglio P, Borghi A, Mangano E, Tritto V, Cesaretti C, et al. A translational approach to spinal neurofibromatosis: clinical and molecular insights from a wide Italian Cohort. Cancers (Basel). 2022;15(1):59.
16. Evans DG, King AT, Bowers NL, Tobi S, Wallace AJ, Perry M, et al. Identifying the deficiencies of current diagnostic criteria for neurofibromatosis 2 using databases of 2777 individuals with molecular testing. Genet Med. 2019;21(7):1525–33.
17. Ho JD, Ho JAS, Ruddock R, Spencer S-A. Isolated, nonsyndromic mucocutaneous plexiform neurofibromas: a systematic review of the clinicopathologic features. Am J Dermatopathol. 2022;44(12):904.
18. Gutmann DH, Ferner RE, Listernick RH, Korf BR, Wolters PL, Johnson KJ. Neurofibromatosis type 1. Nat Rev Dis Primers. 2017;3:17004.
19. Fisher MJ, Blakeley JO, Weiss BD, Dombi E, Ahlawat S, Akshintala S, et al. Management of neurofibromatosis type 1-associated plexiform neurofibromas. Neuro Oncol. 2022;24(11):1827–44.
20. Beert E, Brems H, Daniëls B, De Wever I, Van Calenbergh F, Schoenaers J, et al. Atypical neurofibromas in neurofibromatosis type 1 are premalignant tumors. Genes Chromosomes Cancer. 2011;50(12):1021–32.
21. Higham CS, Dombi E, Rogiers A, Bhaumik S, Pans S, Connor SEJ, et al. The characteristics of 76 atypical neurofibromas as precursors to neurofibromatosis 1 associated malignant peripheral nerve sheath tumors. Neuro Oncol. 2018;20(6):818–25.
22. Miettinen MM, Antonescu CR, Fletcher CDM, Kim A, Lazar AJ, Quezado MM, et al. Histopathologic evaluation of atypical neurofibromatous tumors and their transformation into malignant peripheral nerve sheath tumor in patients with neurofibromatosis 1—a consensus overview. Hum Pathol. 2017;67:1–10.
23. Röhrich M, Koelsche C, Schrimpf D, Capper D, Sahm F, Kratz A, et al. Methylation-based classification of benign and malignant peripheral nerve sheath tumors. Acta Neuropathol. 2016;131(6):877–87.
24. Weiss BD, Wolters PL, Plotkin SR, Widemann BC, Tonsgard JH, Blakeley J, et al. NF106: a neurofibromatosis clinical trials consortium phase II trial of the MEK inhibitor mirdametinib (PD-0325901) in adolescents and adults with NF1-related plexiform neurofibromas. J Clin Oncol. 2021;39(7):797–806.
25. Hornick JL, Bundock EA, Fletcher CDM. Hybrid schwannoma/perineurioma: clinicopathologic analysis of 42 distinctive benign nerve sheath tumors. Am J Surg Pathol. 2009;33(10):1554–61.
26. Harder A, Wesemann M, Hagel C, Schittenhelm J, Fischer S, Tatagiba M, et al. Hybrid neurofibroma/schwannoma is overrepresented among schwannomatosis and neurofibromatosis patients. Am J Surg Pathol. 2012;36(5):702–9.
27. Inatomi Y, Ito T, Nagae K, Yamada Y, Kiyomatsu M, Nakano-Nakamura M, et al. Hybrid perineurioma-neurofibroma in a patient with neurofibromatosis type 1, clinically mimicking malignant peripheral nerve sheath tumor. Eur J Dermatol. 2014;24(3):412–3.
28. Kacerovska D, Michal M, Kuroda N, Tanaka A, Sima R, Denisjuk N, et al. Hybrid peripheral nerve sheath tumors, including a malignant variant in type 1 neurofibromatosis. Am J Dermatopathol. 2013;35(6):641–9.
29. Hornick JL, Fletcher CDM. Soft tissue perineurioma: clinicopathologic analysis of 81 cases including those with atypical histologic features. Am J Surg Pathol. 2005;29(7):845–58.
30. Alkhaili J, Cambon-Binder A, Belkheyar Z. Intraneural perineurioma: a retrospective study of 19 patients. Pan Afr Med J. 2018;30:275.
31. Schaefer I-M, Ströbel P, Thiha A, Sohns JM, Mühlfeld C, Küffer S, et al. Soft tissue perineurioma and other unusual tumors in a patient with neurofibromatosis type 1. Int J Clin Exp Pathol. 2013;6(12):3003–8.
32. Al-Adnani M. Soft tissue perineurioma in a child with neurofibromatosis type 1: a case report and review of the literature. Pediatr Dev Pathol. 2017;20(5):444–8.
33. White B, Belzberg A, Ahlawat S, Blakeley J, Rodriguez FJ. Intraneural perineurioma in neurofibromatosis type 2 with molecular analysis. Clin Neuropathol. 2020;39(4):167–71.

34. Pendleton C, Spinner RJ, Dyck PJB, Mauermann ML, Ladak A, Restrepo CE, et al. Association of intraneural perineurioma with neurofibromatosis type 2. Acta Neurochir (Wien). 2020;162(8):1891–7.
35. Uusitalo E, Rantanen M, Kallionpää RA, Pöyhönen M, Leppävirta J, Ylä-Outinen H, et al. Distinctive cancer associations in patients with neurofibromatosis type 1. J Clin Oncol. 2016;34(17):1978–86.
36. Carney JA. Psammomatous melanotic schwannoma. A distinctive, heritable tumor with special associations, including cardiac myxoma and the Cushing syndrome. Am J Surg Pathol. 1990;14(3):206–22.
37. Bonomo G, Gans A, Mazzapicchi E, Rubiu E, Alimonti P, Eoli M, et al. Sporadic spinal psammomatous malignant melanotic nerve sheath tumor: a case report and literature review. Front Oncol. 2023;13:1100532.
38. Torres-Mora J, Dry S, Li X, Binder S, Amin M, Folpe AL. Malignant melanotic schwannian tumor: a clinicopathologic, immunohistochemical, and gene expression profiling study of 40 cases, with a proposal for the reclassification of "melanotic schwannoma". Am J Surg Pathol. 2014;38(1):94–105.
39. Evans DGR, Huson SM, Birch JM. Malignant peripheral nerve sheath tumours in inherited disease. Clin Sarcoma Res. 2012;2(1):17.
40. Ferner RE, Golding JF, Smith M, Calonje E, Jan W, Sanjayanathan V, et al. [18F]2-fluoro-2-deoxy-D--glucose positron emission tomography (FDG PET) as a diagnostic tool for neurofibromatosis 1 (NF1) associated malignant peripheral nerve sheath tumours (MPNSTs): a long-term clinical study. Ann Oncol. 2008;19(2):390–4.
41. Azizi AA, Slavc I, Theisen BE, Rausch I, Weber M, Happak W, et al. Monitoring of plexiform neurofibroma in children and adolescents with neurofibromatosis type 1 by [18 F]FDG-PET imaging. Is it of value in asymptomatic patients? Pediatr Blood Cancer. 2018;65(1):26733.
42. Suppiah S, Mansouri S, Mamatjan Y, Liu JC, Bhunia MM, Patil V, et al. Multiplatform molecular profiling uncovers two subgroups of malignant peripheral nerve sheath tumors with distinct therapeutic vulnerabilities. Nat Commun. 2023;14(1):2696.
43. Cortes-Ciriano I, Steele CD, Piculell K, Al-Ibraheemi A, Eulo V, Bui MM, et al. Genomic patterns of malignant peripheral nerve sheath tumor (MPNST) evolution correlate with clinical outcome and are detectable in cell-free DNA. Cancer Discov. 2023;13(3):654–71.
44. Szymanski JJ, Sundby RT, Jones PA, Srihari D, Earland N, Harris PK, et al. Cell-free DNA ultra-low-pass whole genome sequencing to distinguish malignant peripheral nerve sheath tumor (MPNST) from its benign precursor lesion: a cross-sectional study. PLoS Med. 2021;18(8):e1003734.
45. Mattox AK, Douville C, Silliman N, Ptak J, Dobbyn L, Schaefer J, et al. Detection of malignant peripheral nerve sheath tumors in patients with neurofibromatosis using aneuploidy and mutation identification in plasma. eLife. 2022;11:e74238.
46. Prudner BC, Ball T, Rathore R, Hirbe AC. Diagnosis and management of malignant peripheral nerve sheath tumors: current practice and future perspectives. Neurooncol Adv. 2020;2(Suppl 1):i40–9.
47. Le Guellec S, Decouvelaere A-V, Filleron T, Valo I, Charon-Barra C, Robin Y-M, et al. Malignant peripheral nerve sheath tumor is a challenging diagnosis: a systematic pathology review, immunohistochemistry, and molecular analysis in 160 patients from the French sarcoma group database. Am J Surg Pathol. 2016;40(7):896–908.
48. Miao R, Wang H, Jacobson A, Lietz AP, Choy E, Raskin KA, et al. Radiation-induced and neurofibromatosis-associated malignant peripheral nerve sheath tumors (MPNST) have worse outcomes than sporadic MPNST. Radiother Oncol. 2019;137:61–70.
49. Okubo T, Saito T, Mitomi H, Takagi T, Torigoe T, Suehara Y, et al. Intraneural lipomatous tumor of the median nerve: Three case reports with a review of literature. Int J Surg Case Rep. 2012;3(9):407–11.
50. Marek T, Mahan MA, Carter JM, Howe BM, Bartos R, Amrami KK, et al. What's known and what's new in adipose lesions of peripheral nerves? Acta Neurochir (Wien). 2021;163(3):835–42.
51. Blackburn PR, Milosevic D, Marek T, Folpe AL, Howe BM, Spinner RJ, et al. PIK3CA mutations in lipomatosis of nerve with or without nerve territory overgrowth. Mod Pathol. 2020;33(3):420–30.
52. Sonneland PR, Scheithauer BW, LeChago J, Crawford BG, Onofrio BM. Paraganglioma of the cauda equina region. Clinicopathologic study of 31 cases with special reference to immunocytology and ultrastructure. Cancer. 1986;58(8):1720–35.
53. Adriani KS, Stenvers DJ, Imanse JG. Pearls & oysters: Lumbar paraganglioma: can you see it in the eyes? Neurology. 2012;78(4):e27–8.
54. Strommer KN, Brandner S, Sarioglu AC, Sure U, Yonekawa Y. Symptomatic cerebellar metastasis and late local recurrence of a cauda equina paraganglioma. Case report. J Neurosurg. 1995;83(1):166–9.
55. McDermott EM, Weiss A-PC. Glomus tumors. J Hand Surg Am. 2006;31(8):1397–400.
56. Yanai K, Tajika T, Kuboi T, Hatori Y, Umeyama T, Arisawa S, et al. A case of solitary digital glomus tumor associated with neurofibromatosis type 1. SAGE Open Med Case Rep. 2023;11:2050313X231193984.
57. Cohen PR. Glomus extradigital tumor: a case report of an extradigital glomus tumor on the wrist and comprehensive review of glomus tumors. Cureus. 2023;15(5):e38737.
58. Legius E, Messiaen L, Wolkenstein P, Pancza P, Avery RA, Berman Y, et al. Revised diagnostic criteria for neurofibromatosis type 1 and Legius syndrome: an international consensus recommendation. Genet Med. 2021;23(8):1506–13.
59. Plotkin SR, Messiaen L, Legius E, Pancza P, Avery RA, Blakeley JO, et al. Updated diagnostic criteria and nomenclature for neurofibromatosis type 2 and schwannomatosis: an international consensus recommendation. Genet Med. 2022;24(9):1967–77.

60. Kallionpää RA, Valtanen M, Auranen K, Uusitalo E, Rinne JO, Peltonen S, et al. Increased risk for dementia in neurofibromatosis type 1. Genet Med. 2021;23(11):2219–22.
61. Eoli M, Saletti V, Finocchiaro G. Neurological malignancies in neurofibromatosis type 1. Curr Opin Oncol. 2019;31(6):554–61.
62. Brems H, Beert E, de Ravel T, Legius E. Mechanisms in the pathogenesis of malignant tumours in neurofibromatosis type 1. Lancet Oncol. 2009;10(5):508–15.
63. Evans DG, Bowers NL, Tobi S, Hartley C, Wallace AJ, King AT, et al. Schwannomatosis: a genetic and epidemiological study. J Neurol Neurosurg Psychiatr. 2018;89(11):1215–9.
64. MacCollin M, Willett C, Heinrich B, Jacoby LB, Acierno JS, Perry A, et al. Familial schwannomatosis: exclusion of the NF2 locus as the germline event. Neurology. 2003;60(12):1968–74.
65. Stratakis CA, Raygada M. Carney complex. In: Adam MP, Ardinger HH, Pagon RA, Wallace SE, Bean LJ, Stephens K, et al., editors. GeneReviews®. Seattle: University of Washington; 1993.
66. Petronzelli F, Sollima D, Coppola G, Martini-Neri ME, Neri G, Genuardi M. CDKN2A germline splicing mutation affecting both p16(ink4) and p14(arf) RNA processing in a melanoma/neurofibroma kindred. Genes Chromosomes Cancer. 2001;31(4):398–401.
67. Sargen MR, Kim J, Potjer TP, Velthuizen ME, Martir-Negron AE, Odia Y, et al. Estimated prevalence, tumor spectrum, and neurofibromatosis type 1-like phenotype of CDKN2A-related melanoma-astrocytoma syndrome. JAMA Dermatol. 2023;159(10):1112–8.

9 Clinical Assessment of Tumor Predisposition Syndromes: Related Nerve Sheath Tumors in Children

Veronica Saletti, Rossana Bucci, Germana Lo Monaco, and Marco Moscatelli

9.1 Introduction

In accordance with the 2021 World Health Organization (WHO) classification of central nervous system (CNS) tumors and the 2020 WHO classification of soft tissue and bone tumors, nerve sheath tumors (NSTs) comprise a heterogeneous group of soft tissue tumors of neuroectodermal origin associated with nerve structures [1–3]. NSTs derive mainly from Schwann cells and perineurial cells, more rarely from specialized neuroendocrine cells of the sympathetic and parasympathetic nervous system. They include benign and malignant tumors such as neurofibroma, plexiform neurofibroma (PN), schwannoma, perineurioma, hybrid peripheral nerve sheath tumor, malignant peripheral nerve sheath tumor (MPNST), epithelioid MPNST or malignant melanotic nerve sheath tumor, and neuroendocrine tumor of the cauda equina (formerly known as paraganglioma of the central nervous system) [1, 2, 4].

In the general population, benign NSTs account for 10–12% of all benign soft tissue tumors [3, 5]. Among benign NSTs, neurofibromas and schwannomas are more common and arise more frequently in adulthood, at which time, they usually develop as single and sporadic lesions. Indeed, in 95% of cases, neurofibromas and schwannomas occur sporadically, while in the remaining cases, they can be part of tumor predisposition syndromes such as neurofibromatosis type 1 (NF1), neurofibromatosis type 2 (NF2), and schwannomatosis (SWN) that are unrelated and genetically distinct syndromes [4, 5]. Malignant NSTs are rare with an incidence from 3 to 10% of soft tissue sarcomas with MPNST having a higher incidence in NF1 patients [3].

Recently, following an advance in phenotypic and genetic characterization, an update of the diagnostic criteria for NF1, NF2, and SWN has been proposed and an updated nomenclature for NF2, and SWN has been recommended by removing the term "neurofibromatosis 2" to emphasize the phenotypic overlap between the two conditions and to minimize misdiagnosis with NF1 [6, 7].

V. Saletti · R. Bucci · G. Lo Monaco
Epileptology and Developmental Neurology Unit, Mariani Foundation Center for Complex Disabilities, Fondazione IRCCS Istituto Neurologico Carlo Besta, Milan, Italy
e-mail: veronica.saletti@istituto-besta.it; rossana.bucci@istituto-besta.it

M. Moscatelli (✉)
Neuroradiology Unit, Fondazione IRCCS Istituto Neurologico Carlo Besta, Milan, Italy

Department of Biomedical Sciences for Health, University of Milan, Milan, Italy
e-mail: marco.moscatelli@istituto-besta.it

I. G. Vetrano, V. Nazzi (eds.), *Peripheral Nerve Sheath Tumors*,
https://doi.org/10.1007/978-3-031-76382-3_9

In case of multiple NSTs, these tumor predisposition syndromes should be suspected at any age; however, in pediatric age, they should be suspected also in case of a single NST, with or without other specific signs and symptoms, as the single lesion can represent the first or most evident sign of NF1 and SWN.

A careful diagnostic approach is therefore necessary not only for the NSTs themselves, in differential diagnosis with not neurogenic lesions, but also for any systemic condition in which they may develop.

The management of NSTs varies depending on the type of lesion, its location, the age at which it develops (childhood versus adulthood), and the genetic aspects (isolated versus syndromic form).

Since an incorrect diagnosis can lead to incorrect management with significant clinical consequences, it is important that the diagnostic approach is addressed in specialized centers with dedicated multidisciplinary teams [8].

NSTs diagnostic workup includes the following: a detailed familial and personal medical history; a careful clinical and neurological examination; magnetic resonance imaging (MRI) studies, which play a central role in NSTs characterization and represent the gold standard imaging modality even if these are not always useful in differentiating between solitary schwannoma and neurofibroma [3–5, 9, 10]; and ultrasound, which could be the first-line imaging procedure in the case of superficial musculoskeletal soft-tissue lesions [4, 11]. This last technique may allow an initial exclusion of mimicking pathologies and, when necessary, it can guide percutaneous biopsy [12]; however, it should be carried out by expert technicians as it is strictly operator-dependent [11]. Occasionally neurophysiological tests may also be necessary in the diagnostic phase [8]. When biopsy or resection is necessary, tumor diagnosis is primarily based on hematoxylin and eosin-stained sections and immunochemistry; molecular analysis is not essential, but it is required on extracted DNA from neurofibromas or schwannomas, to diagnose mosaic forms of NF1 and SWNs [4].

We summarize the main features of NF1 and SWN focusing on the types of NSTs, which characterize these tumor predisposition syndromes. In addition to the main clinical, radiological, and pathological characteristics of the NSTs, we report the recent management recommendations with emphasis in pediatric age.

9.2 Neurofibromatosis Type 1 and Related Nerve Sheath Tumors in Children

NF1 is one of the most common autosomal dominant disorders with an estimated birth incidence of approximately one in 2.000–2.500 without predilection for sex and race [13].

It is due to germline or constitutional heterozygous inactivating pathogenic variants (PVs) of the *NF1* gene located on chromosome 17q11 and encoding for the tumor suppressor protein neurofibromin. Approximately half of the patients inherit the *NF1* PV from an affected parent, while the remaining 50% are carriers of a "de novo" PV. To date, approximately 3000 different PVs have been described [14].

NF1 is considered a neurocutaneous disease since it is characterized by a prevalent cutaneous and neurological involvement; it is, however, a multisystem condition as it can affect different organs/systems with broad and variable manifestations/complications. The pathology is also a cancer-prone syndrome, as neurofibromin has a role in many signaling pathways, among which a negative regulator of the RAS/mitogen-activated protein kinase (MAPK) pathway. The tumors especially affect the central and peripheral nervous system because neurofibromin, although ubiquitously expressed, has the highest level of expression within the nervous system. The neoplasms typically develop in the context of biallelic inactivation of the *NF1* gene due to a second somatic mutation with consequent loss of neurofibromin expression and unregulated cell proliferation and tumorigenesis [6, 15].

NF1 diagnosis can be made when the diagnostic criteria, established by a panel of international experts, are met. Initially developed in 1987 by the National Institute of Health, they were found to be highly specific and sensitive, so much so that they were only revised in 2021 with the formulation of the criteria reported by Legius and coauthors [6].

NF1 penetrance approaches 100% by the end of the first decade of life since NF1 features are typically age-dependent and may present with wide variability, even among affected family members; moreover, genotype-phenotype correlations are very limited [14, 16–18].

In addition to the diagnostic criteria, NF1 patients have an increased risk of arterial hypertension and vasculopathy and may show macrocephaly, short stature, epilepsy, hydrocephalus, neurodevelopmental disorders (such as speech delay, motor coordination deficit, learning disabilities, and autism spectrum disorder), and brain anomalies represented by hyperintense areas on T2-weighted MRI, now known as focal area of signal intensity (FASI) and formerly known as unidentified bright objects (UBOs). FASI are sometimes difficult to distinguish from low-grade glioma, but they typically lack contrast enhancement, do not cause mass effects, and do not require imaging surveillance because they are not related to tumor development or focal neurologic deficits [19–22].

In regard to the diagnostic criteria, in the 2021 revision, the finding of the *NF1* gene PV, which occurs through next-generation sequencing techniques (NGS) and multiplex ligation-dependent probe amplification (MLPA) which is a useful technique for identifying deletions and duplications, has been included. Previously, the diagnosis was based almost exclusively on clinical-instrumental diagnostic criteria because molecular analysis was very complex and reserved for very selected cases.

The other diagnostic criteria are represented by the following: café-au-lait (CAL) spots, skinfold freckling, typical bone lesions, Lisch nodules (or iris hamartomas) and choroidal abnormalities, optic pathways glioma, and neurofibromas.

The first clinical sign, which represents an alarm bell for parents and pediatricians, is represented by the CAL spots, often present from birth or from the first months of life in 99% of affected people; these progressively grow up until puberty both in number and size but have no correlation with the development of more significant complications in later life. Smaller macules, which represent the so-called skinfold freckling, usually follow at the age of 2–3 years.

The typical bone lesions (sphenoid dysplasia, anterolateral curvature of the tibia, or pseudarthrosis of the long bones) are rare (1–2%) but are strongly suggestive of the pathology and appear from the first year of life, sometimes causing significant sequelae.

Lisch nodules (or iris hamartomas) and choroidal abnormalities, frequent from preschool age, don't impact visual acuity. The optic pathway glioma, a benign tumor that can affect any section of the optical pathways from the optic nerves to the optic radiations, develops in approximately 15–20% of cases, typically in preschool age, very rarely after the age of 10 age of life, and in most cases, it has an indolent course without causing visual function disturbances; therefore, it rarely requires pharmacological and sporadically surgical therapy.

9.2.1 *Neurofibromas* are the Hallmark of NF1

They are benign NSTs that arise from Schwann cells but are composed by multiple additional cell types including fibroblasts, mast cells, macrophages, perineural cells and extracellular matrix as collagen [23]. They are centrally located, not encapsulated and not easily separable from the normal nerve [5]. Two main types of neurofibromas can be present simultaneously in NF1 patients: localized/nodular and plexiform neurofibromas.

1. **Localized/nodular neurofibromas** are mainly localized on the skin arising from the terminal ends of small nerves.

   Although they can appear during the first years of life, they usually develop starting from puberty and are present in almost all elderly NF1 subjects; they may appear or increase in number and size during periods of hormonal excess such as puberty and pregnancy.

When they develop on the skin, they initially appear flat with hyperpigmented skin, later as papules, and then pass onto the globular stage, which is a larger painless nodule of 20–30 mm dimensions; in the final phase, they can be pedunculated. They are soft and mobile compared to the deep tissues.

Less frequently, neurofibromas can have a subcutaneous location. In these cases, they may present in young and middle-aged adults as palpable, slow-growing, duroplastic subcutaneous nodules, typically less than 5 cm in diameter and can cause pain on palpation.

More rarely, they can be located at deep nerves arising from major nerves and, unlike cutaneous ones, they can cause sensory or motor deficits. The cranial nerves are always spared [24].

Cutaneous and subcutaneous neurofibromas are not at risk for malignant transformation but may have a significant impact on the quality of life causing aesthetic problems (especially when they are relatively large in size, or very numerous, or in particular locations such as, for example, the face), itching or discomfort if in friction areas. Removal of cutaneous or subcutaneous neurofibromas by plastic surgeons or dermatologists may be recommended if they rub on clothing or cause discomfort [25, 26]. Intraneural deep nodular neurofibromas may occasionally present the morphological precursor of malignant transformation [24].

2. **Plexiform neurofibromas** (PNs) are benign PNTs that involve major peripheral nerves or nerve roots and multiple nerve fascicles. They are vascular and multicellular tumors that typically grow along a nerve and its branches, frequently colliding into large masses, especially in the brachial or lumbosacral plexi.

   PN is a type of peripheral NST arising virtually only in NF1 patients, and it is considered pathognomonic of this syndrome [24]. In NF1 subjects, PNs are clinically visible in up to 30% of cases, but MRI studies have demonstrated a prevalence between 50 and 60% of cases [27–29].

   Unlike cutaneous neurofibromas, they are thought to be congenital or to develop during early childhood, although they are mostly diagnosed in pre-adolescence [30]. Natural history studies showed that PNs growth varies between patients, even between different tumors in the same patient, but younger age is typically associated with more rapid tumor growth [31].

   PNs can arise throughout the entire body even if they are more common in the craniofacial area, neck, pelvis, and lower extremities [28, 29, 32]. They may extend superficially or deeply and may have overlying skin manifestations, such as thickened orange peel skin with purplish-brown discoloration or hairy patches or hyperpigmentation, and may have a heterogeneous texture or nodularity on palpation [26].

   PN-related morbidity is highly variable; asymptomatic and symptomatic PN can coexist in the same individual based on tumor location and size, age, and nerve involvement [30]. Comorbidities include diffuse or localized pain, disfigurement, and motor dysfunction, but they can cause airway or other vital structures compression and visual impairment [29, 33, 34].

   The gold standard diagnostic imaging modality for PN is MRI with short tau inversion recovery sequences which is a fat suppression technique that provides optimal visualization of PN. On MRI, based on the morphology, PN can present as a diffuse type with a homogeneous internal structure, as a diffuse nervous thickening, particularly in the roots of the spinal nerves, and as a complex multinodular mass composed of irregular/serpentine structures (bag of wors appearance). Relative to adjacent structures, boundaries may be well or poorly defined [30].

   PN is usually isointense to muscle on T1-weighted images, with heterogeneous high signal intensity on T2-weighted images. Enhancement is usually moderate and variable after contrast medium injection [35]. The target sign (central hypointensity and peripheral hyperintensity on T2, diffusion, and post-contrast T1-weighted MRI sequences) is suggestive of a benign nature of the lesion [5].

Even if some experts recommend targeted and whole-body MRI to screen for subclinical PNs early in life, currently, screening imaging is not recommended for diagnosis, and there is no standard of care for frequency of surveillance imaging to monitor for tumor growth once diagnosed [36, 37]. Most PNs do not require intervention but clinical and imaging observation.

Historically, medical treatment of PNs is focused on patient education, monitoring, and symptom management. Until recently, surgical resection was the only treatment option with the main goal of removing tumors and reducing disfigurement and pain. However, surgical resection is challenging as PNs are highly vascularized, complete resection is rarely possible, and resection-related morbidity risk is very important [32].

In the past decade, significant advances have been made in the medical management of PNs, and promising therapeutic agents have been identified. Thanks to the therapeutic effects demonstrated by Selumetinib, a MEK-inhibitor that acts by inhibiting the RAS-MAPK pathway hyperactivated by neurofibromin loss of function, the Food and Drug Administration in 2020 and the European Medicine Agency in 2021 approved the drug for the treatment of symptomatic and inoperable PN in children with NF1 [38]. Other MEK inhibitors have also demonstrated efficacy and are under investigation for NF1 children and adults with PNs [39–41]. However, despite these successes, there are still open questions regarding the duration and the long-term toxicities of this type of treatment. Furthermore, since the NF1+/− microenvironment also contributes to the origin of PN, other clinical studies focusing on the microenvironment are ongoing [42].

Currently, treatment is considered when PN causes significant morbidity or when it is in progression with potential morbidity. If treatment is indicated, a multidisciplinary team of expert clinicians and surgeons should select between medical and surgical intervention [30, 37].

9.2.2 Malignant Peripheral Nerve Sheath Tumors

Despite screening imaging is not recommended for PNs diagnosis and despite the lack of standard of care for monitoring diagnosed PNs, they require careful monitoring as they have an estimated lifetime risk of approximately 15% of possible malignant transformation into the aggressive malignant peripheral nervous system tumor (MPNST) [13, 43]. Most NF1 patients develop MPNST in early adulthood, but malignant transformation can also occur in childhood. Malignant transformation of PN to MPNST not only requires biallelic loss of functional *NF1* gene but also the loss of other tumor suppressor genes and/or the amplification of oncogenes. Specifically, loss of *CDKN2A/B* has been suggested as a key factor in the progression of MPNST and has been identified in approximately 50% of atypical neurofibromatous neoplasms of uncertain biological potential (ANNUBP), previously called atypical neurofibromas (AN), which are considered the intermediate state between PN and MPNST. ANNUBPs are characterized by peculiar radiological characteristics as well as peculiar molecular and histological characteristics [44–46].

Rapid growth of PN and worsening pain are indicators of possible malignant transformation and should be induced immediately for further evaluation with advanced MRI techniques (diffusion-weighted and dynamic contrast-enhanced sequences) and fluorodeoxyglucose (FDG) positron emission tomography (PET). This last technique has high diagnostic accuracy as MPNST are FDG-avid. However, AN/

ANNUBP frequently has increased uptake of FDG, and benign PN can have varied FDG uptake. Biopsy is of consequence needed.

While the usefulness of surgically intervening on AN/ANNUBP is discussed, gross total surgical resection of the lesion with wide negative margins is the gold standard intervention for MPNST. The benefits of chemotherapy remain undetermined and the role of adjuvant radiotherapy is debated. Novel targeted therapies and immunotherapies are under investigation. This type of tumor is the leading cause of early death with 5-year overall survival estimates between 20 and 50%. Poor prognosis is associated with tumor size greater than 5 cm, deep location, incomplete resection, and metastatic disease (most commonly to the lung, brain, and bone [37].

9.3 Schwannomatosis and Related Nerve Sheath Tumors in Children

On the basis of the recently revised nomenclature, schwannomatosis represents an umbrella term used to describe genetically distinct tumor predisposition syndromes, NF2 and SWN, that share overlapping phenotypes and are both characterized by the development of typical NSTs that are schwannomas [7]. In addition to the revised nomenclature, updated diagnostic criteria have been proposed by incorporating recent advances in genetics, ophthalmology, neuropathology, and neuroimaging, but it is likely that diagnostic criteria will also be continually refined based on the identification of new genes associated with SWN [7].

NF2 is an autosomal dominant syndrome that predisposes individuals to multiple tumors of the nervous system. It is caused by heterozygous loss of function mutation in the *NF2* gene located on chromosome 22q12. *NF2* mutations result in loss of production of the tumor suppressor protein Merlin that functions in multiple signaling pathways including the PI3K/Akt/mTOR and Raf/MEK/ERK. About half of the cases inherit the mutation, while the remaining cases occur de novo. The disease typically presents in adulthood; however, nearly 20% of patients are diagnosed at less than 15 years of age with earlier presentation associated with a poorer prognosis. More severe phenotypes are also associated with truncating mutations [47, 48]. Bilateral vestibular schwannomas are pathognomonic, but patients can develop other intracranial and peripheral schwannomas as well as meningiomas (up to 80% of patients) and intramedullary ependymomas (up to 35% of patients) and peculiar ophthalmologic signs, such as lenticular opacities, retinal hamartomas, and epiretinal membranes.

The first description of NF2 by Wishart was in 1822, while the first descriptions of NF1 by von Recklinghausen were in 1882. For a long time, the two conditions were confused; however, today the difference is clear to most clinicians. With the exception of café-au-lait (CAL) spots, which is a very small percentage of patients with NF2 can be >6, individuals with NF2 do not develop the other typical and diagnostic features of NF1 [6, 7].

The diagnostic criteria for NF2 were first formulated in 1988 [49], then revised in 1992 [50], and finally in 2019 [51], when the so-called "Manchester criteria" were published with the word "glioma" replaced by "ependymoma," the word "neurofibroma" removed, and the age limit of 70 years for the development of vestibular schwannomas, and molecular criteria, introduced.

The *NF2* gene was cloned in 1993 [52, 53]. Currently, a genetic diagnosis approach based on NGS and MLPA, high-resolution karyotypic studies, allows the identification of the germline mutation in 96% of affected subjects of the second generation with the NF2 diagnostic criteria and both vestibular schwannomas. In de novo NF2 patients, especially if with later onset and asymmetric presentation, the detection rate is lower because approximately 60% of these patients are mosaic [7, 54, 55].

Tumorigenesis in NF2, as in NF1, follows the two-hit model in which somatic inactivation of the second *NF2* allele occurs in addition to *NF2* germline PV.

The first evidence of SWN as a disease separate from NF2 dates back to the late 1990s when isolated and families with non-cranial, non-intradermal, and painful schwannomas were described [56, 57]. However, the first SWN diagnostic criteria were published in 2005 [58]. These required the presence of at least two schwannomas (1 proven histologically) with the exclusion of VS after the age of 18.

After the identification of the *SMARCB1* gene as a cause of SWN and the description of unilateral VS in SWN, the diagnostic criteria were modified with the inclusion of a molecular classification [59–61]. In 2014 another gene, *LZTR1* was identified as causative of SWN [62]. However, germline PVs of *SMARCB1* and *LZTR1* genes, both located centromeric to *NF2* on chromosome 22, also studied by RNA-based testing, do not explain all cases of SWN, but up to 80% of familial cases and only about 30% of sporadic cases [7]. Consequently, it is likely that other genes may cause SWN.

Patients with SWN do not harbor germline *NF2* PV; however, tumors in SWN show somatic *NF2* PV and/or loss of the second *NF2* allele. For tumorigenesis, a three-step/four-hit hypothesis has been proposed with biallelic inactivation (germinal and somatic) of *SMARCB1* or *LZTR1* genes and biallelic *NF2* gene inactivation in tumors [7, 63].

While nowadays there are no more reasons for diagnostic uncertainties between NF1 and NF2, sometimes diagnostic uncertainties exist between NF1 and SWN, but especially between NF2 and SWN. In the first case, some patients with schwannomatosis may be misdiagnosed as having NF1, as schwannomas or hybrid tumors may be confused with neurofibromas [7]. In the second case, the distinction can be very complex. Not only sporadic patients with SWN can result negative for germline *SMARCB1* or *LZTR1* PVs, but they can also be difficult to distinguish from NF2 mosaics which require for the diagnosis the identification of the same *NF2* PV in 2 anatomically unrelated tumors [7, 64].

For these reasons, the umbrella term schwannomatosis and a new classification, by incorporating clinical features and the PV identified on molecular analysis, have been recently proposed [7].

In the revised nomenclature, NF2 has been termed "*NF2*-related schwannomatosis," and SWN has been termed "*SMARCB1*-related SWN," or "*LZTR1*-related SWN," or "22q-related SWN" for patients with multiple schwannomas and loss of chromosome 22q in multiple schwannomas, or "SWN-not otherwise specified" for patients with clinical features of NF2/SWN without molecular analysis, and "SWN-not elsewhere classified" for patients in whom molecular analysis of blood and tumors failed to detect a PV [7]. Revised diagnostic criteria for *NF2*-related SWN, for *SMARCB1* or *LZTR1*-related SWN, and for 22q-related SWN have been reported by Legius and coauthors [7].

9.3.1 Schwannomas

Schwannomas characterize all forms of schwannomatosis. They are benign NSTs derived from Schwann cells of the cranial and peripheral nerves. They may develop at multiple body sites including the spinal roots, but, when intracranial, they mainly involve the vestibular branch of cranial nerve VIII and are referred to as vestibular schwannomas (VS) [3, 4]. Malignant transformation is exceptionally rare [65]. They are usually slowly growing globular well-circumscribed (often <5 cm) masses, located eccentrically to the nerve fibers and encapsulated by the perineurium.

At MRI, schwannomas are hypo-intense or iso-intense on T1-weighted images and hyperintense, heterogeneous, or not on T2-weighted images. Sometimes a hypo-intense pseudocapsule corresponding to the perineurium is visible [35]. Schwannomas appear well-defined, readily surrounded by a halo of fat, the so-called split fat sign, characteristic of schwannomas [66]. Enhancement intensity is often diffuse, sometimes predominantly peripheral.

Multiple schwannomas are typical of *NF2*-related SWN and non-*NF2*-related SWN. While bilateral vestibular schwannomas (VSs) are a criterion for *NF2*-related SWN, patients with non-

NF2-related SWN generally have a preponderance of peripheral schwannomas involving spinal roots, plexi, and peripheral nerves, less frequently intracranial tumors including vestibular usually unilateral.

Unlike the *NF2*-related form, in non-*NF2*-related SWN, the main clinical manifestation is pain which often requires multiple surgical procedures.

9.3.1.1 Vestibular Schwannomas

The most common tumor in *NF2*-related SWN patients is the VS, which arises from the vestibular branches of the vestibulocochlear nerve in over 95% of patients. VS can cause hearing loss and tinnitus, vertigo, unsteadiness, facial neuropathy, and brainstem compression; hydrochepalus may occur in case of large cerebellopontine angle tumors. At MRI, VS typically appear isointense on T1-weighted and heterogeneously intense on T2-weighted imaging and show contrast-enhancement on T1 sequences post-gadolinium.

Management of pediatric VS includes observation, surgery, and chemotherapy such as bevacizumab, while radiotherapy should be postponed if feasible because, like other tumor predisposition syndromes, it carries significant risks of secondary malignancies. Regarding surgery, to date recommendations based on tumor size have not been proposed for pediatric patients. Surgery however is considered for patients with tumor progression and prior loss of hearing and for patients with brainstem compression, and it is indicated for patients with obstructive hydrocephalus or brainstem dysfunction due to mass effect regardless of tumor size. Surgery is considered also a hearing preservation therapy, but the risks include further hearing loss or deafness, tinnitus or exacerbations of preexisting symptoms, and facial nerve damage. The evaluation for surgery, consequently, must incorporate patient age, hearing, and neurological status, and tumor size [37, 67].

Effective systemic chemotherapies for VS are lacking. Bevacizumab is a non-FDA and EMA-approved therapy that has resulted in VS volume reduction in 32–43% and hearing improvement in 36–45% of patients with *NF2*-related SWN, but these improvements appear to be less common in children than adults [68, 69]. Despite these results, some clinical benefits, such as improvement in tinnitus, quality of life, and disease stabilization, may also occur in children [37, 69]. Other targeted therapies are under evaluation in clinical trials [37].

9.3.1.2 Non-vestibular Schwannomas

Non-vestibular schwannomas are common in *NF2*-related and not *NF2*-related SWN. They can arise from the cranial nerves, peripheral nerves, and nerve roots. Clinically, they can present as cutaneous or subcutaneous, asymptomatic, or painful nodules; when they develop on the spinal roots, they can grow through the neural foramina, causing radicular pain, sensory, or less frequently, motor symptoms. Observation is the initial preferred management, but if schwannomas are progressive or symptomatic, surgery is the primary therapeutic intervention. As with VS, effective systemic therapies are lacking, but platform-basket trial designs (such as the current NCT04374305 trial) should facilitate more rapid evaluation of new treatments for NF2-related SWN [37].

9.3.2 Perineuriomas

Perineuriomas are benign, rare, but probably under-recognized, peripheral nerve tumor composed of differentiated neoplastic perineurial cells that arise from the perineurium of peripheral nerves [70, 71]. Perineuriomas are mainly sporadic tumors but a few cases have been reported to be associated with NF1 or, more frequently, with NF2-related SWN [24].

Perineuriomas are commonly classified in intraneural subtype, extraneural or soft tissue subtype, and sclerosing subtype; the former appears as solid and pale masses that cause a localized cylindrical thickening of the affected nerve, typically presented during childhood and adolescence before the age of 20 years; the soft tissue subtype presents as well-defined rubbery nodules, white or light brown in color typically in middle-aged patients; the latter is characterized as small painless subcutaneous masses occurring almost exclusively on the palms and fingers of young patients [71–74].

Intraneural perineuriomas (IP) are considered tumors of childhood and adolescence as they occur more often before the age of 20 years [70, 71, 73]. They usually present as a slowly progressive painless mononeuropathy with muscle weakness, which can progress to complete loss of function, and atrophy, rarely with contractures, pain or sensory loss, and limb-length discrepancy. More often they present as a mononeuropathy, but a plexopathy can occur [70, 71, 73]. Electromyography is useful for localizing and determining the degree and distribution of axonal loss in nerve branches.

Ultrasound and MRI studies, especially with slices on 3-Tesla MRI, are important tools for localizing the lesions and for permitting differentiation from other nerve lesions. At MRI, studies IP can be focal and restricted to a short segment or more elongated and fusiform encompassing most of the nerve or even a single plexus. The sciatic nerve or its branches are commonly affected, followed by brachial plexus, radial nerve, and ulnar nerve. On MRI, classic features of IP have been identified. If combined with some basic clinical data, they have been considered as the Perineurioma Diagnostic Criteria [71]. They include a lack of cancer history, unifocal disease, moderate to severe T2-hyperintensity, moderate to severe contrast enhancement, fusiform shape, enlargement of the involved nerves, and age less than or equal to 40 years. These characteristics are considered highly specific and, if present, may obviate the need for tissue diagnosis. However, rarely IP have been described without contrast enhancement and multifocal; in addition, rare histologic variants including hybrid peripheral nerve sheath tumors (with histologic features of perineuriomas in combination with neurofibromas or schwannomas), which do not present typically clinical and radiological features of IP, have been described [71].

Advancements in genetic testing have revealed that many IP, but not soft tissue subtypes, harbor somatic mutations in the *TRAF7* gene (typically associated with the development of meningiomas) [75], while in soft tissue, perineurioma macrodeletions of *NF1* and *NF2* genes and rearrangement of chr22 regions containing the *NF2* gene have not always been described in patients with systemic features of NF2 syndrome [73, 76, 77].

In addition in some cases of IP, chromosomal abnormalities at 22q12 are present [75] and four cases of IP in patients with NF2 are described [74, 78]. These evidences suggest a potential link between the *NF2* gene and the development of perineuriomas, which probably are more common than currently recognized in NF2 syndrome. It is well-known that a polyneuropathy, unrelated to the expected distribution of known tumors, can develop in NF2 patients. Consequently, NF2 patients presenting with weakness in a peripheral nerve distribution should raise suspicion for an alternate diagnosis, among which perineurioma [74]. Furthermore, the advancement in genetic testing is opening the possibility of targeted therapy in the future [74, 78].

Despite the increase in knowledge regarding IP, the clinical management still remains the most challenging issue as surgical treatment with nerve grafts and nerve transfer is controversial. Treatment has historically been quite variable. Nonoperative management with a watch-and-wait approach is still the most widely used, given the benign nature of the lesions and their tendency to remain radiologically mostly stable or slowly progressive, despite the relentless worsening of clinical symptoms. Neurolysis of the affected nerve and surgical resection with or without nerve graft repair are the most commonly reported surgeries with variable and not well-defined results.

Rather than treating the underlying tumor, one approach might be to try to focus on restoring function with tendon transfer, nerve grafting, or nerve transfer. As is now known, IPs generally do not grow in length but in thickness; therefore, tendon or nerve transfers may become increasingly used even if the optimal timing for these treatments remains unclear [71].

9.3.3 Hybrid Peripheral Nerve Sheath Tumors

They constitute a subtype of benign NSTs which has been described and added to the WHO classification of tumors of soft tissue and bone in 2013. They are neoplasms that are difficult to categorize into one specific diagnostic category as they present as a combination of neurofibroma/schwannoma, schwannoma/perineurioma, and neurofibroma/perineurioma. As MRI features of hybrid NSTs are not delineated, the definitive diagnosis is histological [3]. The diagnosis of a hybrid tumor should always raise suspicions of a tumor predisposition syndrome.

Most tumors with a variable combination of neurofibromatous and schwannomatous areas are associated with the different types of schwannomatosis and are associated with chromosome 22 monosomy and less frequently NF1. While hybrid tumors with both neurofibroma and perineurioma features have been described, there is increasing evidence of a tumors composed of perineurioma and schwannoma areas [24].

9.4 Conclusions

The most common NSTs in pediatric age are schwannomas, neurofibromas, and plexiform neurofibromas and rarely perineuriomas; malignant NSTs such as MPNST are significantly rarer.

The interest in NSTs is mainly due to the fact that these tumors can be diagnosed in the context of tumor-predisposing syndromes. In children, neurofibromas and plexiform neurofibromas are usually associated with NF1 that is characterized by typical clinical features not found in other pathologies; plexiform neurofibromas are pathognomonic of NF1 and have a potential risk of malignant transformation to MPNST, sometimes through an intermediate called ANNUBP. The ANNUBP diagnostic category was added recently and is only used for patients with NF1.

Schwannomas and hybrid tumors instead characterize schwannomatoses, which can pose important problems of differential diagnosis, especially in sporadic cases in pediatric age. Perineuriomas are mostly sporadic but may be associated specifically with *NF2*-related schwannomatosis.

Consequently, knowledge of the spectrum of these tumors and their development in the context of tumor predisposing syndromes is essential to provide correct information on prognosis and plan the appropriate management of patients.

References

1. Louis DN, Perry A, Wesseling P, et al. The 2021 WHO classification of tumors of the central nervous system: a summary. Neuro-Oncology. 2021;23:1231–51.
2. Anderson WJ, Doyle LA. Updates from the 2020 World Health Organization classification of soft tissue and bone tumours. Histopathology. 2021;78:644–57.
3. Debs P, Fayad LM, Ahlawat S. MR neurography of peripheral nerve tumors and tumor-mimics. Semin Roentgenol. 2022;57:232–40.
4. Pellerino A, Verdijk RM, Nichelli L, Andratschke NH, Idbaih A, Goldbrunner R. Diagnosis and treatment of peripheral and cranial nerve tumors with expert recommendations: an EUropean network for RAre CANcers (EURACAN) initiative. Cancers (Basel). 2023;15(7):1930. https://doi.org/10.3390/cancers15071930.
5. Ahlawat S, Chhabra A, Blakely J. Magnetic resonance neurography of peripheral nerve tumors and tumorlike conditions. Neuroimaging Clin N Am. 2014;24:171–92.
6. Legius E, Messiaen L, Wolkenstein P, et al. Revised diagnostic criteria for neurofibromatosis type 1 and Legius syndrome: an international consensus recommendation. Genet Med. 2021;23:1506–13.
7. Plotkin SR, Messiaen L, Legius E, et al. Updated diagnostic criteria and nomenclature for neurofibromatosis type 2 and schwannomatosis: an international consensus recommendation. Genet Med. 2022;24:1967–77.
8. Uerschels A-K, Dengler NF, Chihi M, Lenkeit A, Dinger TF, Jabbarli R, Sure U, Hagenacker T, Wrede KH, Gembruch O. Benign peripheral nerve sheath tumors: an interdisciplinary diagnostic and therapeutic challenge. Neurosurg Rev. 2023;46:205.
9. Soldatos T, Fisher S, Karri S, Ramzi A, Sharma R, Chhabra A. Advanced MR imaging of peripheral nerve sheath tumors including diffusion imaging. Semin Musculoskelet Radiol. 2015;19:179–90.
10. Agarwal A, Chandra A, Jaipal U, Bagarhatta M, Mendiratta K, Goyal A, Kumar R, Mangalhara N. Can imaging be the new yardstick for diagnosing peripheral neuropathy?-a comparison between high resolu-

tion ultrasound and MR neurography with an approach to diagnosis. Insights Imaging. 2019;10:104.
11. Tøttrup M, Eriksen JD, Hellfritzsch MB, Sørensen FB, Baad-Hansen T. Diagnostic accuracy of ultrasound-guided core biopsy of peripheral nerve sheath tumors. J Clin Ultrasound. 2020;48:134–8.
12. Pendleton C, Spinner RJ. Image-guided percutaneous biopsy of peripheral nerve tumors of indeterminate nature: risks and benefits. Acta Neurochir. 2020;162:1425–9.
13. Uusitalo E, Leppävirta J, Koffert A, Suominen S, Vahtera J, Vahlberg T, Pöyhönen M, Peltonen J, Peltonen S. Incidence and mortality of neurofibromatosis: a total population study in Finland. J Invest Dermatol. 2015;135:904–6.
14. Koczkowska M, Chen Y, Callens T, et al. Genotype–phenotype correlation in NF1: evidence for a more severe phenotype associated with missense mutations affecting NF1 codons 844-848. Am J Hum Genet. 2018;102:69–87.
15. Knudson AGJ. Mutation and cancer: statistical study of retinoblastoma. Proc Natl Acad Sci USA. 1971;68:820–3.
16. Koczkowska M, Callens T, Gomes A, et al. Expanding the clinical phenotype of individuals with a 3-bp in-frame deletion of the NF1 gene (c.2970_2972del): an update of genotype–phenotype correlation. Genet Med. 2019;21:867–76.
17. Koczkowska M, Callens T, Chen Y, et al. Clinical spectrum of individuals with pathogenic NF1 missense variants affecting p.Met1149, p.Arg1276, and p.Lys1423: genotype–phenotype study in neurofibromatosis type 1. Hum Mutat. 2020;41:299–315.
18. Mo J, Moye SL, McKay RM, Le LQ. Neurofibromin and suppression of tumorigenesis: beyond the GAP. Oncogene. 2022;41:1235–51.
19. Ferner RE, Chaudhuri R, Bingham J, Cox T, Hughes RA. MRI in neurofibromatosis 1. The nature and evolution of increased intensity T2 weighted lesions and their relationship to intellectual impairment. J Neurol Neurosurg Psychiatry. 1993;56:492–5.
20. Ferner RE, Huson SM, Thomas N, et al. Guidelines for the diagnosis and management of individuals with neurofibromatosis 1. J Med Genet. 2007;44:81–8.
21. Gutmann DH, Ferner RE, Listernick RH, Korf BR, Wolters PL, Johnson KJ. Neurofibromatosis type 1. Nat Rev Dis Prim. 2017;3:17004.
22. Nix JS, Blakeley J, Rodriguez FJ. An update on the central nervous system manifestations of neurofibromatosis type 1. Acta Neuropathol. 2020;139:625–41.
23. Le LQ, Liu C, Shipman T, Chen Z, Suter U, Parada LF. Susceptible stages in Schwann cells for NF1-associated plexiform neurofibroma development. Cancer Res. 2011;71:4686–95.
24. Magro G, Broggi G, Angelico G, Puzzo L, Vecchio GM, Virzì V, Salvatorelli L, Ruggieri M. Practical approach to histological diagnosis of peripheral nerve sheath tumors: an update. Diagnostics (Basel, Switzerland). 2022;12(6):1463. https://doi.org/10.3390/diagnostics12061463.
25. Chamseddin BH, Le LQ. Management of cutaneous neurofibroma: current therapy and future directions. Neuro-oncol Adv. 2020;2:i107–16.
26. Miller DT, Freedenberg D, Schorry E, Ullrich NJ, Viskochil D, Korf BR. Health supervision for children with neurofibromatosis type 1. Pediatrics. 2019;143(5):e20190660. https://doi.org/10.1542/peds.2019-0660.
27. Huson SM, Harper PS, Compston DA. Von Recklinghausen neurofibromatosis. A clinical and population study in south-East Wales. Brain. 1988;111(Pt 6):1355–81.
28. Mautner V-F, Asuagbor FA, Dombi E, Fünsterer C, Kluwe L, Wenzel R, Widemann BC, Friedman JM. Assessment of benign tumor burden by whole-body MRI in patients with neurofibromatosis 1. Neuro-Oncology. 2008;10:593–8.
29. Nguyen R, Kluwe L, Fuensterer C, Kentsch M, Friedrich RE, Mautner V-F. Plexiform neurofibromas in children with neurofibromatosis type 1: frequency and associated clinical deficits. J Pediatr. 2011;159:652–5.e2.
30. Fisher MJ, Blakeley JO, Weiss BD, et al. Management of neurofibromatosis type 1-associated plexiform neurofibromas. Neuro-Oncology. 2022;24(11):1827–44. https://doi.org/10.1093/neuonc/noac146.
31. Dombi E, Solomon J, Gillespie AJ, et al. NF1 plexiform neurofibroma growth rate by volumetric MRI: relationship to age and body weight. Neurology. 2007;68:643–7.
32. Prada CE, Rangwala FA, Martin LJ, Lovell AM, Saal HM, Schorry EK, Hopkin RJ. Pediatric plexiform neurofibromas: impact on morbidity and mortality in neurofibromatosis type 1. J Pediatr. 2012;160:461–7.
33. Gross AM, Singh G, Akshintala S, Baldwin A, Dombi E, Ukwuani S, Goodwin A, Liewehr DJ, Steinberg SM, Widemann BC. Association of plexiform neurofibroma volume changes and development of clinical morbidities in neurofibromatosis 1. Neuro-Oncology. 2018;20:1643–51.
34. Iheanacho I, Yoo HK, Yang X, Dodman S, Hughes R, Amin S. Epidemiological and clinical burden associated with plexiform neurofibromas in pediatric neurofibromatosis type-1 (NF-1): a systematic literature review. Neurol Sci. 2022;43:1281–93.
35. Chick G, Hollevoet N, Victor J, Bianchi S. The role of imaging in isolated benign peripheral nerve tumors: a practical review for surgeons. Hand Surg Rehabil. 2016;35:320–9.
36. Carton C, Evans DG, Blanco I, et al. ERN GENTURIS tumour surveillance guidelines for individuals with neurofibromatosis type 1. EClinicalMedicine. 2023;56:101818.
37. Kotch C, Brosius SN, De Raedt T, Fisher MJ. Updates in the management of central and peripheral nervous system tumors among patients with neurofibromatosis type 1 and neurofibromatosis type 2. Pediatr Neurosurg. 2023;58(5):267–80. https://doi.org/10.1159/000529507.

38. Gross AM, Wolters PL, Dombi E, et al. Selumetinib in children with inoperable plexiform neurofibromas. N Engl J Med. 2020;382:1430–42.
39. Mueller S, Reddy AT, Dombi E, et al. NFB-17. MEK INHIBITOR binimetinib shows clinical activity in children with neurofibromatosis type 1- associated plexiform neurofibromas: a report from PNOC and the NF clinical trials consortium. Neuro Oncol. 2020;22:iii420–1.
40. Weiss BD, Wolters PL, Plotkin SR, et al. NF106: a Neurofibromatosis clinical trials Consortium phase II trial of the MEK inhibitor Mirdametinib (PD-0325901) in adolescents and adults with NF1-related plexiform neurofibromas. J Clin Oncol. 2021;39:797–806.
41. Ronsley R, Hounjet CD, Cheng S, et al. Trametinib therapy for children with neurofibromatosis type 1 and life-threatening plexiform neurofibroma or treatment-refractory low-grade glioma. Cancer Med. 2021;10:3556–64.
42. Fisher MJ, Shih C-S, Rhodes SD, et al. Cabozantinib for neurofibromatosis type 1-related plexiform neurofibromas: a phase 2 trial. Nat Med. 2021;27:165–73.
43. Evans DGR, Baser ME, McGaughran J, Sharif S, Howard E, Moran A. Malignant peripheral nerve sheath tumours in neurofibromatosis. J Med Genet. 2002;39(5):311–4. https://doi.org/10.1136/jmg.39.5.311.
44. Beert E, Brems H, Daniëls B, et al. Atypical neurofibromas in neurofibromatosis type 1 are premalignant tumors. Genes Chromosomes Cancer. 2011;50:1021–32.
45. Miettinen MM, Antonescu CR, Fletcher CDM, et al. Histopathologic evaluation of atypical neurofibromatous tumors and their transformation into malignant peripheral nerve sheath tumor in patients with neurofibromatosis 1—a consensus overview. Hum Pathol. 2017;67:1–10.
46. Pemov A, Hansen NF, Sindiri S, et al. Low mutation burden and frequent loss of CDKN2A/B and SMARCA2, but not PRC2, define premalignant neurofibromatosis type 1-associated atypical neurofibromas. Neuro-Oncology. 2019;21:981–92.
47. Baser ME, Friedman JM, Aeschliman D, Joe H, Wallace AJ, Ramsden RT, Evans DGR. Predictors of the risk of mortality in neurofibromatosis 2. Am J Hum Genet. 2002;71:715–23.
48. Selvanathan SK, Shenton A, Ferner R, Wallace AJ, Huson SM, Ramsden RT, Evans DG. Further genotype–phenotype correlations in neurofibromatosis 2. Clin Genet. 2010;77:163–70.
49. NIH. Neurofibromatosis. Conference statement. National Institutes of Health Consensus Development Conference. Arch Neurol. 1988;45:575–8.
50. Evans DG, Huson SM, Donnai D, Neary W, Blair V, Newton V, Harris R. A clinical study of type 2 neurofibromatosis. Q J Med. 1992;84:603–18.
51. Evans DG, King AT, Bowers NL, et al. Identifying the deficiencies of current diagnostic criteria for neurofibromatosis 2 using databases of 2777 individuals with molecular testing. Genet Med. 2019;21:1525–33.
52. Trofatter JA, MacCollin MM, Rutter JL, Murrell JR, Duyao MP, Parry DM, Eldridge R, Kley N, Menon AG, Pulaski K. A novel moesin-, ezrin-, radixin-like gene is a candidate for the neurofibromatosis 2 tumor suppressor. Cell. 1993;72:791–800.
53. Rouleau GA, Merel P, Lutchman M, Sanson M, Zucman J, Marineau C, Hoang-Xuan K, Demczuk S, Desmaze C, Plougastel B. Alteration in a new gene encoding a putative membrane-organizing protein causes neuro-fibromatosis type 2. Nature. 1993;363:515–21.
54. Evans DG, Wallace AJ, Wu CL, Trueman L, Ramsden RT, Strachan T. Somatic mosaicism: a common cause of classic disease in tumor-prone syndromes? Lessons from type 2 neurofibromatosis. Am J Hum Genet. 1998;63:727–36.
55. Evans DG, Hartley CL, Smith PT, et al. Incidence of mosaicism in 1055 de novo NF2 cases: much higher than previous estimates with high utility of next-generation sequencing. Genet Med. 2020;22:53–9.
56. MacCollin M, Woodfin W, Kronn D, Short MP. Schwannomatosis: a clinical and pathologic study. Neurology. 1996;46:1072–9.
57. Wolkenstein P, Benchikhi H, Zeller J, Wechsler J, Revuz J. Schwannomatosis: a clinical entity distinct from neurofibromatosis type 2. Dermatology. 1997;195:228–31.
58. MacCollin M, Chiocca EA, Evans DG, et al. Diagnostic criteria for schwannomatosis. Neurology. 2005;64:1838–45.
59. Hulsebos TJM, Plomp AS, Wolterman RA, Robanus-Maandag EC, Baas F, Wesseling P. Germline mutation of INI1/SMARCB1 in familial schwannomatosis. Am J Hum Genet. 2007;80:805–10.
60. Smith MJ, Kulkarni A, Rustad C, Bowers NL, Wallace AJ, Holder SE, Heiberg A, Ramsden RT, Evans DG. Vestibular schwannomas occur in schwannomatosis and should not be considered an exclusion criterion for clinical diagnosis. Am J Med Genet A. 2012;158A:215–9.
61. Plotkin SR, Blakeley JO, Evans DG, et al. Update from the 2011 international Schwannomatosis workshop: from genetics to diagnostic criteria. Am J Med Genet A. 2013;161A:405–16.
62. Piotrowski A, Xie J, Liu YF, et al. Germline loss-of-function mutations in LZTR1 predispose to an inherited disorder of multiple schwannomas. Nat Genet. 2014;46:182–7.
63. Sestini R, Bacci C, Provenzano A, Genuardi M, Papi L. Evidence of a four-hit mechanism involving SMARCB1 and NF2 in schwannomatosis-associated schwannomas. Hum Mutat. 2008;29:227–31.
64. Kehrer-Sawatzki H, Kluwe L, Friedrich RE, et al. Phenotypic and genotypic overlap between mosaic NF2 and schwannomatosis in patients with multiple non-intradermal schwannomas. Hum Genet. 2018;137:543–52.
65. Carter JM, O'Hara C, Dundas G, Gilchrist D, Collins MS, Eaton K, Judkins AR, Biegel JA, Folpe AL. Epithelioid malignant peripheral nerve sheath

tumor arising in a schwannoma, in a patient with "neuroblastoma-like" schwannomatosis and a novel germline SMARCB1 mutation. Am J Surg Pathol. 2012;36:154–60.
66. Woertler K. Tumors and tumor-like lesions of peripheral nerves. Semin Musculoskelet Radiol. 2010;14:547–58.
67. Blakeley JO, Evans DG, Adler J, et al. Consensus recommendations for current treatments and accelerating clinical trials for patients with neurofibromatosis type 2. Am J Med Genet A. 2012;158A:24–41.
68. Blakeley JO, Ye X, Duda DG, et al. Efficacy and biomarker study of bevacizumab for hearing loss resulting from neurofibromatosis type 2-associated vestibular schwannomas. J Clin Oncol. 2016;34:1669–75.
69. Plotkin SR, Duda DG, Muzikansky A, et al. Multicenter, prospective, phase II and biomarker study of high-dose bevacizumab as induction therapy in patients with neurofibromatosis type 2 and progressive vestibular schwannoma. J Clin Oncol. 2019;37:3446–54.
70. McMillan HJ, Torres C, Michaud J, Ying Y, Boyd KU, Bourque PR. Diagnosis and outcome of childhood perineurioma. Childs Nerv Syst. 2016;32:1555–60.
71. Lenartowicz KA, Smith BW, Jack MM, Wilson TJ, Klein CJ, Amrami KK, Spinner RJ. What is new in intraneural perineurioma? Acta Neurochir. 2023;165:3539–47.
72. Skovronsky DM, Oberholtzer JC. Pathologic classification of peripheral nerve tumors. Neurosurg Clin N Am. 2004;15:157–66.
73. Mauermann ML, Amrami KK, Kuntz NL, Spinner RJ, Dyck PJ, Bosch EP, Engelstad J, Felmlee JP, Dyck PJB. Longitudinal study of intraneural perineurioma—a benign, focal hypertrophic neuropathy of youth. Brain. 2009;132:2265–76.
74. Pendleton C, Lenartowicz KA, Howe BM, Spinner RJ. Limb undergrowth in intraneural perineuriomas: an under-recognized association. World Neurosurg. 2020;141:e670–6.
75. Klein CJ, Wu Y, Jentoft ME, Mer G, Spinner RJ, Dyck PJB, Dyck PJ, Mauermann ML. Genomic analysis reveals frequent TRAF7 mutations in intraneural perineuriomas. Ann Neurol. 2017;81:316–21.
76. Lasota J, Fetsch JF, Wozniak A, Wasag B, Sciot R, Miettinen M. The neurofibromatosis type 2 gene is mutated in perineurial cell tumors: a molecular genetic study of eight cases. Am J Pathol. 2001;158:1223–9.
77. Pitchford CW, Schwartz HS, Atkinson JB, Cates JMM. Soft tissue perineurioma in a patient with neurofibromatosis type 2: a tumor not previously associated with the NF2 syndrome. Am J Surg Pathol. 2006;30:1624–9.
78. White B, Belzberg A, Ahlawat S, Blakeley J, Rodriguez FJ. Intraneural perineurioma in neurofibromatosis type 2 with molecular analysis. Clin Neuropathol. 2020;39:167–71.

IOUS and CEUS in PNS Tumors

10

Giovanni Carone, Massimiliano Del Bene,
and Francesco Prada

10.1 Introduction

Surgery of peripheral nervous system tumors recognizes intraoperative ultrasound (ioUS) as one of the most valuable aids. IoUS is particularly indicated for small parts surgeries, as it can provide high-resolution, real-time images able to guide surgical dissection, lesion localization, and orientation [1]. Interestingly, US is part of the routine presurgical imaging in peripheral nervous system tumors, and for this reason, it is a well-known modality, to which every neurosurgeon is accustomed. As a consequence, the surgeon can rely on a standardized real-time imaging modality to acquire scans every time is deemed necessary [2]. Furthermore, the significant technological developments of recent years have made ioUS real multi-modal imaging able to characterize both anatomical and functional features. B-mode is the leading modality, allowing the acquisition of extremely defined real-time images of lesions and relative relationships. Doppler imaging (power, color, spectral, and submodalities) provides information on vascular tree, the pattern of vascularization, and relationships. Contrast-enhanced ultrasound can depict in detail the vascularization of a lesion: different phases, and anatomy in real time, together with the modifications that can take place during surgery. Elastosonography (shear wave and strain elastography) can highlight the mechanical features of lesions and surrounding tissues, thus providing relevant information to guide the surgical approach [3].

G. Carone
Department of Neurosurgery, Fondazione IRCCS Istituto Neurologico Carlo Besta, Milan, Italy
e-mail: giovanni.carone@unimi.it

M. Del Bene
Department of Neurosurgery, Fondazione IRCCS Istituto Neurologico Carlo Besta, Milan, Italy

Department of Experimental Oncology, IEO, European Institute of Oncology IRCCS, Milan, Italy

Department of Pharmacological and Biomolecular Sciences, University of Milan, Milan, Italy
e-mail: massimiliano.delbene@istituto-besta.it

F. Prada (✉)
Department of Neurosurgery, Fondazione IRCCS Istituto Neurologico Carlo Besta, Milan, Italy

Department of Neurological Surgery, University of Virginia Health System, Charlottesville, VA, USA

Acoustic Neuroimaging and Therapy Laboratory, Fondazione IRCCS Istituto Neurologico Carlo Besta, Milan, Italy

Focused Ultrasound Foundation, Charlottesville, VA, USA
e-mail: francesco.prada@istituto-besta.it

10.2 US Modalities

Several modalities can be used to assist during surgery. Each of these has specific characteristics that, when combined, can aid in surgical decision-making.

In this chapter, the different types of ultrasound modalities are presented, before reviewing the relative practical applications.

I. G. Vetrano, V. Nazzi (eds.), *Peripheral Nerve Sheath Tumors*,
https://doi.org/10.1007/978-3-031-76382-3_10

10.2.1 B-mode

B-mode represents the mainstay for ioUS. In most situations, B-mode is acquired using a linear array transducer with an upper frequency of 15–18 MHz to assess the anatomy, shape, echogenicity, and cross-sectional area (CSA) of the nerve and relative lesion [4]. Typically, a distal to proximal approach is used to identify any abnormalities along the course of the nerve, with structures assessed in both orthogonal planes (cross-sectional and longitudinal). Comparison with the contralateral nerve may be useful in the assessment of unilateral disease, while the mobility of the nerve, particularly around joints, can be dynamically assessed where relevant [5].

In B-mode images, nerves are identified by the honeycomb-like appearance in the transverse plane and parallel hyperechoic fascicular borders in the longitudinal plane (perineurium) [6]. In pathological conditions, increased intraneural oedema may result in decreased echogenicity and loss of the fascicular pattern. Since nerve enlargement is the primary diagnostic marker, accurate measurement of nerve size is crucial for the diagnosis of abnormal nerves [7]. To quantify nerve size, cross-sectional area (CSA) and swelling ratio (ratio between the CSA of the pathological nerve and that of an unaffected site) can be measured on transverse images, while diameter can be measured on longitudinal images [8]. To ensure accurate measurement, it's important to hold the transducer perpendicular to the nerve with minimal pressure to obtain accurate measurements. The site of maximum magnification should be selected for measurement, but it's equally important to find the minimum cross-sectional area at that site to ensure proper perpendicular scanning. Higher frequency transducers improve image resolution, and B-mode echogenicity carries information about nerve structure, with measurement of the hypoechoic fractional area of nerves on US B-mode images, or the nerve density index, proposed as an objective measure of overall nerve echogenicity, capable of distinguishing between normal and abnormal nerves.

Nerves differ from tendons, displaying a dense fibrillar structure in B-mode, and are distinguished from blood vessels using Doppler US [9] (Fig. 10.1a).

10.2.2 Doppler Imaging

Vascularity assessment is an important aspect of nerve imaging and can be performed using power Doppler or color Doppler. The power Doppler box is placed over the nerve, and the gain is slowly increased to detect non-specific color signals or focal color flow signals that indicate arterial pulsation and blood flow within the nerve tissue. No color Doppler signal, either intraneural or epineural, is observed in the normal nerve [10]. Therefore, if there is vascularization around the nerve, there is either a pathological picture at the base or it is not a nerve. However, it should be noted that at the highest gain levels, nonspecific color signals may appear as noise; only focal color flow signals synchronized with arterial pulsation should be interpreted as evidence of blood. Causes for the presence of blood are vascular congestion—presumably due to compression or inflammatory response in various neuropathies—tumor lesions, and neurolymphomatosis. To assess this, current quantitative measures of intraneural blood flow include manual counting of Doppler signals or analysis of Doppler waveforms to assess velocity, but these measures are likely to underestimate intraneural blood flow due to detection thresholds for flow velocity [11]. Ultrasensitive power Doppler modalities can further aid in identifying pathological microcirculation, perfusion patterns, and very slow-flow capillaries. These modalities can approach the information provided by contrast-enhanced ultrasound (Fig. 10.1b).

10.2.3 Contrast-Enhanced Ultrasound

Contrast-enhanced ultrasound (CEUS) is a promising technique for the assessment of peripheral

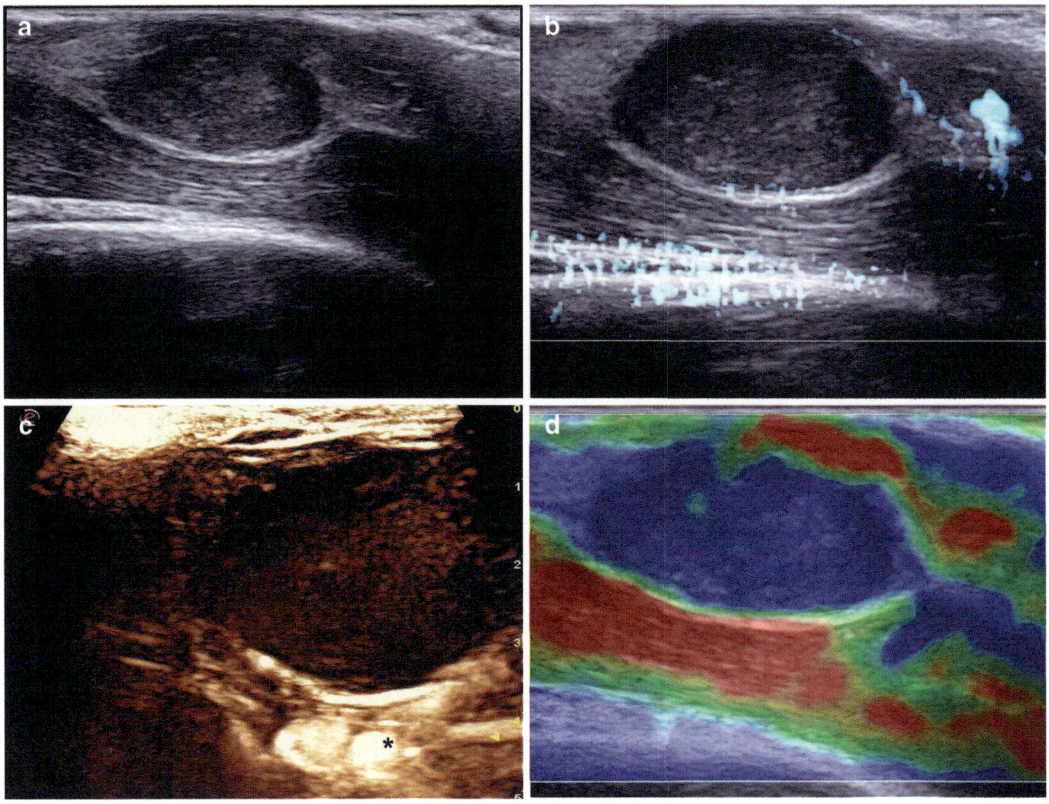

Fig. 10.1 Illustrative image of the main intraoperative ultrasound scans routinely employed for peripheral nerve tumors. (**a**) B-mode scan of a schwannoma in which the loss of the typical honeycomb-like aspect and the hyperechogenic contour can be observed. (**b**) Doppler US, with the typical absence of intralesional vascular patterns. (**c**) CEUS, with the lesion capturing inhomogeneous contrast according to the contrastographic phase. The surrounding vascular structures can be observed (*). (**d**) elastosonography to show the density of the lesion in relation to the surrounding structures

nerve tumors. Using last-generation US contrast agents, which are gas-filled microbubbles with purely endovascular distribution, CEUS can dynamically image the entire microcirculation, including the tiny vascular branches and the associated capillary network supplying nerve lesions [12]. This allows the noninvasive quantification of early angiogenesis in tumors, providing valuable insights into the relationships between tumors and nerves further enhancing lesion recognition and characterization [13]. CEUS can detect low-flow intraneural perfusion patterns with high sensitivity, thus directly guiding surgical decision-making; however, further studies are needed to evaluate its usefulness in assessing disease prognosis and outcome. Potential evolutions of this methodology could be related to the introduction of targeted microbubbles, to differentiate between benign and malignant tumors and to implement targeted therapies [14] (Fig. 10.1c).

10.2.4 Elastosonography

Elastosonography is a US-based noninvasive imaging technique aimed at measuring the stiffness of tissues to identify underlying pathological changes in tissue composition and architecture. In detail, the technique uses two major methods—strain (SE) and shear-wave elastography (SWE)—to measure tissue mechanical properties in different ways. SE measures stiffness by recording the displacement of a tissue induced through a mechanical stimulus provided with the probe. SWE uses a known high-intensity acoustic stimulation to cause tissue displacement [15]. As a consequence, SE can provide only qualitative, or semiquantitative measures, while SWE can quantify the stiffness of the tissue. Both methods are now routinely used to study peripheral nerve neuropathies, as in the case of the median nerve at the wrist to detect carpal tunnel syndrome [16]. On this line, sonoelastography has also been introduced in PNS tumor surgery, where it can provide mechanical information, which can help in identifying and characterizing pathological tissues, especially in those cases where tumors have undergone degenerative change or present atypical appearances (e.g., cystic degeneration) [16] (Fig. 10.1d).

10.3 Applications

Peripheral nerve sheath (PNS) tumors are a heterogeneous subset of soft tissue tumors that may originate from a peripheral nerve [17]. Different histology subtypes may present different characteristics on ultrasound [6]. The main features supporting PNS diagnosis are either a direct association with a neural structure or their location along a typical nerve distribution. The first step is to distinguish a true tumor from a nonneoplastic nerve condition, such as a neuroma, peripheral nerve ganglion, intraneural venous malformation, nerve lipomatosis, or focal nerve hypertrophy. By combining clinical features with conventional and advanced imaging, it is usually possible to identify neurogenic tumors with confidence. As in other areas, the use of intraoperative ultrasound has several advantages. First of all, the ease of the imaging and the possibility of effortless repeating the imaging with low economic impact and low time expenditure [18]. In the following paragraphs, the major applications of ioUS in PNS tumor surgery are presented, according to the most relevant histology subtypes.

10.3.1 Schwannoma

Conventional schwannoma is an encapsulated spindle cell tumor composed predominantly of well-differentiated Schwann cells [19]. On ultrasound (US), it typically appears as a round or oval hypoechoic mass with well-defined margins and may have slight posterior acoustic enhancement with loss of the typical honeycomb appearance. In most cases, the tumor is in continuity with the nerve at its proximal and distal poles. Relying on high-frequency probes, it is possible to visualize the association of the schwannoma with a swollen fascicle, while unaffected fascicles may appear displaced in the periphery. A positive perfusion pattern may be seen on CEUS and ultrasensitive power Doppler [20]. The actual usefulness of advanced US techniques such as elastography and contrast in the diagnosis of peripheral nerve sheath tumors (PNSTs) requires further investigation. However, as the different tumor subtypes have partially different histological characteristics, it is important to consider that different types of schwannomas may have unique imaging characteristics that differ from conventional schwannomas. For example, the "ancient" variant of schwannoma and epithelioid schwannoma may show intralesional calcification and degenerative changes such as hemorrhage and cyst formation, resulting in extensive signal heterogeneity [21]. In addition, plexiform schwannoma in deep soft tissue may present as a multinodular inhomogeneous mass. Finally, cellular schwannomas in paravertebral locations may show extensive bone erosion, extension into the spinal canal, and compression of spinal roots, which may lead to a misdiagnosis of malignancy [22].

10.3.2 Neurofibroma

To understand US imaging, it is important to know that histologically neurofibroma is composed of Schwann cells, perineurial or perineurial-like cells, fibroblasts, mast cells, and interspersed myelinated and unmyelinated axons surrounded by a myxoid and collagenous extracellular matrix [23]. In US, the fascicles are markedly enlarged and thickened in the B-mode nerve cross sections, with normal-appearing epineuria. The localized intraneural neurofibroma typically presents as a fusiform hypoechoic mass with continuity to the nerve at its proximal and distal ends. Unlike schwannomas, neurofibromas lack a capsule and tend to infiltrate the affected nerve [24]. CEUS and ultrasensitive power Doppler can depict an intralesional capillary tree and a pathological perfusion pattern. Different subtypes of neurofibroma can have different imaging characteristics [12]. For example, diffuse neurofibroma involving soft tissue can vary in shape from uniform regional soft tissue enlargement to pendulous, sac-like masses. Plexiform neurofibroma typically presents as a multinodular mass involving a large nerve or plexus, with a bag-like or ropy appearance [24]. ioUS can play a pivotal role in differentiating neurofibroma (which can evolve in malignant peripheral nerve sheath tumors) and schwannoma tumors, thus owing to important implications for surgical strategy.

10.3.3 Intraneural Perineurioma

Histologically, the affected nerve in this condition shows concentric layers of perineurial cells surrounding nerve fibers, resembling a pseudo-onion bulb [25]. The degree of myelin and axonal loss is variable. In ultrasound, it presents as a long segment of fusiform enlargement with tapering proximal and distal ends. Crucially, the normal nerve fascicle architecture is preserved, and hypoechogenic enlarged fascicles are present throughout the cross-section of each nerve [26].

10.3.4 B-cell Lymphoma

In B-cell lymphoma, the ultrasound pattern is particularly striking [27]. The pathological nerve shows significant enlargement in its cross-sectional area, while the epineurium appears normal. The fascicles within the nerve were markedly enlarged with a dense and hyperechogenic appearance. Within these fascicle groups, there are usually hypoechogenic septa, dividing them into smaller sections [6].

References

1. Pedro MT, Antoniadis G, Scheuerle A, Pham M, Wirtz CR, Koenig RW. Intraoperative high-resolution ultrasound and contrast-enhanced ultrasound of peripheral nerve tumors and tumorlike lesions. Neurosurg Focus. 2015;39(3):E5.
2. Haldeman CL, Baggott CD, Hanna AS. Intraoperative ultrasound-assisted peripheral nerve surgery. Neurosurg Focus. 2015;39(3):E4.
3. Prada F, Ciocca R, Corradino N, et al. Multiparametric intraoperative ultrasound in oncological neurosurgery: a pictorial essay. Front Neurosci. 2022;16:881661.
4. Cartwright MS, Passmore LV, Yoon J-S, Brown ME, Caress JB, Walker FO. Cross-sectional area reference values for nerve ultrasonography. Muscle Nerve. 2008;37(5):566–71.
5. Koenig RW, Schmidt TE, Heinen CPG, et al. Intraoperative high-resolution ultrasound: a new technique in the management of peripheral nerve disorders. J Neurosurg. 2011;114(2):514–21.
6. Lawande AD, Warrier SS, Joshi MS. Role of ultrasound in evaluation of peripheral nerves. Indian J Radiol Imaging. 2014;24(3):254–8.
7. Suk JI, Walker FO, Cartwright MS. Ultrasonography of peripheral nerves. Curr Neurol Neurosci Rep. 2013;13(2):328.
8. Chen I-J, Chang K-V, Wu W-T, Özçakar L. Ultrasound parameters other than the direct measurement of ulnar nerve size for diagnosing cubital tunnel syndrome: a systemic review and meta-analysis. Arch Phys Med Rehabil. 2019;100(6):1114–30. https://doi.org/10.1016/j.apmr.2018.06.021.
9. Carroll AS, Simon NG. Current and future applications of ultrasound imaging in peripheral nerve disorders. World J Radiol. 2020;12(6):101–29.
10. Ghasemi-Esfe AR, Khalilzadeh O, Vaziri-Bozorg SM, et al. Color and power Doppler US for diagnosing carpal tunnel syndrome and determining its severity: a quantitative image processing method. Radiology. 2011;261(2):499–506.
11. Ghasemi-Esfe AR, Khalilzadeh O, Mazloumi M, et al. Combination of high-resolution and color Doppler ultrasound in diagnosis of carpal tunnel syndrome. Acta Radiol. 2011;52(2):191–7.
12. Kearns KN, Sokolowski JD, Chadwell K, et al. The role of contrast-enhanced ultrasound in neurosurgical disease. Neurosurg Focus. 2019;47(6):E8.
13. Wang Y, Tang P, Zhang L, Guo Y, Wan W. Quantitative evaluation of the peripheral nerve blood perfusion with high frequency contrast-enhanced ultrasound. Acad Radiol. 2010;17(12):1492–7.

14. Ajmal S. Contrast-enhanced ultrasonography: review and applications. Cureus. 2021;13(9):e18243.
15. Battaglia PJ, Carbone-Hobbs V, Guebert GM, Mackinnon SE, Kettner NW. High-resolution ultrasonography and shear-wave sonoelastography of a cystic radial nerve Schwannoma. J Ultrasound. 2017;20(3):261–6.
16. Wee TC, Simon NG. Ultrasound elastography for the evaluation of peripheral nerves: a systematic review. Muscle Nerve. 2019;60(5):501–12.
17. Woertler K. Tumors and tumor-like lesions of peripheral nerves. Semin Musculoskelet Radiol. 2010;14(5):547–58.
18. Ganau M, Ligarotti GK, Apostolopoulos V. Real-time intraoperative ultrasound in brain surgery: neuronavigation and use of contrast-enhanced image fusion. Quant Imaging Med Surg. 2019;9(3):350–8.
19. Hilton DA, Hanemann CO. Schwannomas and their pathogenesis. Brain Pathol. 2014;24(3):205–20.
20. Lee SJ, Yoon ST. Ultrasonographic and clinical characteristics of schwannoma of the hand. Clin Orthop Surg. 2017;9(1):91–5.
21. Lefebvre G, Le Corroller T. Ultrasound and MR imaging of peripheral nerve tumors: the state of the art. Skeletal Radiol. 2023;52(3):405–19.
22. Reynolds DLJ, Jacobson JA, Inampudi P, Jamadar DA, Ebrahim FS, Hayes CW. Sonographic characteristics of peripheral nerve sheath tumors. AJR Am J Roentgenol. 2004;182(3):741–4.
23. Magro G, Broggi G, Angelico G, et al. Practical approach to histological diagnosis of peripheral nerve sheath tumors: an update. Diagnostics (Basel). 2022;12(6):1463.
24. Tagliafico AS, Isaac A, Bignotti B, Rossi F, Zaottini F, Martinoli C. Nerve tumors: what the MSK radiologist should know. Semin Musculoskelet Radiol. 2019;23(1):76–84.
25. Kleihues P, Louis DN, Scheithauer BW, et al. The WHO classification of tumors of the nervous system. J Neuropathol Exp Neurol. 2002;61(3):215–25. https://doi.org/10.1093/jnen/61.3.215.
26. Wang L-M, Zhong Y-F, Zheng D-F, et al. Intraneural perineurioma affecting multiple nerves: a case report and literature review. Int J Clin Exp Pathol. 2014;7(6):3347–54.
27. Misdraji J, Ino Y, Louis DN, Rosenberg AE, Chiocca EA, Harris NL. Primary lymphoma of peripheral nerve: report of four cases. Am J Surg Pathol. 2000;24(9):1257–65.

11 Neurophysiological Evaluation and Intraoperative Monitoring in Peripheral Nervous System Tumors

Grazia Devigili and Valentina Catanzaro

Abbreviations

BPNST	Benign peripheral nerve sheath tumors
cMAP	Compound motor amplitude potential
SAP	Sensory action potential
EDX	Electrodiagnostic (examination)
EMG	Electromyography
IONM	Intraoperative neurophysiological monitoring
MEP	Motor evoked potential
MPNST	Malign peripheral nerve sheath tumors
NAP	Nerve action potential
NF 1	Neurofibromatosis type 1
NF 2	Neurofibromatosis type 2
PNSTs	Peripheral nerve sheath tumors
SEP	Somatosensory evoked potential
US	Ultrasonography

G. Devigili (✉)
Neurophysiology Unit, Fondazione IRCCS Istituto Neurologico Carlo Besta, Milan, Italy

Parkinson and Movement Disorders Unit, Department of Clinical Neurosciences, Fondazione IRCCS Istituto Neurologico Carlo Besta, Milan, Italy
e-mail: grazia.devigili@istituto-besta.it

V. Catanzaro
Neurophysiology Unit, Fondazione IRCCS Istituto Neurologico Carlo Besta, Milan, Italy
e-mail: valentina.catanzaro@istituto-besta.it

11.1 Introduction and Background

Over the last 30 years, intraoperative neurophysiological monitoring (IONM) has grown into a widely used method for several surgical procedures, including peripheral nerve surgery. Significant advancements in neurophysiological technology improved the reliability of acquired data during surgery and, therefore, expanded the use and application of IONM over recent years [1, 2]. IONM techniques encompass different electrophysiological methodologies to continuously evaluate the functional status of different neural pathways during surgery. In particular, the IONM comprises two main functions: monitoring and mapping. The monitoring goals are to detect possible damage to the nervous system while they are still *reversible,* minimize the damage to the neural structures, and maximize tumor resection. Conversely, intraoperative mapping aims to identify functional neurologic structures, especially when the anatomy is significantly modified by lesions or trauma; for this reason, it is performed at the beginning of exploration but is usually done also several times during surgery as a guide for a wider resection, concerning the confidence of neurosurgeon in the neurophysiological support [3–5].

In peripheral neurosurgery of tumors, the primary technique is represented by mapping; however, depending on the location of the lesion and

I. G. Vetrano, V. Nazzi (eds.), *Peripheral Nerve Sheath Tumors*,
https://doi.org/10.1007/978-3-031-76382-3_11

different surgical approaches, several techniques are available, allowing ever more of a tailored multimodality IONM planning. During surgery, the peripheral nervous system (PNS) structures undergo risks of damage due not only to manipulation and resection extension but also to stretching, compression, and heating from electrocauterization or ischemic stress by local blood vessel clamping [1]. The IONM *monitoring* during all surgical times, using motor and/or somatosensory evoked potentials and F-waves techniques, gives real-time data on nervous system function, also concerning proximal structures or segments not accessible by surface electrodes or surgical mapping. Additionally, neurophysiological monitoring can also guide the process of decision-making during exploration and surgery giving [6–8], with a principal role in surgery of trauma in which the visual inspection together with the neurophysiological findings suggests if a lesion in continuity should be repaired or might regenerate without repaired or, conversely, should be resected and a nerve graft or transfer should be considered [5, 7, 9, 10]. In IONM applied to surgery of tumors, the main goal is represented by tumor resection. However, all the strategies and efforts must aim to preserve nerve function. For surgery applied to the PNS, the more crucial role is the intraoperative recordings of compound nerve action potentials (CNAP) generated by stimulation and recording electrodes applied directly to the nerve trunk, which can provide timely information about the integrity of peripheral nerves during surgery. With the different levels of complexity, the IONM is not always necessary, as in the case of subcutaneous schwannoma, not in continuity with the main nervous trunk; otherwise, lesions of larger size or in continuity with the principal nervous trunks, plexus, or spinal root are highly recommended.

In this chapter, we described the neurological and neurophysiological preoperative evaluation, including the ultrasonography (US) and electrodiagnostic (EDX) combined, as well as different IONM approaches, and their role in guiding surgical plan and intraoperative decision, including for the entity of resection and, in some cases, grafting or neurotization. The main neurophysiological techniques, comprising free-running electromyography (EMG), stimulated EMG, direct nerve stimulation, NAP, somatosensory, and motor-evoked potentials, are briefly described. Finally, it is useful to mention that the success of preoperative and intraoperative neurophysiological studies depends on the mutual understanding of the surgical strategy and the neurophysiological findings and their significance by the neurophysiologist and the neurosurgeon.

11.2 Preoperative Neurophysiological Evaluation

Benign nerve sheath tumors (BNST) usually complain of mild sensory symptoms, either local or in the peripheral nerve or dermatomeric distribution. Sometimes the compression of the lesion can induce positive symptoms like paresthesia or electric shock paroxysms. The presence of motor involvement and weakness or hipotrophy is relatively uncommon, and EDX showed mild findings [11]. Other intravenous lesions include intraneural ganglion cysts, lipomas, hemangiomas, granular cell tumors, perineuriomas that cause sensory-motor symptoms and signs, and significant abnormalities at the EDX [12].

Otherwise, in the case of patients with early onset of pain, worsening in a short time, and rapid progression of neurological deficit, the presence of a malignant peripheral nerve sheath tumor should be considered, and a timely diagnostic workup is recommended.

Clinical and neurophysiological evaluation is crucial before surgery to define the peripheral nervous system functionality, and additional data are relevant to define the surgical plan and tailored IONM. The aims of the electrodiagnostic studies include the careful diagnosis of the site and understanding of PNS injury related or not to the PNS tumor, the clarification of which nerve or nerves are involved and their residual function, the grading of damage according to the current neurophysiological scoring, and the presence of situation or pre-existing PNS pathology that could mislead the IONM interpretation.

EDX studies include nerve conduction studies (NCS), needle EMG, and other complementary techniques (i.e., H reflex, F-waves). Finally, the ultrasonography combined with EDX studies, which represent a different diagnostic approach introduced in the last decade in clinical neurophysiological units, gives complementary data regarding morphological details, anatomic location and extension of nerve involvement, and relationship to other structures, including vessels or other nerves. All these data are fundamental to defining the surgical plan. Interestingly, there are peculiar clinical and EDX findings suggestive of malignant lesions. All these data are reasonably crucial, but an effective preoperative interaction between the neurosurgeon and neurophysiologist and the knowledge of the significance of neurophysiological findings is mandatory.

11.2.1 Peripheral Nerve Ultrasound

In recent years, after the improvement of US technology, including the development of transducers with high frequency (15–22 MHz) as well as the image processing software enhancement, high-resolution US imaging is become an ideal screening and diagnostic tool in clinical neurophysiological studies for several reasons: it is a noninvasive, painless, and relatively not expansive technique that allows dynamic assessment of the nerve anatomy in real time with a good resolution [12, 13]. Although the MRI is still considered the gold standard in diagnosis, US should be considered a complementary method with the advantages of being portably and quickly adopted in several clinical settings, including during surgery with intraoperative nerve ultrasound [14]. However, the US also has some limitations; it requires specific training and experience and has a poor resolution of nerves located deeply or close to the bone. Thorough knowledge of anatomy and pathology and significant experience are necessary, considering that US is an operator-dependent technique.

Nerve US is performed in B or 2D Mode using a linear transducer with a frequency usually ranging from 12 to 18–22 MHz. The nervous trunks and brachial plexus are easily investigated, particularly regarding the shape, echogenicity, and cross-sectional area determination, in two orthogonal planes (cross-sectional and longitudinal views). The US appearance of a normal peripheral nerve is similar to nerve histology. In the section, nerves display a honeycomb fashion due to the hypoechoic fascicles surrounded by hyperechoic epineurium and perineurium (Fig. 11.1). High-frequency probes can quantify the number and size of fascicles and follow the fascicular pattern along the entire nerve trunk, especially for the principal nerves not too deep. In the longitudinal view, the nerve appears like a train track with a linear hypoechoic fascicle with hyperechoic edges. Also, the mobility is tested with dynamic maneuvers, not only for the most common entrapment neuropathies like the ulnar nerve at the elbow but also in PNSTs that can induce functional entrapment in other nerves close to the lesion. Additionally, the B-mode can be used with color or power Doppler to assess and quantify the presence of blood flow, which can indicate a hypervascularity or increase in blood volume within the nerve.

Recently, the contrast-enhanced ultrasound (CEUS) able to measure intraneural blood flow improved the accuracy of vascularity determination. CEUS has been applied for neuropathy entrapments and intraoperative differentiation of peripheral nerve tumors [15–17]. US, as a complementary technique of EDX, can differentiate not tumoral mass causing focal neuropathy like intraneural ganglion cysts or may identify nerve infiltration by lymphoma that determines focal mononeuropathy or plexopathy and increase of CSA and hypervascularity [18].

PNSTs, including schwannomas and neurofibromas, appear as hypoechoic lesions in continuity with the nerve fascicules. Neurofibroma may usually occur in superficial nerves or, less often, in large and proximal trunks or nerve plexus and presents sensory-motor symptoms and signs related to the distribution. Schwannomas may be

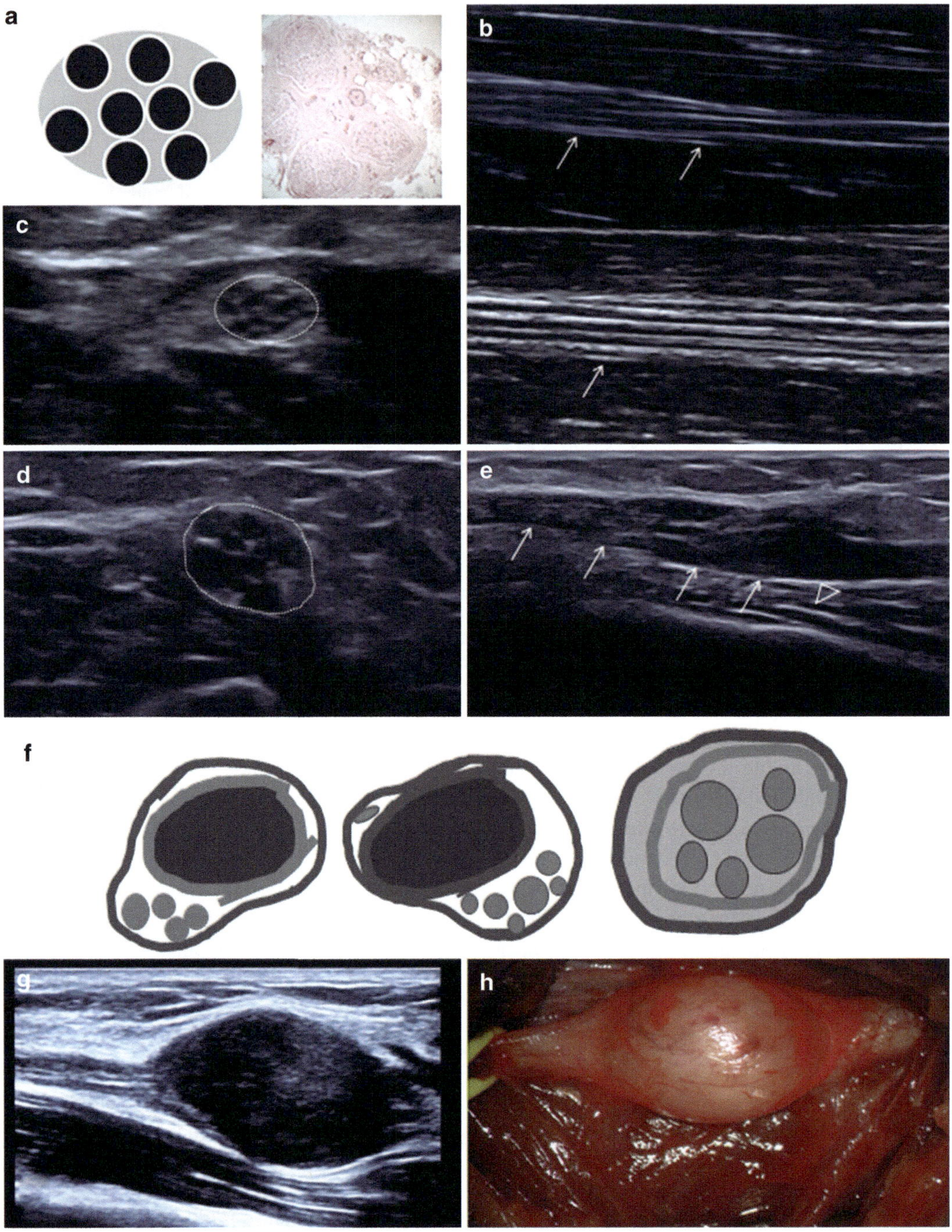

Fig. 11.1 In (**a**), representative scheme of normal nerve ultrasound in axial view and hematoxylin-eosin staining of a sural nerve. Histology is characterized by individual fascicles, densely packed with myelinated axons and surrounded by endoneurium, perifascicular connective tissue, and epineurium. The nerve ultrasound image (**b**, **c**), in axial (**d**) and longitudinal (**e**) view, shows hypoechoic fascicles surrounded by hyperechoic connective tissue and epineurium. US imaging of a normal median nerve demonstrating a typical fascicular pattern (**c**) compared with the median nerve at the elbow in a patient with schwannomatosis shows high variability of fascicular pattern with a significant increase of fascicular diameter and fusiform fascicular lesion. In (**f**), a scheme of different US patterns of schwannoma and neurofibroma is depicted in axial view. In (**g**, **h**), US and intraoperative view of schwannoma

localized anywhere on a peripheral nerve, usually along the flexor surfaces of the upper and lower extremities, including the skin and subcutaneous tissues of the neck and head, as well as in spinal intradural and extramedullary sites and cranial nerves. Although the distinction between different forms of PNSTs is usually challenging, several US characteristics give valuable support in diagnosis. Schwannomas show an eccentrical site between preserved fascicules rather than obliterating them (Fig. 11.1), while neurofibromas appear as concentric lesions with disrupted fascicules [19, 20]. At US, schwannomas usually appear as an oval, hypoechoic lesion with a well-defined margin and a slight posterior acoustic enhancement (Fig. 11.1).

Concerning the subtypes, schwannomas can show intralesional signs of hemorrhage, calcification, and cystic mass. Furthermore, the lesion is clearly in continuity with the nerve trunk, and sometimes, it is possible to recognize the arising hypoechoic and swollen fascicle with the other healthy fascicles displaced at the periphery.

Intraneural neurofibromas are usually fusiform and hypoechoic lesions, in continuity with the nerve trunk, without a capsule, with the tendency to infiltrate the nerve involved (Fig. 11.2). The presence of cystic or hemorrhagic signs should be considered with suspicion of malignant transformation.

Plexiform neurofibroma is a multinodular lesion involving entire nerves or plexus with intra- and extra-fascicular growth, and they do not exhibit a capsule. Moreover, plexiform neurofibromas involve main nerve trunks showing a classical bag-of-worm fashion.

Schwannomas, neurofibromas, plexiform neurofibromas, and malignant transformation of a benign PNST may be associated with neurofibromatosis type 1 [33].

In other types of PNST, US is less informative, as in intraneural perineurioma, where the affected nerve shows a long segment of fusiform enlargement with a thinning at the poles and relative preservation of the fascicular architecture [34]. Otherwise, hybrid nerve sheath tumors present the coexistence of different features of the other PNST (Fig. 11.7).

Malignant PNSts are fast-growing lesions, classified as soft tissue sarcomas, characterized

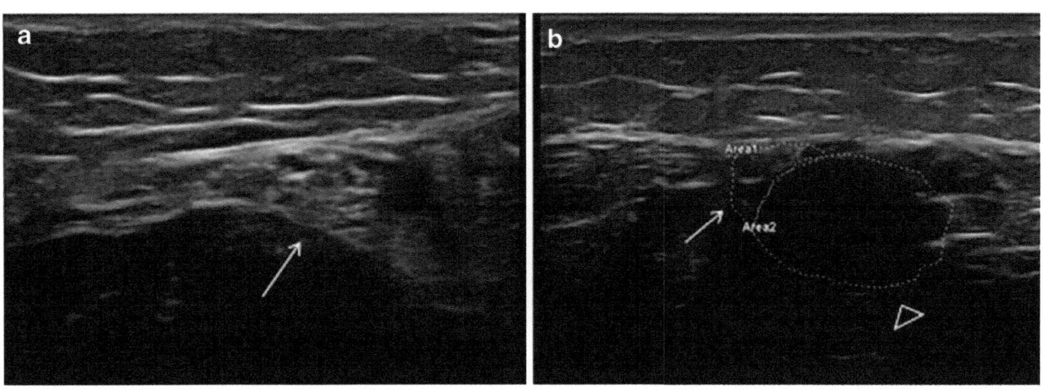

Fig. 11.2 (**a**–**d**) Nerve US of neurofibroma of the tibial nerve at the distal third of the leg. US (**a**, **b**) showing tibial nerve immediately proximal to the neurofibroma, and (**b**) the neurofibroma arising from a single fascicle and dislocation of the healthy fascicles. In (**c**), the longitudinal view with low flux signals. In (**d**), the CT scans images. (**e**) Nerve conduction studies at the preoperative evaluation. The top traces show the motor nerve conduction studies at the peroneal and tibial nerves. The time base for the recordings is 5 ms/division. The bottom shows sensory nerve conduction studies at the sural, superficial peroneal, medial plantar, and lateral plantar nerves. The time base for the recordings is 1 ms/division. In (**f**), an example of the IONM. On the left, triggered EMG recordings demonstrate large amplitude polyphasic compound muscle potentials (CMAPs) in the AH muscle during nerve stimulation, sweeping the surgical field. The most significant response corresponds to the location closest to the nerve. The bottom trace shows the corresponding free-running EMGs with superimposed stimulation artifacts. On the right is an example of an F-wave recording from the tibial nerve during surgery

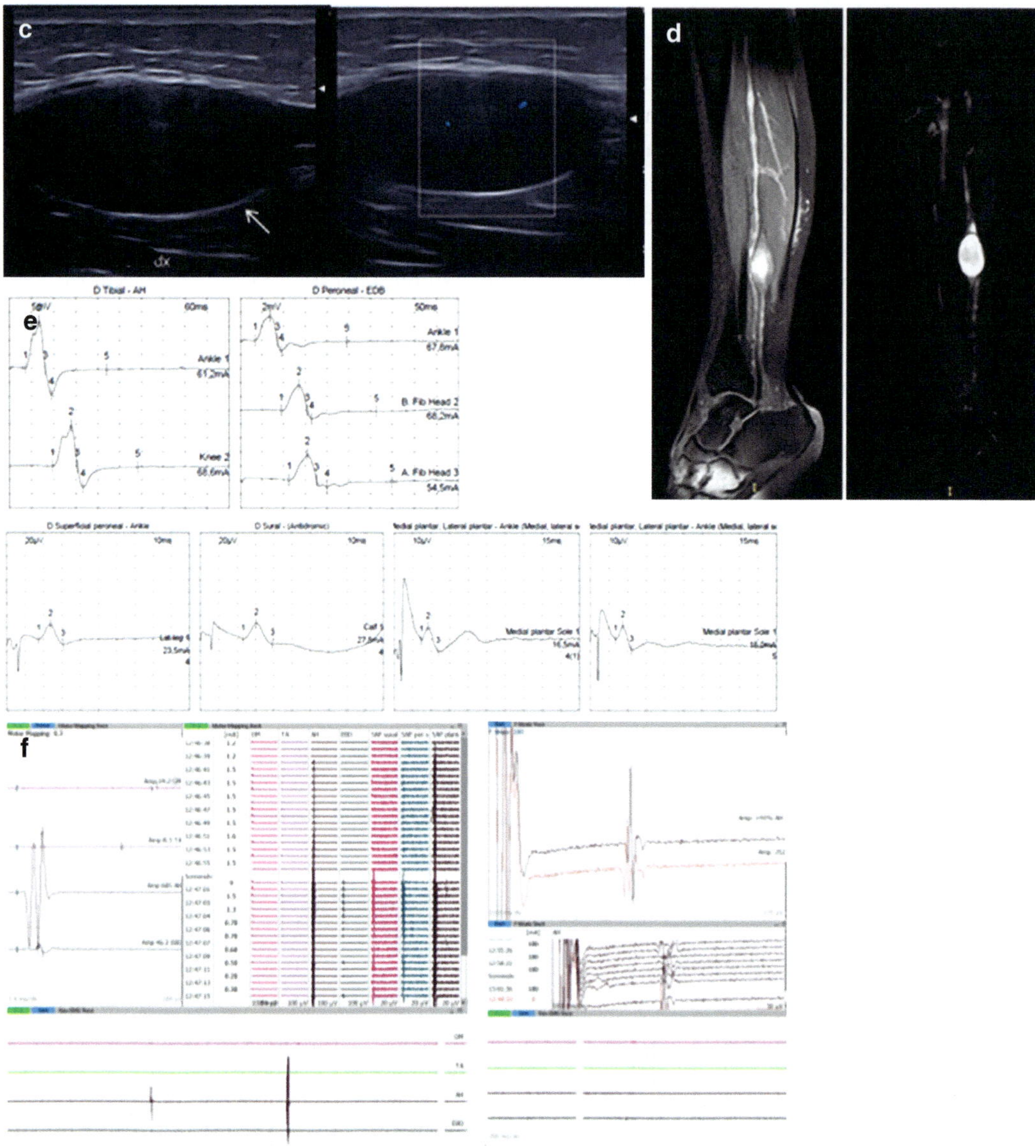

Fig. 11.2 (continued)

by pain and nervous deficit rapidly worsening, that arise from large peripheral nerves or the brachial or lumbosacral plexus, often following radiation treatment or in NF1. In US, MPNST usually shows peculiar characteristics, including larger diameter (>5 cm), unclear boundaries, cystic degeneration, intratumor necrosis or hemorrhage, edema, and calcification within the lesion.

11.3 Intraoperative Monitoring

Intraoperative neurophysiologic monitoring to preserve nervous function, during exposure, dissection, and tumor removal, is applied using both mapping and monitoring techniques, including free-running EMG, triggered EMG via direct nerve stimulation and muscle recordings, nerve action potentials (NAPs), and motor and somatosensory evoked potentials (MEPs and SEPs).

11.3.1 Neurophysiological Techniques

11.3.1.1 EMG in Free Run

EMG in free run can be continuously monitored intraoperatively in any muscle accessible using surface or needle electrodes that provide a better signal-to-noise ratio. The muscle plan is related to the neuroanatomical distribution of nerve structures interested in PNS tumors and must be done during preoperative planning since placing additional EMG recordings after surgery starts is impossible. Usually, a couple of muscles should be chosen for each root level and peripheral nerve to correctly localize the nervous source of irritation or damage. Moreover, regional muscles outside of the nerve distribution should also be monitored to exclude volume conduction for excessive stimulation and protection. Free run EMG are recorded continuously during the surgery; it can also be broadcast to timely hear any change in muscle activity.

Recording EMG can identify mechanical and thermal irritation by recording neurotonic discharges characterized by high-frequency bursts or trains of motor unit action potentials firing rapidly and irregularly. Bursts are generally <200 ms in duration, while the repetitive firing of one or more motor units lasts more than 200 ms producing EMG trains, shown in multiple short bursts or continuous firing [21]. Trains have been further defined, particularly for cranial nerve surgery, by their characteristics and supposed ability to predict postoperative outcomes [22]. Notably, the presence of neurotonic discharges does not necessarily imply nerve damage, and their lack does not entirely exclude nerve injury, as shown by a sharp laceration may not induce any neurotonic discharges but only electric silence [23]. Temporal and anatomical correlations are essential to evaluate EMG changes during surgery correctly. Furthermore, EMG activity alteration can result from surgical maneuvers and other factors, including temperature, ambient electrical noise, or neuromuscular blockade. Therefore, careful discrimination of all possible confounding factors is continuously performed by the IONM team during all the surgical phases.

11.3.1.2 Triggered Electromyography and Nerve Action Potentials

Two main methods exist to establish the presence or absence of electrical transmission across peripheral nerve lesions: direct nerve stimulation or triggered electromyography (EMG) and nerve action potentials (NAPs). Despite both these techniques not allowing proper continuous monitoring of the peripheral nerve activity, the information obtained is diagnostic and essential for the surgery [32].

11.3.1.3 Triggered Electromyography

Direct nerve stimulation is performed with a small bipolar or monopolar handheld stimulator probe within the operative field by the surgeon (Fig. 11.3). Usually, a monopolar stimulator is used during dissection for stimulation through tissue, such as a tumor, to identify if a neural structure is nearby; otherwise, a small bipolar stimulator is preferred when the neural structure is exposed because it minimizes current spread to adjacent nerves and muscles. This is particularly important when complex structures or multiple nerves are nearby, such as in the brachial plexus. Direct motor stimulation may identify the nerve and its function and allows for the evaluation of different thresholds during nerve exposure, tumor dissection, and resection; for this reason, direct stimulation is performed several times during surgery. The cMAP amplitude recorded by applying a supramaximal stimulation is related to the number of functioning axons between the stimulation site and the muscle recordings. An absent cMAP can result in the absence of motor nerve function, neuromuscular blockade, or technical problems. In the case of a significant reduction or complete disappearance of the cMAP, a significant post-operatory neural deficit is expected. Otherwise, the preservation of cMAP amplitude is predictive of a good outcome. Moreover, the threshold of excitability gives additional information to the surgeon and requires a careful stimulation strategy: firstly, the stimulation of a healthy distal site is performed to determine the threshold level; then, the stimulation is performed at a slight increase of intensity to obtain reliable cMAPs by proximal and distal stimuli, and pre-resection thresholds are finally recorded.

An example of an IONM strategy during schwannoma resection starts with evaluating the nerve trunk proximally and distally to identify functional fascicles streaming around the tumor in the pseudocapsule (Fig. 11.3). Once a safety entry zone (fascicle-free) is identified, the surgeon assesses the pseudocapsule layers and identifies the true capsule. At the poles of the lesion, a single nonfunctional fascicle that courses into the tumor is usually found, tested for function, and finally cut.

11.3.1.4 NAP Recording and Significance

Intraoperative recordings of nerve action potential (NAP) are a valuable technique that provides quick and reliable information on the status of peripheral nerves during surgical exploration and nerve lesion resection (Fig. 11.4). The information provided by NAP recording is handy in peripheral nerve tumors as well as peripheral nerve injury can predict the clinical course and help the decision-making process during surgery [32]. NAP recordings are particularly helpful in identifying functioning sensory and motor nerves or fascicles across the nerve tumor. Individual fascicles can be stimulated while recording either NAP or triggered EMG to map which axons are nonfunctional.

The nerve is directly stimulated on the exposed nerve segment by a tripolar or bipolar probe, and the recording probe is generally 3–6 cm apart from the stimulating probe. A tripolar three-pronged electrode consisting of a single cathode between two anodes should be preferred over the bipolar probe to focus further on the site of stimulation [24, 25].

Notably, there are several rules to obtain reliable and high-quality recordings. First, the distance between stimulating and recording electrodes should be more than 4–5 cm. Thus, with a shorter distance, the stimulus artifact can obscure the NAP, while longer lengths of the nerve (8–10 cm) will facilitate recording. Alternatively, in case of a shorter length (<4 cm), it could be considered to stimulate or record percutaneously by using needle electrodes at some point down the length of the nerve. Using very short-duration stimulus pulses reduces stimulus artifacts and can discriminate the type of fibers; for instance, short-duration stimuli can elicit large fibers. Furthermore, the operating field should be kept dry, and the nerve should be very slightly lifted out of the fluid.

The stimulus is applied at the distal site with the recording electrodes proximal along the nerve or vice versa (Fig. 11.5). Stimulus intensities are usually only 1–10 mA but should be increased to 20–25 mA if there is no response [6, 25]. NAP recording is sometimes technically challenging; averaging 3–5 responses could be necessary. A response larger than 2 mV and/or longer duration than 2 ms is more likely a muscle action potential. The response should exhibit threshold behavior as the stimulus intensity is raised and lowered. Despite

Fig. 11.3 Schwannoma in the posterior tibial nerve at the popliteal fossa, causing increasingly paroxysmal pain and hyperalgesia. In (**a**), nerve US shows a fusiform lesion in continuity with the tibial nerve (24.3 × 18 × 16 mm). The nerve conduction studies were regular (**b**). In (**c**), mapping before the intracapsular resection of schwannoma of the tibial nerve. In (**d**), on the left, triggered EMG recordings demonstrating a large amplitude CMAP in the abductor hallucis muscle during nerve stimulation using a monopolar probe. On the right, a smaller cMAP was recorded on the same muscle after stimulation far from healthy fascicules

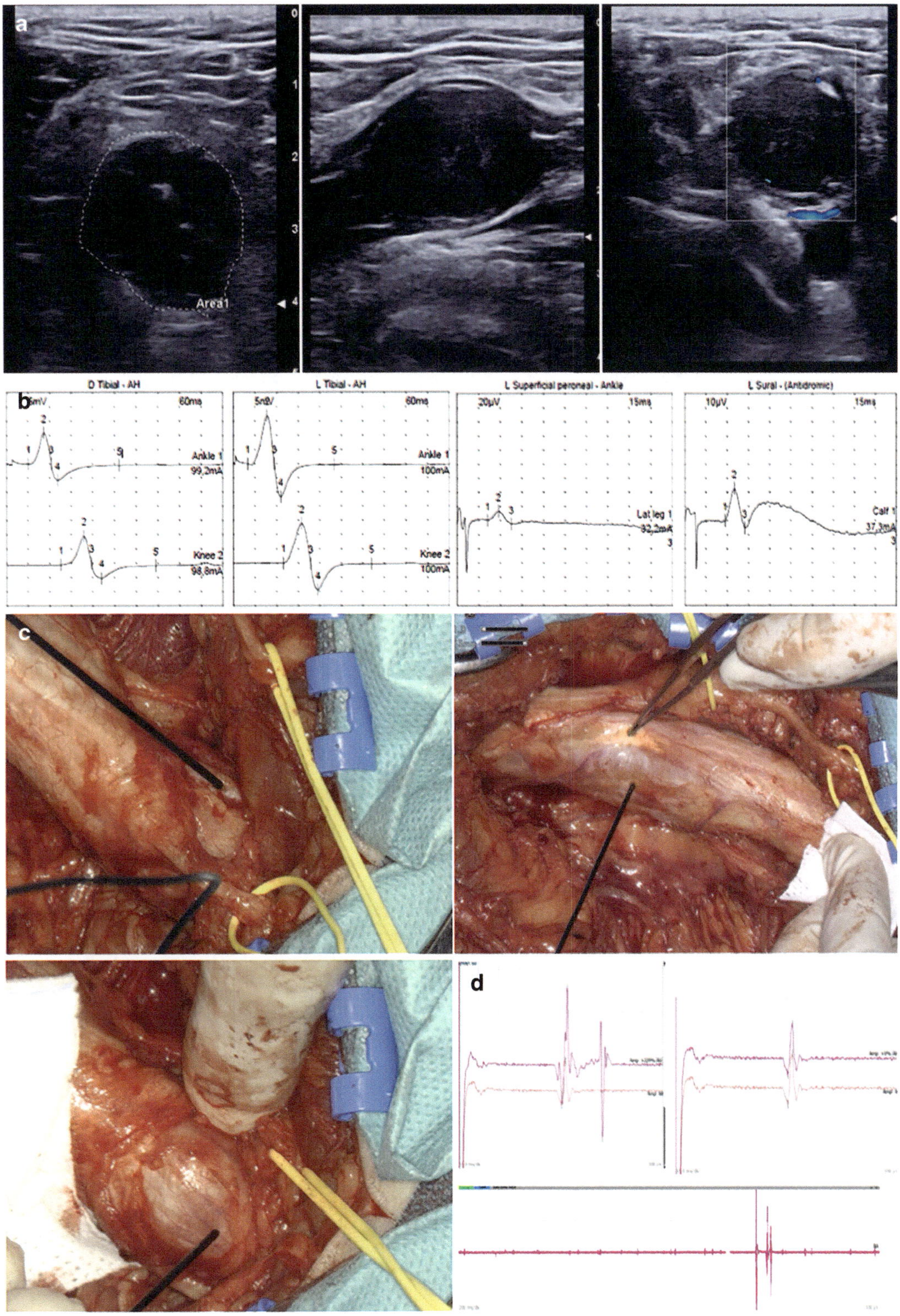
a
Area1
b
D Tibial - AH
L Tibial - AH
L Superficial peroneal - Ankle
L Sural - (Antidromic)
5mV
60ms
Ankle 1
99,2mA
Knee 2
98,8mA
100mA
20µV
15ms
Lat leg 1
82,2mA
10µV
Calf 1
37,3mA
c
d

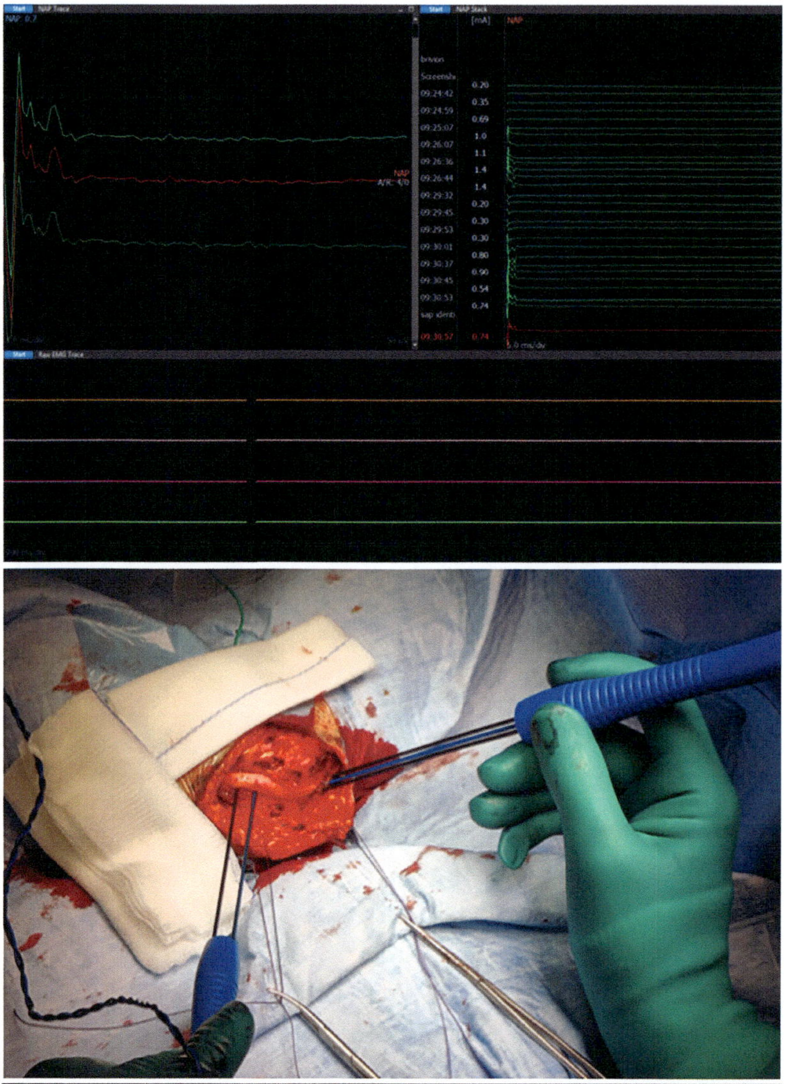

NERVE ACTION POTENTIALS: USUAL SETTINGS			
recording settings		**stimulus settings**	
frequency filter setting	HFF 2500-3000 Hz - LFF 10-30 Hz	duration	0.1–03msec
time base	0.5–5msec/div	intensity	3–5mA
ampification	50uV	frequency	1Hz
sweeps	3–10		

Fig. 11.4 Stimulating and recording technique for nerve action potentials (NAPs). The surgeon uses a stimulating electrode with three contacts that can better localize the stimulus to the region of electrode application, compared with bipolar stimulation

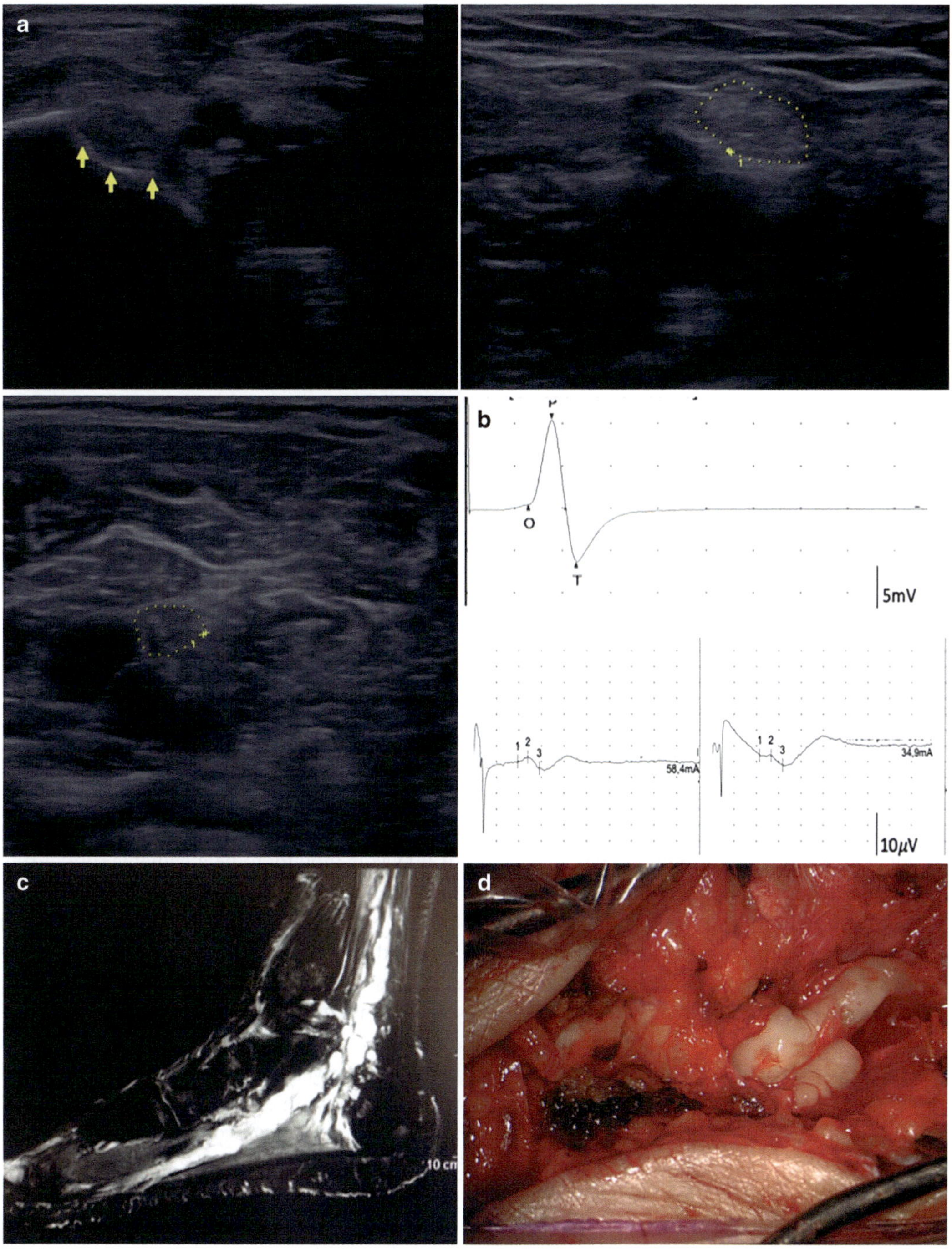

Fig. 11.5 Plexiform neurofibroma of the tibial nerve in a 19-year-old male with swelling of the medial ankle, severe pain, and sensory symptoms evoked by light touch, standing, and walking. In (**a**), nerve US. The nerve conduction studies (NCSs) were substantially unremarkable; in (**b**), motor conduction study of the tibial nerve and sensory conduction study of medial and lateral plantar nerves. The time base for the recordings is 1 ms/division for sensory NCS and 5 ms/div for motor NCS. In (**c**), MRI findings. In (**d**), the intraoperative view. After surgery, the patient showed significant pain relief and improvement in the other symptoms and signs

the cMAP recording, the NAP may also be obtained with neuromuscular blockade, since there is no neuromuscular junction in the pathway.

NAP recording could be performed for fascicules too, which is particularly useful in some cases of PNST resection. In particular, for nerve injuries due to trauma, the presence of NAP suggests a conservative surgery (neurolysis, split repair), while in the absence of NAP, other strategies like a suture or nerve graft should be considered. Nerve action potentials can estimate the number of axons in continuity and predict the likelihood of spontaneous recovery [30]. Otherwise, the absence of NAP has been correlated to a lack of functional continuity and poor clinical recovery.

In schwannomas, the lesion rises from a single fascicle entering and exiting from the lesion, whereas the other fascicles are usually normal and functionally intact; conversely, in neurofibromas, there are several fascicles or one large exiting fascicle [5]. Using a small bipolar probe is possible to record a single NAP from each fascicle to verify the functional continuity. During surgery, sequential NAP or CMAP recordings are then performed to identify functional versus nonfunctional axons. If a single nonfunctional fascicle entering the tumor can be identified, a safe resection can be completed without a neurologic deficit. Otherwise, when multiple fascicles are involved, NAP recordings can identify functioning fascicles that can be protected, thereby limiting postoperative neurologic deficit or helping the decision-making process to define which fascicles should be sacrificed.

11.3.1.5 Somatosensory and Motor Evoked Potential

In the case of proximal root or spinal nerve lesions, it may become necessary to stimulate spinal nerves and record from nerve trunks, stimulate the exposed nerve and roots in the neck, and record from the sensory cortex, using the somatosensory evoked potential (SEP) techniques [29, 31].

The absence of SEP is related to proximal damage, and the site of stimulation can deduce the topographical map of damage: in case of a proximal stimulation at the nerve root, proximal damage should be considered, while for peripheral nerve trunk, the lesion occurred in a segment between cortex and the site of stimulation.

The motor evoked potentials (MEP) are elicited by transcortical anodal stimulation performed over the motor cortex using brief trains of four to seven stimuli with a 1–4 ms interstimulus interval (ISI) and intensities ranging up to 500 V. The recording is performed from peripheral muscles, but for proximal lesions also directly from the root close to the intervertebral foramen using paired hook electrodes. The presence of a response indicates ventral root continuity [5].

11.4 IONM in PNST Surgery and Outcome

Resection of benign PNSTs may determine, in some cases, a worsening of neuropathic pain or other positive symptoms and/or weakness. In the literature, the rate of postoperative deficits reported is about 12–30% [26–28]. The support of IONM is thought to be a valuable tool during the resection of PNSTs, in particular in neurofibromas with high risks of motor deficits and with wider lesions (>3 cm) [28]. However, some surgeons are more confident with a handheld nerve stimulator to map and assess motor fascicles, examining visibly for movement. Only a few studies investigated the risks or benefits of using IONM, with contrasting findings [26, 28], and further studies are needed to clarify these issues better. In addition, the most limiting factor is the availability of a team of neurophysiologists experienced in IOMN, which is still limited to a few centers (Figs. 11.6, 11.7, 11.8, and 11.9).

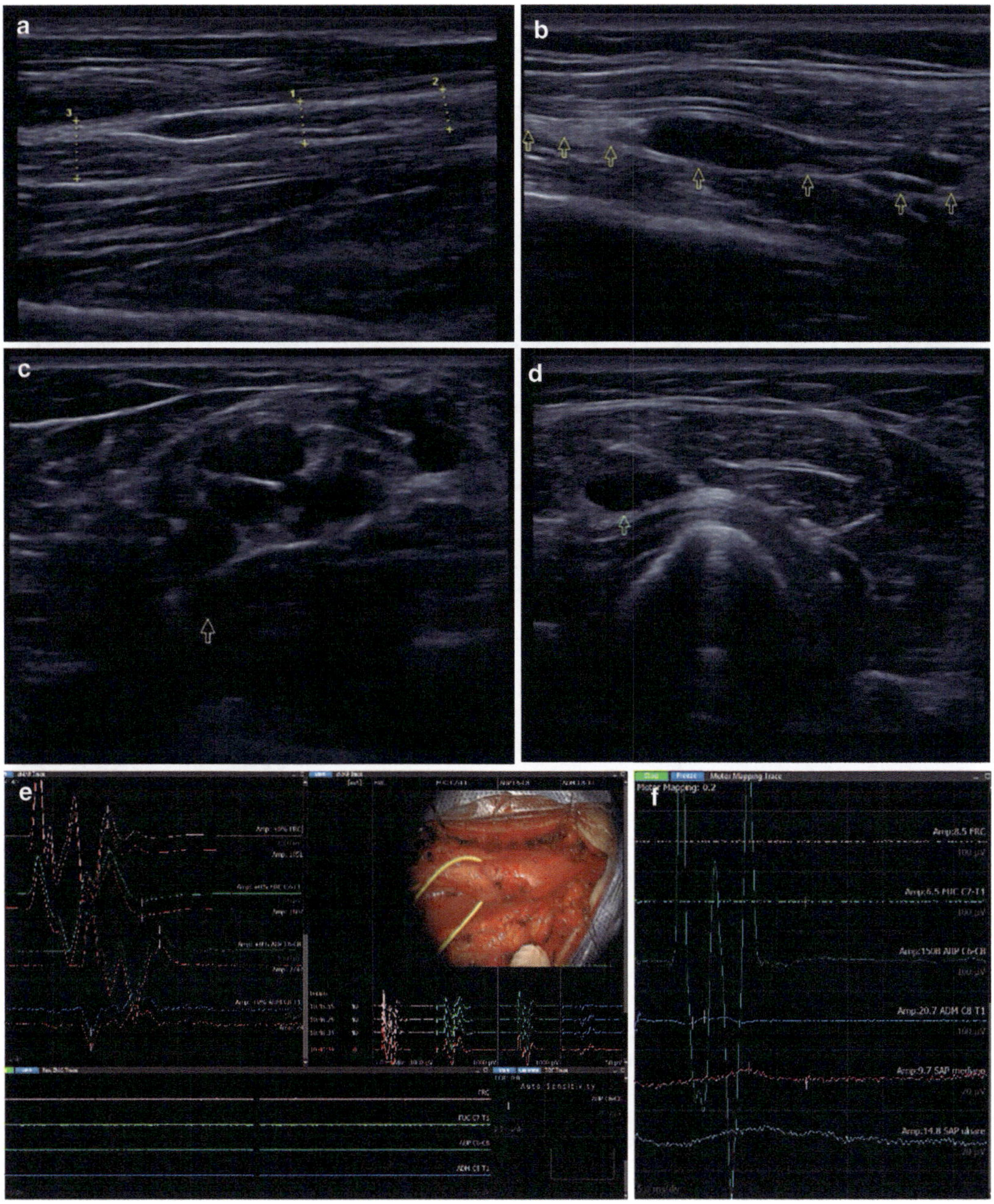

Fig. 11.6 In (**a–d**), nerve US in a 37-year-old female with NF1. In (**a**) and (**b**), longitudinal views of the median nerve and radial nerve, and in (**c**, **d**), axial views of the median nerve at the elbow (**c**) and radial nerve at the middle third of the arm. In (**e**), an example of IONM, in the top, motor evoked potentials, in the bottom free-running EMGs; on the right (**f**), triggered EMG showing large amplitude polyphasic compound muscle potentials (CMAPs) of the ABP muscle during nerve stimulation

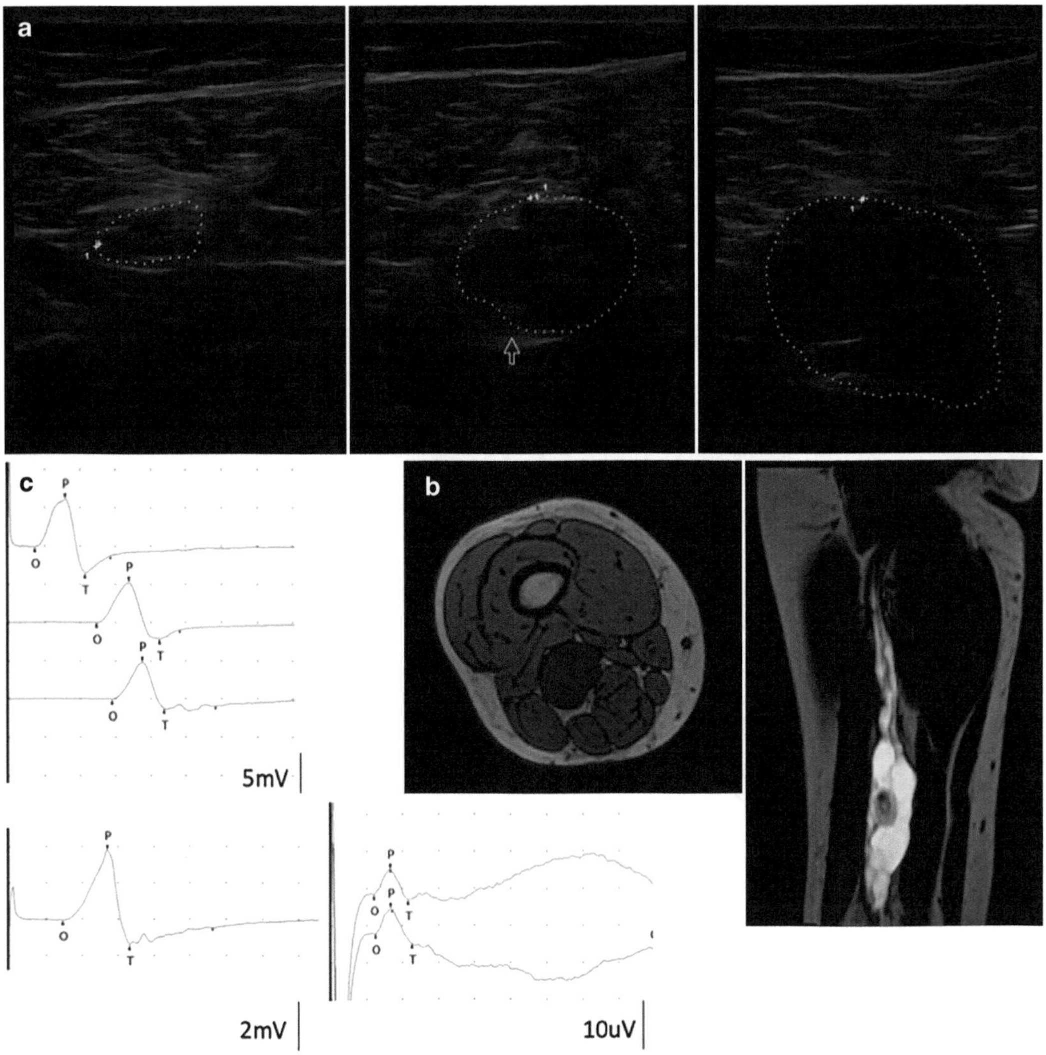

Fig. 11.7 Hybrid nerve sheath tumor presenting in a 45-year-old male with swelling of the right thigh and only sensory symptoms evoked by sitting and leg extension. In (**a**), axial US imaging; in (**b**), the CT scans. In (**c**), peroneal and tibial nerve conduction studies (the time base for the recordings is 5 ms/division), on the left sural nerve conduction study. Notably, the sensory and motor nerve conduction studies were normal (bottom), without significant differences with the contralateral side. EMG findings (not shown) were normal

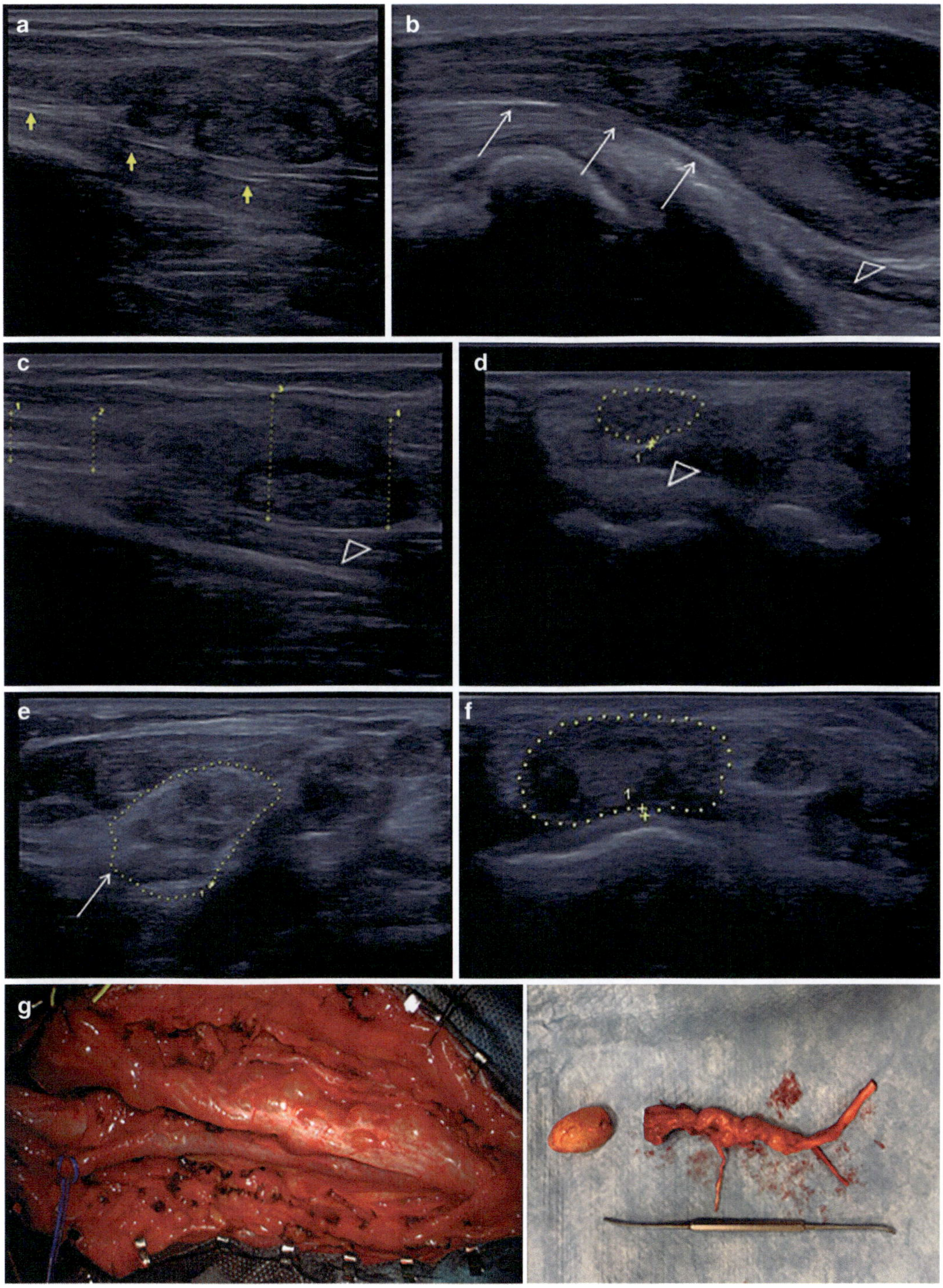

Fig. 11.8 Different US imaging of ulnar (**a**–**d**) and median (**e**–**f**) nerves in a 48-year-old female with schwannomatosis. At the bottom (**g**), the intraoperative view shows a large hybrid nerve sheath tumor originating from the ulnar nerve, along with the result of the resection. The histological examination revealed a prevalence of neurofibroma tissues, with a less pronounced schwannoma part

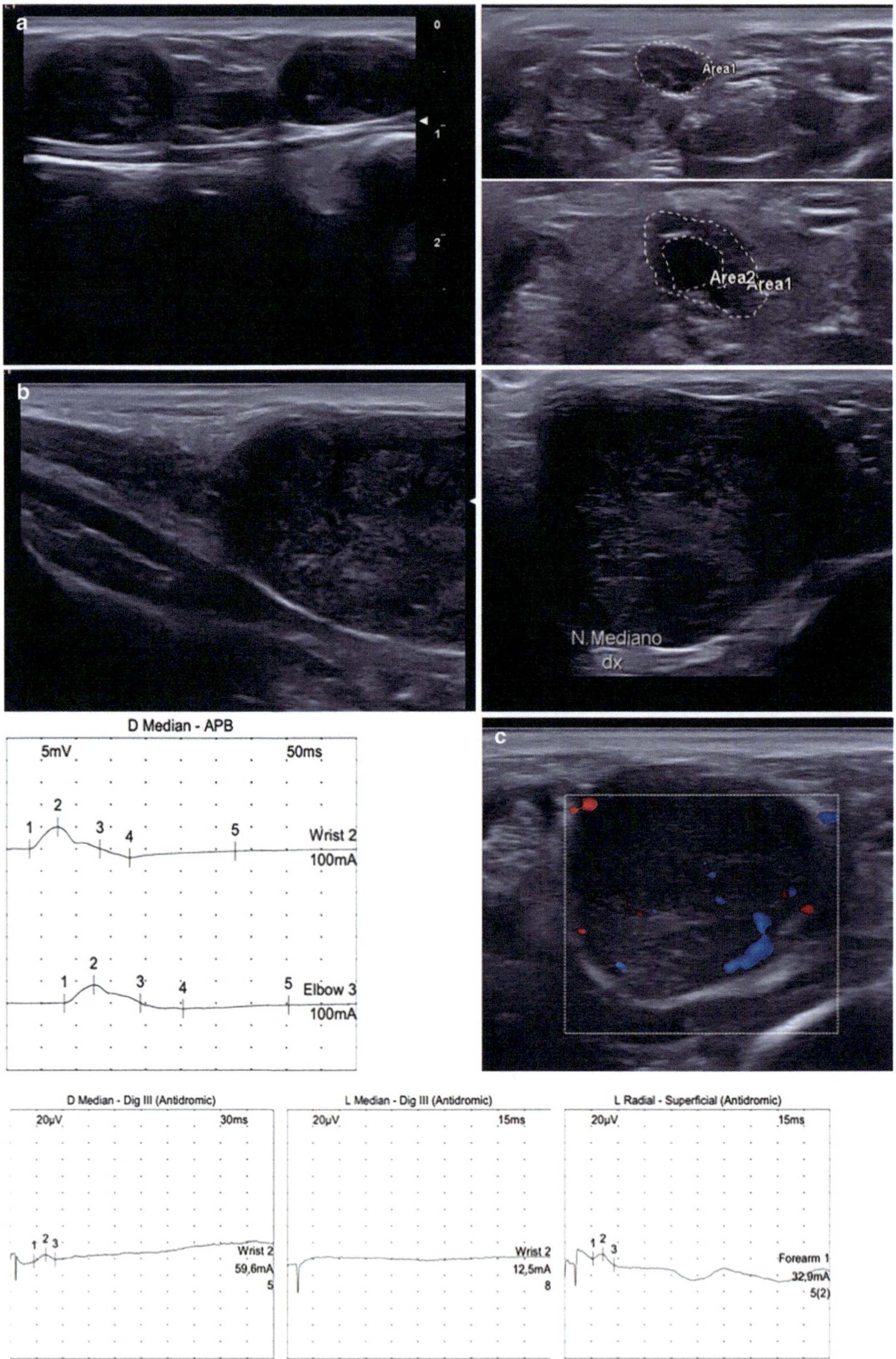

Fig. 11.9 Multiple nerve sheath tumors at all levels in a 38-year-old female with NF-2. In (**a**), a longitudinal view of the right superficial radial nerve. On the right, a sequence of images of the right median nerve scanning from the carpus to the distal third of the forearm, in (**b**), the same lesion in a longitudinal view. In (**c**), the color Doppler examination depicted low flow within the lesion. In this case, NCSs showed sensory-motor axonal neuropathy

References

1. Happle L, Kline D. Intraoperative neurophysiology of the peripheral nervous system. In: Deletis V, Shils J, Sala F, Seidel K, editors. Neurophysiology and neurosurgery. A modern approach. 2nd ed; 2020.
2. Strommen JA, Skinner S, Crum BA. Neurophysiology during peripheral nerve surgery. Handb Clin Neurol. 2022;186:295–318. https://doi.org/10.1016/B978-0-12-819826-1.00022-3.
3. Simon NG, Talbott J, Chin CT, Kliot M. Peripheral nerve imaging. Handb Clin Neurol. 2016a;136:811–26. https://doi.org/10.1016/B978-0-444-53486-6.00040-5.
4. Simon NG, Spinner RJ, Kline DG, Kliot M. Advances in the neurological and neurosurgical management of peripheral nerve trauma. J Neurol Neurosurg Psychiatry. 2016b;87(2):198–208. https://doi.org/10.1136/jnnp-2014-310175.
5. Spinner RJ, Kline DG. Surgery for peripheral nerve and brachial plexus injuries or other nerve lesions. Muscle Nerve. 2000;23(5):680–95. https://doi.org/10.1002/(sici)1097-4598(200005)23:5<680::aid-mus4>3.0.co;2-h.
6. Kline DG, Hackett ER, May PR. Evaluation of nerve injuries by evoked potentials and electromyography. J Neurosurg. 1969;31(2):128–36.
7. Kline DG. Surgical repair of peripheral nerve injury. Muscle Nerve. 1990;13(9):843–52. https://doi.org/10.1002/mus.880130911.
8. Simon MV. Intraoperative neurophysiology: a comprehensive guide to monitoring and mapping. Springer; 2018. p. 540–3.
9. Grant GA, Goodkin R, Kliot M. Evaluation and surgical management of peripheral nerve problems. Neurosurgery. 1999;44(4):825–39; discussion 839–40. https://doi.org/10.1097/00006123-199904000-00077.
10. Korus L, Ross DC, Doherty CD, Miller TA. Nerve transfers and neurotization in peripheral nerve injury, from surgery to rehabilitation. J Neurol Neurosurg Psychiatry. 2016;87(2):188–97. https://doi.org/10.1136/jnnp-2015-310420.
11. Spinner RJ, Hanna AS, Maldonado AA, Wilson TJ. Peripheral nerve. Oper Neurosurg (Hagerstown). 2019;17(Suppl 2):S229–55. https://doi.org/10.1093/ons/opz072.
12. Lee FC, Singh H, Nazarian LN, Ratliff JK. High-resolution ultrasonography in the diagnosis and intraoperative management of peripheral nerve lesions. J Neurosurg. 2011;114(1):206–11. https://doi.org/10.3171/2010.2.JNS091324.
13. Lefebvre G, Le Corroller T. Ultrasound and MR imaging of peripheral nerve tumors: the state of the art. Skeletal Radiol. 2023;52(3):405–19. https://doi.org/10.1007/s00256-022-04087-5.
14. Koenig RW, Schmidt TE, Heinen CP, et al. Intraoperative high-resolution ultrasound: a new technique in the management of peripheral nerve disorders. J Neurosurg. 2011;114(2):514–21. https://doi.org/10.3171/2010.9.JNS10464.
15. Cantisani V, Orsogna N, Porfri A, Fioravanti C, D'Ambrosio F. Elastographic and contrast-enhanced ultrasound features of a benign schwannoma of the common fibular nerve. J Ultrasound. 2013;16(3):135–8.
16. Pedro MT, Antoniadis G, Scheuerle A, Pham M, Wirtz CR, Koenig RW. Intraoperative high-resolution ultrasound and contrast-enhanced ultrasound of peripheral nerve tumors and tumorlike lesions. Neurosurg Focus. 2015;39:E5.
17. Wang Y, Tang P, Zhang L, Wan W, He C, Tang J. Gray-scale contrast-enhanced ultrasonography for quantitative evaluation of the blood perfusion of the sciatic nerves with crush injury. Acad Radiol. 2011;18:1285–91.
18. Vijayan J, Chan YC, Therimadasamy A, Wilder-Smith EP. Role of combined B-mode and Doppler sonography in evaluating neurolymphomatosis. Neurology. 2015;85:752–5.
19. Toia F, Gagliardo A, D'Arpa S, Gagliardo C, Gagliardo G, Cordova A. Preoperative evaluation of peripheral nerve injuries: what is the place for ultrasound? J Neurosurg. 2016;125(3):603–14. https://doi.org/10.3171/2015.6.JNS151001.
20. Simon NG, Cage T, Narvid J, Noss R, Chin C, Kliot M. High-resolution ultrasonography and diffusion tensor tractography map normal nerve fascicles in relation to schwannoma tissue prior to resection. J Neurosurg. 2014;120(5):1113–7. https://doi.org/10.3171/2014.2.JNS131975.
21. Harper CM. Intraoperative cranial nerve monitoring. Muscle Nerve. 2004;29:339–51.
22. Romstöck J, Strauss C, Fahlbusch R. Continuous electromyography monitoring of motor cranial nerves during cerebellopontine angle surgery. J Neurosurg. 2000;93(4):586–93.
23. Nelson KR, Vasconez HC. Nerve transection without neurotonic discharges intraoperative electromyographic monitoring. Muscle Nerve. 1995;18(2):236–8. https://doi.org/10.1002/mus.880180215.
24. Ohsawa I, Inui K. Use of tripolar electrodes for minimization of current spread in uncut peripheral nerve stimulation. Neurosci Res. 2009;64(1):63–6.
25. Tiel RL, Happel LT Jr, Kline DG. Nerve action potential recording method and equipment. Neurosurgery. 1996;39(1):103–8; discussion 108–9. https://doi.org/10.1097/00006123-199607000-00019.
26. Levi AD, Ross AL, Cuartas E, et al. The surgical management of symptomatic peripheral nerve sheath tumors. Neurosurgery. 2010;66(4):833–40. https://doi.org/10.1227/01.NEU.0000367636.91555.70.
27. Siqueira MG, Socolovsky M, Martins RS, et al. Surgical treatment of typical peripheral schwannomas: the risk of new postoperative deficits. Acta Neurochir. 2013;155(9):1745–9.
28. Wilson TJ, Hamrick F, Alzahrani S, et al. Analysis of the effect of intraoperative neuromonitoring during resection of benign nerve sheath tumors on gross-total resection and neurological complications. J Neurosurg. 2021;135(4):1231–40. https://doi.org/10.3171/2020.8.JNS202885.

29. Burkholder LM, Houlden DA, Midha R, Weiss E, Vennettilli M. Neurogenic motor evoked potentials: role in brachial plexus surgery. Case report. J Neurosurg. 2003;98(3):607–10. https://doi.org/10.3171/jns.2003.98.3.0607.
30. Kim DH, Kam AC, Chandika P, Tiel RL, Kline DG. Surgical management and outcomes in patients with median nerve lesions. J Neurosurg. 2001;95:584–94.
31. Sugioka H. Evoked potentials in the investigation of traumatic lesions of the peripheral nerve and the brachial plexus. Clin Orthop Relat Res. 1984;184:85–92.
32. Zelenski NA, Oishi T, Shin AY. Intraoperative neuromonitoring for peripheral nerve surgery. J Hand Surg Am. 2023;48(4):396–401. https://doi.org/10.1016/j.jhsa.2022.11.022.
33. Abreu E, Aubert S, Wavreille G, Gheno R, Canella C, Cotten A. Peripheral tumor and tumor-like neurogenic lesions. Eur J Radiol. 2013;82(1):38–50. https://doi.org/10.1016/j.ejrad.2011.04.036.
34. Mauermann ML, Amrami KK, Kuntz NL, et al. Longitudinal study of intraneural perineurioma-a benign, focal hypertrophic neuropathy of youth. Brain. 2009;132:2265–76.

12 Fluorescein-Guided Surgery of Peripheral Nerve Sheath Tumors

Jacopo Falco, Niccolò Innocenti,
Ignazio Gaspare Vetrano, Marica Eoli,
Morgan Broggi, Vittoria Nazzi, and Francesco Acerbi

12.1 Background

Peripheral nerve sheath tumors (PNST) constitute a miscellaneous group that counts for 10–15% of all soft tissue primary tumors. Among the histopathologic entities within the PNST, benign lesions, i.e., schwannomas and neurofibromas, are the most common overall and more frequently affect adults. PNST can be sporadic lesions or associated with neurocutaneous syndromes characterized by multisystemic manifestations, prominently affecting structures primarily derived from the ectoderm. Syndromic PNST affect young adults and children and are generally present in multiple localizations, including several subcutaneous lesions; although it is a rare subgroup, malignant PNST are more frequent in syndromic diseases than sporadic conditions.

The mainstay of treatment is constituted by surgical enucleation, which should not lead to neurological impairment; gross-total resection is a pivotal predictor of local control and recurrence. In the case of malignant lesions, characterized by aggressive infiltrative growth, postoperative radiation is recommended as complementary therapy, whereas the efficacy of chemotherapy remains unclear. The most important innovation to reduce the risk of neurological deterioration is the use of intraoperative neurological monitoring (IONM), which plays a fundamental role in identifying functional nerves, in localizing the safest entry point inside the tumor capsule and in recognizing anatomical structures in case of large and destructive tumors.

This field has been largely explored in last years for central nervous system tumors (CNS), where different dyes, like 5-aminolevulinic acid (5-ALA) and, more recently, sodium fluorescein (SF) and indocyanine green (ICG) have demonstrated to have a positive impact on intra-axial tumor resection [1–6]. However, their potential role in PNST surgery is still unexplored. In particular, 5-ALA has been preliminarily reported in a few cases of spinal neurinomas, but none of them presented a positive fluorescence pattern [6,

J. Falco · N. Innocenti · M. Broggi · V. Nazzi
F. Acerbi (✉)
Department of Neurosurgery, Fondazione IRCCS Istituto Neurologico Carlo Besta, Milan, Italy
e-mail: morgan.broggi@istituto-besta.it; vittoria.nazzi@istituto-besta.it; francesco.acerbi@istituto-besta.it

I. G. Vetrano
Department of Neurosurgery, Oncological Neurosurgery Unit, Fondazione IRCCS Istituto Neurologico Carlo Besta, Milano, Italy
e-mail: ignazio.vetrano@istituto-besta.it

M. Eoli
Molecular Neuro-Oncology Unit, Fondazione IRCCS Istituto Neurologico Carlo Besta, Milan, Italy
e-mail: marica.eoli@istituto-besta.it

I. G. Vetrano, V. Nazzi (eds.), *Peripheral Nerve Sheath Tumors*,
https://doi.org/10.1007/978-3-031-76382-3_12

7]. Furthermore, the role of ICG is at the moment experimental in this field [8].

The sodium salt of fluorescein, a green fluorescent synthetic water-soluble dye, acts as a vascular fluorophore, in a similar way to the paramagnetic contrast agents in MRI, improving visualization and consequently the extent of resection of several histotypes of CNS tumors in which the blood-brain barrier (BBB) is disrupted [2, 9]. In addition, in an experimental model, it has been demonstrated that fluorescein presents also an endoneurial extravasation, in a similar way to BBB extravasation [10]. At the moment, only a few studies have reported on the use of fluorescein during the resection of PNST [11–13].

12.2 Intraoperative Imaging Technique and Sodium Fluorescein

Although the outstanding advantages provided by the IONM, distinguishing between functional intact nerves and the silent fibers from whence the PNST arose is not always easy to perform; therefore, the development of visualization tools represents a valuable method to better visualize pathological tissue and possibly improve tumor resection. Intraoperative ultrasound is a standardized, operator-dependent method, which allows a visualization superimposable to preoperative diagnostics. It can play a fundamental role in tumor localization and identification; however, surgical subversion can limit its use in the final phase of resection, thus not allowing for recognizing small tumor remnants.

As already studied in central nervous system (CNS) neuro-oncological surgery, the use of intraoperative fluorophores could represent an alternative method to directly recognize pathological tissue during tumor resection. In this research field, 5-aminolevulenic acid has demonstrated inconsistent results due to the lack of a positive fluorescence pattern, probably in relation to its specific mechanism of action.

Otherwise, the potential role of sodium fluorescein (SF) in PNST surgery, although is still little explored, seems to be promising. Fluorescein is a green fluorescent synthetic organic compound, which has countless medical applications. It represents a water-soluble dye with a major blue excitation peak in the region of 460–500 nm and a major green emission peak in the region of 540–690 nm. Although initially its use has been limited to ophthalmology, since the introduction of a dedicated and integrated filter in the surgical microscope, SF applications in neurosurgery have increased exponentially. Experimental models have demonstrated that SF causes an endoneurial extravasation, in a way like blood-brain barrier extravasation, confirming the behavior of fluorescein as an unspecific, vascular dye.

12.3 Surgical Technique and SF Assistance

The safest surgical technique forecasts a sufficiently large tumor exposure, followed by inspection and pseudocapsule or true capsule direct electrophysiological stimulation, to identify possible functional fascicles running inside. To guarantee an accurate IONM, general anesthesia should be administered in a totally intravenous fashion, possibly with a targeted controlled infusion (TIVA-TCI) of a combination of propofol and remifentanil, maintained at a constant plasma concentration due to pharmacokinetic models; muscle relaxants are only used during intubation and, when necessary, a curare antagonist is administered before starting the electrostimulation. The subsequent surgical step is therefore the incision of a safe, silent entry zone, followed by intraneural and longitudinal dissection allowing a progressive isolation and tumor removal via en bloc enucleation. In case of huge tumors or neurofibromas, when an en bloc enucleation is not feasible, neoplastic tissue is usually removed piecemeal. Current neurophysiological protocol for peripheral nerve surgery comprises continuous free-running and stimulus-triggered electromyography, to identify the best and safest entry point inside the pseudocapsule or true capsule and to verify during and at the end of the procedure the functional integrity of the nerve.

The standardized surgical protocol of fluorescein-guided technique is based on intravenous (i.v.) SF injection at standard dose of 0.5 or 1 mg/kg, immediately upon completion of the induction of general anesthesia. Surgery was performed with the aid of a surgical microscope equipped with an integrated fluorescent filter tailored to the excitation and emission wavelength of sodium fluorescein. During resection, the microscope could be switched alternatively from fluorescent to white-light illumination.

No patients, in our experience, presented fluorescein-related side effects or adverse reactions to SF administration. The only visible manifestation of intraoperative fluorescein administration is the onset of a transient and harmless yellowish stain of the urine that rapidly disappears after 24 h. The lack of any side effect, in particular any allergic reaction, was predominantly related to the low dosage used in this study, thanks to the use of a dedicated filter in the microscope, that allowed a more accurate identification of fluorescent tissue.

12.4 SF Role in Different Histotypes

In our wide experience, the schwannomas present a bright and homogeneous fluorescence. Before tumor capsule incision, in yellow-light, the difference between the fluorescent aspect of nerves and the underlying nodule was not particularly evident. Interestingly, as soon as the pseudocapsule had been opened, the schwannomas showed a distinctly brighter fluorescent appearance compared to the surrounding nerves. This difference became even more evident while proceeding with resection (Fig. 12.1).

All neurofibromas demonstrated an intense and homogeneous yellow-green fluorescence (Fig. 12.2). Differently than in schwannomas, due to the presence of various entering and exiting fascicles, resection started by proximal or distal ends of the tumor, with a progressive dissection and piecemeal removal. In all neurofibromas, SF was considered useful to better delineate the difference between the tumor and the origi-

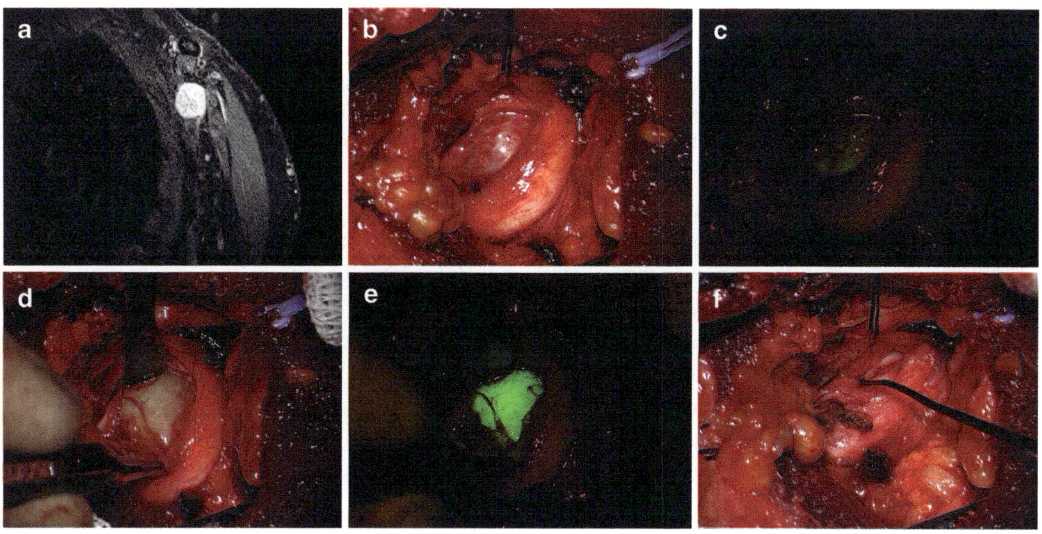

Fig. 12.1 Left brachial plexus schwannoma (coronal MRI scan in **a**). After skin incision and subcutaneous tissues dissection, the tumor appeared surrounded by the nerve (**b**). The first examination under YELLOW 560 filter (**c**) showed a slight and diffuse fluorescence both of the tumor and the nerve covering the tumor. However, after pseudocapsule incision (**d**), the tumor showed an intense and homogeneous fluorescence (**e**), whereas the nerve was only slightly fluorescent. Tumor removal was performed under intraoperative neurophysiological monitoring (**f**). (Figure caption from Nazzi et al. [14] licensed under CC-BY 4.0)

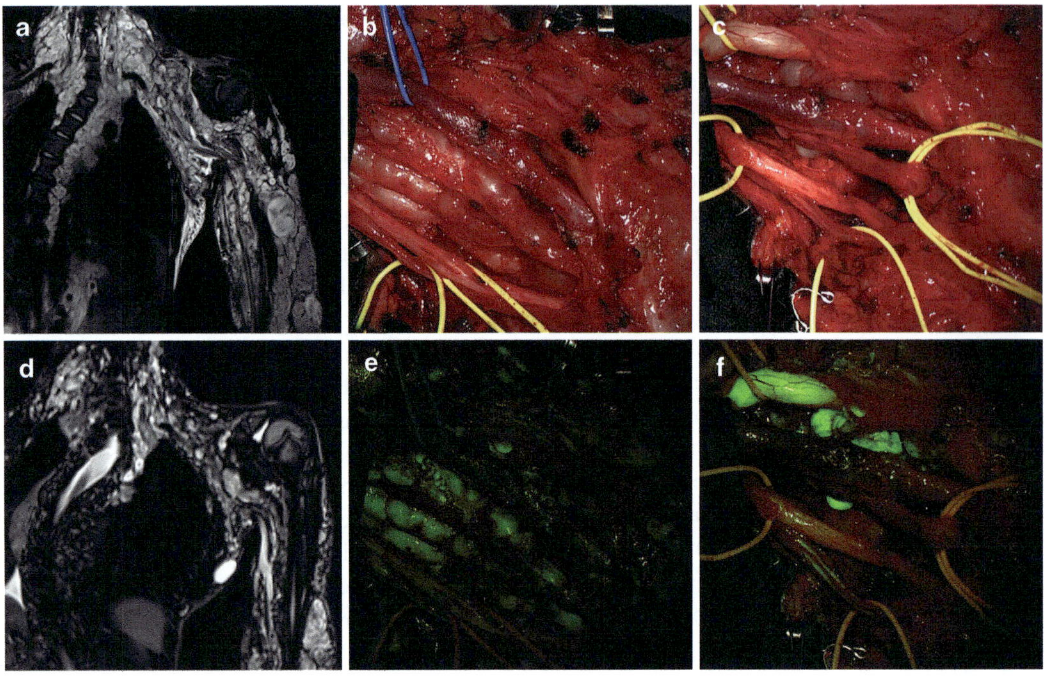

Fig. 12.2 Coronal view in a huge plexiform neurofibroma of the whole upper left arm. Due to the rapid increase of a nodular lesion between radial and median nerves, the patients underwent partial surgery (coronal view in (**a**)—preoperative and (**d**)—postoperative), with progressive exposure of the nerves by alternating the white light (**b**, **c**) and the YELLOW560 filter activation (**e**, **f**). The tumor showed intense fluorescence if compared to the moderate fluorescein uptake of the nerve, whereas the neurofibroma partially substituted the radial nerve (**f**). (Figure caption from Nazzi et al. [14] licensed under CC-BY 4.0)

nating fibers, showing the eventual presence of small remnants that were not visible under white-light illumination and that could be removed.

Perineurioma and hybrid nerve sheath tumors were characterized by a faint fluorescence enhancement, although the rarity of these histotypes cannot allow a strong generalization. Other tumors and lesions such as B-cell lymphomas, solitary fibrous tumor, or other malignant entities appeared highly fluorescent with an inhomogeneous pattern due to sporadic necrotic regions.

In MPNST, fluorescein enhancement was slight and inhomogeneous (Fig. 12.3), and it was not considered useful to remove the lesion, as also reported by Pedro et al. [15]. This issue still does not show a specific explanation, considering that malignant central nervous system tumors have the highest degree of fluorescein uptake, differently from malignant forms of PNS tumors.

Regarding the rate of tumor resection, in our opinion, after a comparison of our historical series, we found a significant improvement in the complete removal rate in plexiform neurofibromas [14]; furthermore, we hypothesized the role of SF in sacrificing yellow nervous fibers, already silent at direct stimulation probably due to tumor infiltration, to potentially increase surgical radicality.

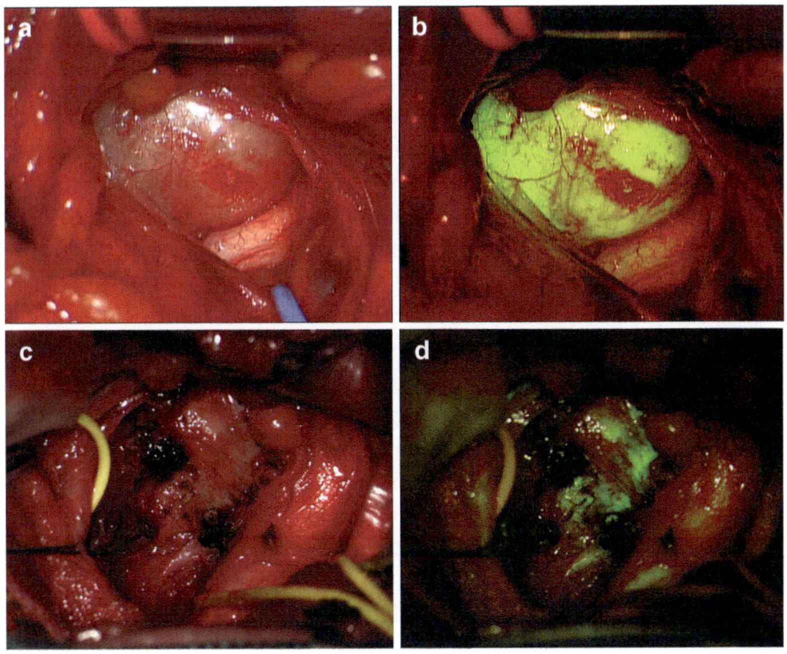

Fig. 12.3 Comparison among the intense and homogeneous fluorescein uptake of a brachial plexus schwannoma (**a**, **b**) and a MPNST of the upper limb (**c**), which showed inhomogeneous and slightly fluorescence (**d**), compared to the surrounding nerve. (Figure caption from Nazzi et al. [14] licensed under CC-BY 4.0)

12.5 Discussion

The standard of care of PNST is surgical resection [16, 17], whereas adjuvant therapies are usually applied as second-line treatments in case of tumor recurrence and for those cases in which surgery is contraindicated or unsuccessful. During the last years, SF has emerged as an intraoperative tracer able to improve brain-tumor visualization, due to its nonspecific, vascular mechanism of action related to the accumulation in brain regions with blood-brain barrier (BBB) disruption, as it happens with MRI contrast enhancement [18–22]. The availability of integrated and specific filters in the surgical microscope has contributed to the wide diffusion of fluorescein. As the reimbursement of SF as a fluorescent tracer in neuro-oncology was approved by AIFA in July 2015 (determination 905/2015, Gazette n.168, 22 July, 2015) and based on our extensive experience with CNS tumors, we empirically started this experience of fluorescent PNST [20, 23].

Regarding the dose used, before starting our published study, we progressively reduced the standard dose used in CNS surgery (from 5 mg/kg to 1 mg/kg) in order to have an optimal discrimination between the tumor and normal surrounding nerves [24]. Few other experiences were available; concomitant groups, and in particular Pedro and coworkers, started using a low dosage of 0.5 or 1 mg/kg of sodium fluorescein, i.v. injected at the time of patient intubation [12, 15].

In our preliminary analysis described in 2019, we discussed that fluorescein plays a useful role during surgical resection of schwannomas by means of highlighting the pathological tissue with a brighter fluorescent appearance compared to the surrounding nerves [24]. Nevertheless, we noted instead that the most important advantage of SF was obtained mainly in neurofibromas helping in identifying diffuse tumor remnants [13, 25].

Similar results were found by the German group led by Pedro: the authors examined overall 21 cases of PNST during surgery under SF [12]. An optimal distinction between tumor and surrounding nerves was observed in all 17 schwannomas scheduled for fluorescein-guided surgery, in 1 neurofibroma, and in 1 MPNST. The authors also experienced the role of SF in fascicular biopsy of two lesions involving the whole nerve segment, finding usefulness in a case of B-cell lymphoma, whereas the fluorescein uptake in the MPNST was so widespread that the SF-contrast was judged useless [12]. Moreover, they stressed the role of ImageJ, an open platform for digital video analysis, in objectifying SF enhancement, confirming the intraoperative impression of increased fluorescence of the PNST in contrast with healthy tissue [12, 15]. In contrast to these findings, Kalamarides and coworkers examined five schwannomas scheduled for fluorescein-guided surgery with fluorescein dosage administration of 0.5 mg/kg, without noticing any benefit deriving from filter activation in tumor visualization or in pathological tissue discrimination from en passant nerve fascicles, except for only one case [26].

In our patients' cohort, a bright fluorescence was present in all cases of schwannomas and neurofibromas, although with a less homogeneous pattern, whereas it was significantly less evident for malignant PNST [24]. Perineurioma and hybrid nerve sheath tumors were characterized by a faint fluorescence enhancement, although the rarity of these histotypes cannot allow a strong generalization.

These findings correlate to the characteristics enhancement of tumors at preoperative MRI; in fact, as already discussed in our previous case series, the concentration of SF depends on an endoneurial extravasation: the key point is the disruption of the blood-nerve barrier (BNB), which shares some structural features with the BBB [10]. The interaction between blood and nerve sheath justifies our clinical findings [27], in particular, the increase of fluorescent enhancement after pseudocapsule opening and during tumor resection: this evidence can be explained considering that tumor growth disrupts the tight junctions, a network of transmembrane and peripheral proteins, altering BNB structure and leading to fluorescein extravasation and accumulation inside neoplastic nodules.

In particular, a stratified analysis of our results revealed that the contribution of fluorescein has been considered more significant for schwannomas and neurofibromas: the former presented a bright fluorescence with intense delineation of tumor borders which made easier surgery, especially for large lesions, which need a piecemeal removal [15, 24]. In the case of neurofibromas, especially for the plexiform variant, the role of SF can be judged fundamental in increasing the rate of resection. Differently than in schwannomas, due to the presence of various entering and exiting fascicles in neurofibromas, the progressive dissection and piecemeal resection are facilitated by fluorescein, which allows furthermore the identification of small remnants, not visible under white-light illumination. On the contrary, the use of fluorescein remains questionable for other tumors, such as MPNST, showing a minimal or heterogeneous pattern of SF uptake and widespread involvement of healthy and functional structures.

Currently, neurological complications did not reduce in the fluorescent group: this data can be explained by the fundamental role played by IONM. We want to stress that the need of intraoperative fascicular mapping and monitoring is still a surgical milestone; the use of fluorescein is imagined as a surgical adjunct without rejecting the standard principles of PNST microsurgical resection to preserve neurological function.

12.6 Conclusion

Fluorescein-guided surgery is a safe and effective technique that can be used during the surgical resection of PNST to better identify and distinguish the most frequent subtypes of PNST from intact functional nerves. Prospective multicentric studies with long-term follow-up could provide major insights into the effects of fluorescein application on the outcome of this cohort of patients and further analyze its role in rarer histotypes. We believe that the institution of an international registry, as reported for the general population affected by PNST [28], of neurosurgical procedures based on fluorescent-guided techniques could help increase the surgical management of patients affected by

PNST. Nevertheless, SF demonstrated to be a safe, effective, and not expensive tool that could help to optimize the management of peripheral nerve tumors.

References

1. Stummer W, Pichlmeier U, Meinel T, Wiestler OD, Zanella F, Reulen HJ. Fluorescence-guided surgery with 5-aminolevulinic acid for resection of malignant glioma: a randomised controlled multicentre phase III trial. Lancet Oncol. 2006;7(5):392–401.
2. Acerbi F, Broggi M, Schebesch KM, et al. Fluorescein-guided surgery for resection of high-grade gliomas: a multicentric prospective phase II study (FLUOGLIO). Clin Cancer Res. 2018;24(1):52–61.
3. Cavallo C, De Laurentis C, Vetrano IG, et al. The utilization of fluorescein in brain tumor surgery: a systematic review. J Neurosurg Sci. 2018;62:690–703.
4. Cho SS, Salinas R, Lee JYK. Indocyanine-green for fluorescence-guided surgery of brain tumors: evidence, techniques, and practical experience. Front Surg. 2019;6:11.
5. Acerbi F, Vetrano IG, Sattin T, et al. The role of indocyanine green videoangiography with FLOW 800 analysis for the surgical management of central nervous system tumors: an update. Neurosurg Focus. 2018;44(6):E6.
6. Wainwright JV, Endo T, Cooper JB, Tominaga T, Schmidt MH. The role of 5-aminolevulinic acid in spinal tumor surgery: a review. J Neuro-Oncol. 2019;141:575–84.
7. Millesi M, Kiesel B, Woehrer A, et al. Analysis of 5-aminolevulinic acid-induced fluorescence in 55 different spinal tumors. Neurosurg Focus. 2014;36(2):E11.
8. Mahjoub A, Morales-Restrepo A, Fourman MS, et al. Tumor resection guided by intraoperative indocyanine green dye fluorescence angiography results in negative surgical margins and decreased local recurrence in an orthotopic mouse model of osteosarcoma. Ann Surg Oncol. 2019;26(3):894–8.
9. Da Silva CE, Da Silva VD, Da Silva JLB. Skull base meningiomas and cranial nerves contrast using sodium fluorescein: a new application of an old tool. J Neurol Surg B Skull Base. 2014;75(4):255–60.
10. Malmgren LT, Olsson Y. Differences between the peripheral and the central nervous system in permeability to sodium fluorescein. J Comp Neurol. 1980;191(1):103–17.
11. Stone JJ, Graffeo CS, de Ruiter GCW, Rock MG, Spinner RJ. Intraoperative intravenous fluorescein as an adjunct during surgery for peroneal intraneural ganglion cysts. Acta Neurochir. 2018;160(3):651–4.
12. Pedro MT, Eissler A, Schmidberger J, et al. Sodium fluorescein–guided surgery in peripheral nerve sheath tumors: first experience in 10 cases of schwannoma. World Neurosurg. 2019;124:e724–32.
13. Vetrano IG, Saletti V, Nazzi V. Fluorescein-guided resection of plexiform neurofibromas: how I do it. Acta Neurochir. 2019;161(10):2141–5.
14. Nazzi V, Innocenti N, Castelli N, et al. Assessing the role of sodium fluorescein in peripheral nerve sheath tumors and mimicking lesions surgery: an update after 142 cases. Front Oncol. 2023;12:1070878. https://doi.org/10.3389/fonc.2022.1070878.
15. Pedro MT, Eissler A, Scheuerle A, et al. Sodium fluorescein as intraoperative visualization tool during peripheral nerve biopsies. World Neurosurg. 2020;133:e513–21.
16. Tiel R, Kline D. Peripheral nerve tumors: surgical principles, approaches, and techniques. Neurosurg Clin N Am. 2004;15(2):167–75.
17. Kim DH, Murovic JA, Tiel RL, Moes G, Kline DG. A series of 397 peripheral neural sheath tumors: 30-year experience at Louisiana State University Health Sciences Center. J Neurosurg. 2005;102:246–55.
18. Acerbi F, Restelli F, de Laurentis C, et al. Fluorescent tracers in neurosurgical procedures: a European survey. J Neurosurg Sci. 2021;65(6):618–25.
19. Falco J, Broggi M, Schiariti M, et al. The role of sodium fluorescein in pediatric supratentorial intra-axial tumor resection: new insights from a monocentric series of 33 consecutive patients. Childs Nerv Syst. 2023;39(6):1463–71.
20. Acerbi F, Broggi M, Eoli M, et al. Fluorescein-guided surgery for grade IV gliomas with a dedicated filter on the surgical microscope: preliminary results in 12 cases. Acta Neurochir. 2013;155(7):1277–86.
21. Falco J, Broggi M, Vetrano IG, et al. Fluorescein sodium in the surgical treatment of pleomorphic xanthoastrocytomas: results from a retrospective study. Front Oncol. 2022;12:1009796. https://doi.org/10.3389/fonc.2022.1009796.
22. Ferroli P, Nakaji P, Acerbi F, Albanese E, Broggi G. Indocyanine green (ICG) temporary clipping test to assess collateral circulation before venous sacrifice. World Neurosurg. 2011;75:122–5.
23. Falco J, Cavallo C, Vetrano IG, et al. Fluorescein application in cranial and spinal tumors enhancing at preoperative MRI and operated with a dedicated filter on the surgical microscope: preliminary results in 279 patients enrolled in the FLUOCERTUM prospective study. Front Surg. 2019;6:49.
24. Vetrano IG, Acerbi F, Falco J, et al. Fluorescein-guided removal of peripheral nerve sheath tumors: a preliminary analysis of 20 cases. J Neurosurg. 2019;134(1):260–9.
25. Vetrano IG, Nazzi V, Acerbi F. What is the advantage of using sodium fluorescein during resection

of peripheral nerve tumors? Acta Neurochir (Wien). 2020;162:1153–5.
26. Kalamarides M, Bernat I, Peyre M. Extracapsular dissection in peripheral nerve schwannoma surgery using bright light and fluorescein sodium visualization: case series. Acta Neurochir. 2019;161(12):2447–52.
27. Ouyang X, Dong C, Ubogu EE. In situ molecular characterization of endoneurial microvessels that form the blood-nerve barrier in normal human adult peripheral nerves. J Peripher Nerv Syst. 2019;24(2):195–206.
28. Dengler NF, Scholz C, Beck J, et al. Rationale and design of the peripheral nerve tumor registry: an observational cohort study. Neurol Res. 2023;45(1):81–5.

Exoscope in PNS Tumor Surgery

13

Niccolò Innocenti, Jacopo Falco, Vittoria Cojazzi, Marica Eoli, Francesco Acerbi, Ignazio Gaspare Vetrano, and Vittoria Nazzi

13.1 Magnifying Vision Instruments in Neurosurgery

Compared to other neurosurgery branches, peripheral nerve tumor surgery can be considered a pretty modern subspecialty. It was not until the late twentieth century that we consistently found scientific works regarding peripheral nerve sheath tumors (PNST) surgery. The first detailed descriptions of such tumors were seen in the eighteenth and nineteenth centuries, thanks to a better understanding of the anatomy of peripheral nerves. However, the assumption that manipulating nerves could be fatal hampered the development of surgical techniques for these tumors. Tumor dissection from the nerve was considered too dangerous, and the standard of care consisted of either an en bloc removal or amputation of the affected limb. The advent of microscopes and the birth of histology in the twentieth century led to an improved comprehension of the distinction between schwannomas and neurofibromas; this advancement rendered the practice of dissecting PNST from the parent nerve the gold-standard treatment. Modern neurosurgery started with the introduction of the binocular optical microscope (OM) by Yaşargil [1, 2]. OM paved the way for the micro-neurosurgery era, providing a much better magnification and light than traditional surgical approaches. The OM allowed recognition and preservation of neurovascular microstructures that were impossible with nonmagnified stereoscopic vision [3]. Since the first experience, OM has been routinely used worldwide for cranial, spinal, and peripheral nerve surgery. Even if it represented a breakthrough in the neurosurgical field, and despite the progressive technological improvement, it is not free from downsides such as limited focus range, shorth-depth magnification, illumination problems, difficulty in looking around corners, and possible uncomfortable positions for surgeons. It must also be noted that the assistant vision is usually worse than the first operator. Over the years, new technologies came into place to overcome these limitations, such as the endoscope (EN), conceived as a tool to obtain a wider surgical field view in narrow spaces and surgical corridors and let the surgeon

N. Innocenti · J. Falco · V. Cojazzi · F. Acerbi
V. Nazzi
Department of Neurosurgery, Fondazione IRCCS Istituto Neurologico Carlo Besta, Milan, Italy
e-mail: vittoria.cojazzi@istituto-besta.it; francesco.acerbi@istituto-besta.it; vittoria.nazzi@istituto-besta.it

M. Eoli
Neuro-Oncology Unit, Fondazione IRCCS Istituto Neurologico Carlo Besta, Milan, Italy
e-mail: marica.eoli@istituto-besta.it

I. G. Vetrano (✉)
Department of Neurosurgery, Oncological Neurosurgery Unit, Fondazione IRCCS Istituto Neurologico Carlo Besta, Milano, Italy
e-mail: ignazio.vetrano@istituto-besta.it

I. G. Vetrano, V. Nazzi (eds.), *Peripheral Nerve Sheath Tumors*,
https://doi.org/10.1007/978-3-031-76382-3_13

use more minor approaches for the same kind of surgery. However, the EN also has some disadvantages, including stereo-blindness, small corridors for instrument maneuverability, a bimanual technique, and the limit of needing an air/fluid-filled cavity. Endoscopes became popular in limited subspecialties of peripheral nervous system (PNS) surgery, such as carpal tunnel syndrome [4]. However, its use has not been widespread in PNS tumors due to anatomical limitations of these pathologies that would not extensively benefit from an endoscopic approach.

Recently, surgeon ergonomics and comfortableness have become more and more a matter of discussion. Different studies have documented that musculoskeletal disease rates among surgeons who perform open surgery are between 66% and 94% [5–8]. Neurosurgeons report a high degree of musculoskeletal work-related symptoms during their career [9, 10], and ameliorating surgeon ergonomics has been a recent focus in developing technologies. In recent years, the exoscope (EX) was developed as a new hybrid optical instrument between the OM and the EN [11, 12]. Although it is conceptually similar in some aspects to an EN, EX is positioned outside ("exo") the body cavity. Surgical EX grants a longer working distance, greater magnification, and a broader field of view compared to standard OM. EX has a light-emitting diode (LED) source that results in less thermal tissue damage and tissue glare. It permits neutral body positioning for both first and second operators during surgery regardless of viewing angle, which would not be possible with conventional microscopes [13, 14]. A multicentric study by Schupper and coauthors found that, when using the robotic EX, surgeons reported a statistically significant reduction of neck and back pain [5]. Moreover, images are presented in 4K high-definition (HD) monitors, and the newest EX have also a three-dimensional (3D) vision on specific monitors, so it is necessary to wear specific eyeglasses. EX was found to be helpful also in teaching settings, having big 3D 4K-HD screens; also, residents and medical students can simultaneously follow gestures and have the same vision as the first operator [15, 16].

13.2 Exoscope in PNST Surgery

In peripheral nerve tumors, depth of magnification and visualization details are crucial to identify and dissect the different layers and tumor-nerve interface to preserve nerves and vessels to reduce the risk of postoperative neurological deficits. In complex PNTs, such as plexiform neurofibromas and MPNSTs, surgeries can last hours, and surgeon ergonomics is also a key topic in this subspecialty. The traditional microscope can present some ergonomic limitations, especially in the abovementioned cases. The structure of the microscope itself can limit the surgical pathway and reduce the comfort of the operating surgeon. Moreover, certain OMs do not offer the same quality of vision to the assisting surgeon and to the medical students, trainees, residents, and operating room staff who usually follow the procedure on small 2D screens. Due to these drawbacks, some authors recently presented their experiences when using new exoscopes (EX) systems in several procedures, such as neuro-oncological surgeries [14, 17–19], neurovascular surgeries such as cerebral bypasses [20] or for aneurysms and arteriovenous fistulas [21, 22], and degenerative spinal procedures (cervical discectomy or corpectomy, laminectomies, osteotomies) [23–28].

Being EX systems also promising in the peripheral nerve surgery subspecialty, some authors started to describe their use of EX in peripheral nerve surgery in general and PNST surgery in particular, despite usually being as small series or case reports [14, 29–33]. It must be noted that EX is a novel technology, and PNST surgery is a particular branch primarily done in specialized tertiary centers; therefore, up to now, only anecdotal reports of EX in PNST surgery are available. Miura and coauthors described the use of EX in tarsal tunnel release [34]. Rodriguez-Unda and coworkers verified the role of EX in recurrent carpal tunnel release surgery [30]. Krishnan et al. were the first ones, in 2017, to document EX use in the removal of PNST: in their series, they described the resection of two intraneural schwannomas of the ulnar nerve. The authors found EX very useful to discern further nerve fascicles surrounding the lesion and minimize neurological deficits. The only limitation that they reported was the cumbersomeness of the repositioning and refocusing of the EX; they used an EX with a fixed mechanical arm to hold the system, but new and different models have pneumatic or electronic arms that can be easily moved and redirected with one-hand or using a specific foot pedal. In 2020, Vetrano et al. presented a pilot study focusing on the use of EX in removing a schwannoma of the deep peroneal nerve (Fig. 13.1) and a schwannoma of the sciatic nerve. They described the two different intraoperative settings that were adopted for the two cases: In setting 1, the surgeon and the assistant are working side by side and are looking at the same large 55-inches 4K-3D monitor (as depicted in Fig. 13.2a). In setting 2, the surgeon and the assistant were positioned opposite to each other and aligned differently to have a clear view of the surgical monitors (Fig. 13.2b). The authors noted that both procedures were performed in a very comfortable position for both surgeons especially in "Setting 1" when the 2 surgeons worked side by side, a setting that the authors think it's ideal in PNST surgery. For these reasons, the authors stress the importance of having a preoperative outline of the surgical setup involving surgeons, anesthesiologists, and operating room staff to find the best operative setting.

During the whole procedure, surgeons maintained a subjectively comfortable position, described as more natural than the position using the traditional OM. The physical obstruction on the surgical field is limited if compared to the microscope, allowing vision on a huge 3D-4K

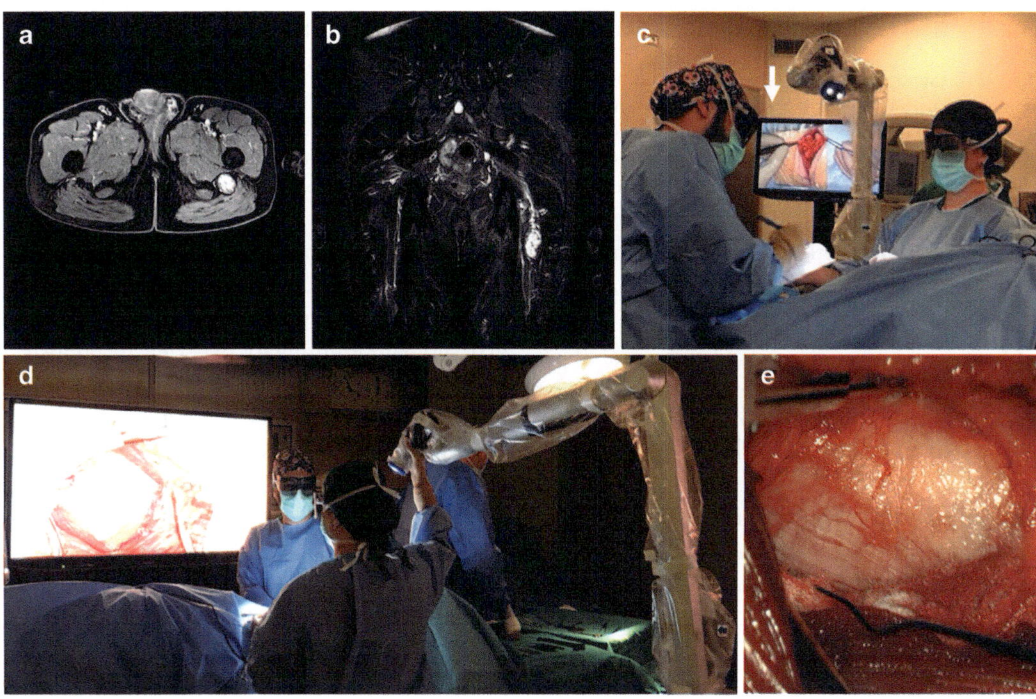

Fig. 13.1 Axial (**a**) and coronal (**b**) MRI show depicted a tumor (arrow) of the left sciatic nerve close to the ischial tuberosity, presenting a homogeneous contrast enhancement. Both the surgeons stood slightly staggered to each other to avoid view impediments; the second surgeon, with this dual monitor setup, had not the same stereoscopic experience as the first one (**c**). In this case, the camera was far from the surgical field, whereas easily maneuverable (**d**). The intraoperative image under exoscopic view (**e**) clearly depicts the difference between nerve fascicles (dotted arrows) and the tumor, even if the schwannoma was strictly adherent to the main nerve
With permission from Vetrano et al. [29]

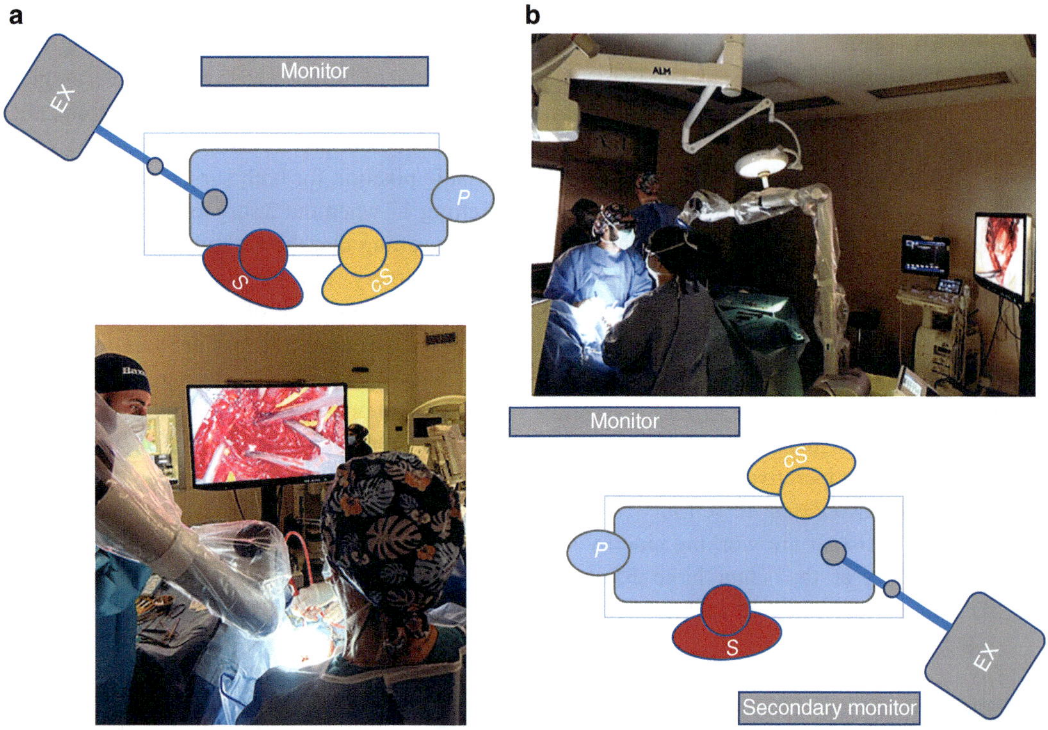

Fig. 13.2 Schematic representation of the operative room setting using an exoscope, with the different positions occupied by the surgical team, the anesthesiologist, and the neurophysiologists. In (**a**), both surgeons and the whole team can visualize the same images on the 55-inch 4K monitor. However, also if the first surgeon and the co-surgeon are positioned opposite to each other and aligned differently, a clear view of the surgical field is possible by using a secondary monitor, with slight discomfort for the second surgeon (**b**). *S* surgeon, *cS* co-surgeon, *P* patient

monitor. Surgeon comfort and the excellent quality of the images represent the main advantages that they found using EX.

13.3 Advantages and Limitations of Exoscope in PNST

Surgeon comfort is reported as one of the principal advantages of EX use in neurosurgery. Murai and coauthors reported a rate of 85.7% satisfaction with the comfort of their position, especially among expert surgeons [18]. In the same series, the authors assessed that 60% of the residents judged the EX as being superior to the microscope, although these results in residents could also be related to their lesser experience in the management of surgical microscopes compared to more expert surgeons. Takahashi et al. reported that the nine neurosurgeons involved in their study greatly appreciated the ability to perform procedures in a comfortable position, even if the assistant could have some problems in visualization when his/her position was opposite to the first surgeon. Preliminary experiences published in the literature show that the EX is considered at least equivalent to the OM in terms of image quality, magnification, lighting, and focal and field depth in most cases. No differences were found between EX and OM regarding efficacy and surgical outcomes. Vetrano et al. reported that with the use of EX, fascicles running into tumor pseudocapsule is clearly discernible before direct stimulation, more than they are with a traditional microscope, denoting a very potential advantage of the exoscopic view (Fig. 13.3). It must be noted that specific OM models do not provide the same vision to the assistant eyepiece, particularly in terms of resolution and stereopsis. These limitations are present for trainees and OR staff, who watch the procedure on a 2D screen. On the other hand, assistants, trainees, and OR

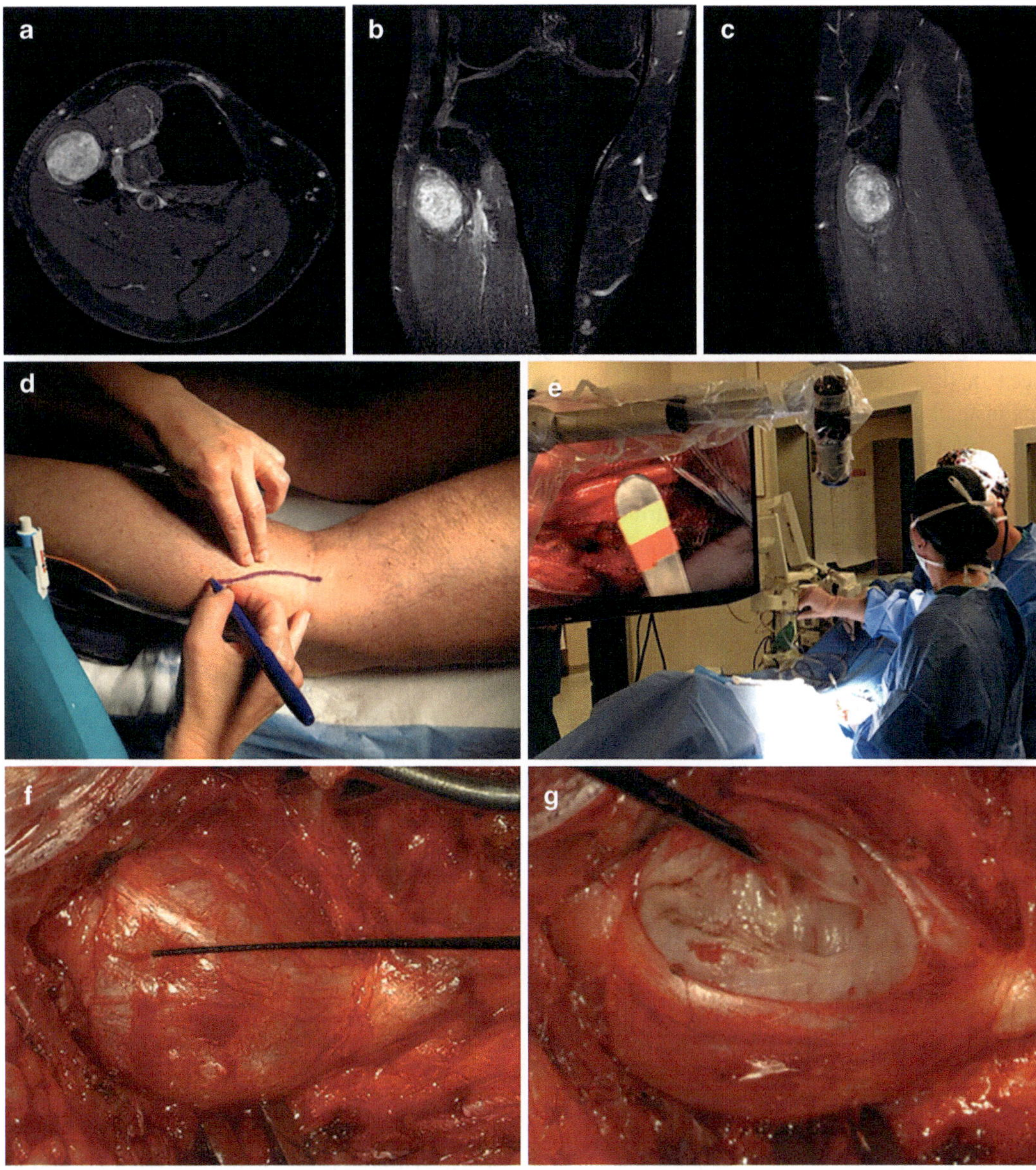

Fig. 13.3 The axial (**a**), coronal (**b**), and sagittal (**c**) MRI after contrast show a lesion (arrow) anterior to the proximal metaphyseal portion of the right fibula with relatively homogeneous contrast enhancement, suggesting the presence of a schwannoma. The patients underwent surgery in the prone position, with the lower limbs slightly flexed at 20 degrees; a slightly curvilinear skin incision distally prolonged to the lateral surface of the right leg was performed to reach the popliteal region and the head of the fibula (**d**). During surgery, the 3D-4K monitor is in front of both surgeons, while the EX-camera is not obstructive (**e**). The intraoperative images under exoscopic view (**f**, **g**) showed, at the bifurcation between the deep and the superficial peroneal nerve, an encapsulated lesion of 3 cm (**f**). It is possible to appreciate the different nerve fascicles (dotted arrows) running just above the tumor; they appear more visible than usual concerning the microscopic binocular view. However, to confirm these findings, the pseudocapsule was incised through the silent zone after direct stimulation, revealing (**g**) a yellow capsulated tumor (arrow) that was progressively enucleated and then removed. With permission from Vetrano et al. [29]

staff can appreciate the same quality of images as those of the surgeon through the wide high-resolution screens providing educational and training opportunities for residents, fellows, and students. The 3D technology also provides a high perception of depth and surgical dissection technique to assistants, who watch the surgery on the screen, and trainees may thus more quickly gain confidence with the anatomic structures and microsurgical techniques. In the first models, wearing 3D glasses for a prolonged time could lead to headache, eyestrain, and nasal pain. New EX has addressed this aspect, and it has not been described anymore.

One of the principal drawbacks is the continuous repositioning/focus recalibration needed during surgery. In fact, this was more commonly seen in first-generation devices. Newer devices are instead equipped with a foot-pedal controller and a pilot unit, which make it easier and quicker to reposition the camera with high precision.

As with all new technologies, a learning curve is required, but most studies published did not report surgeons, finding it challenging to adapt. Takahashi et al. assess that surgeons who have performed more than five procedures with the EX provided higher scores for characteristics of the exoscope system when administered the evaluation questionnaire, compared to surgeons with lesser experience. As previously noted, a specific OR setting is necessary when using EX, especially regarding the positioning of output screens that may be cumbersome and conflict with other OR instruments. A collaborative effort among all different components of the team is needed, especially in the first cases, to overcome these potential problems. Exoscope systems can integrate neuronavigation, diffusion tensor imaging tractography, and fluorescence-guided surgery, allowing surgeons to exclusively use the exoscope without switching back and forth between OM and EX. However, EX models include filters for indocyanine and 5-ALA but not sodium fluorescence. Even though it is not the gold standard, various groups worldwide are implementing SF use in PNST removal, especially in complex cases. Compared to conventional OM or surgical loupes, a common impression is that the EX allows better ergonomics, and surgeons usually report significantly less neck and back pain, at least in spinal procedures [5]. Also, economic issues must be taken into account: several EXs are cheaper than OMs in terms of purchase and management, and OM costs range from $250,000 to >$750,000. Whereas this aspect may play an important role in developing countries [25], some newer EXs could be more expensive than OMs. Cost analysis is a relevant aspect that should be considered when comparing products to implement new devices, especially in centers where loupes magnification (and not OM) is routinely used for PNST surgery. In conclusion, EX systems are a novelty in the neurosurgical field, and further studies are desirable to understand better the role of novel devices in this surgical "niche." Given the characteristics of these technologies and the reports available up to now, EX look like very competitive tools to be used in PNST surgery.

References

1. Yaşargil MG, Krayenbühl H. The use of the binocular microscope in neurosurgery. Bibl Ophthalmol. 1970;81:62–5.
2. Yaşargil MG. A legacy of microneurosurgery: memoirs, lessons, and axioms. Neurosurgery. 1999;45(5):1025–92.
3. Jannetta PJ. The surgical binocular microscope in neurological surgery. Am Surg. 1968;34(1):31–4.
4. Li Y, Luo W, Wu G, Cui S, Zhang Z, Gu X. Open versus endoscopic carpal tunnel release: a systematic review and meta-analysis of randomized controlled trials. BMC Musculoskelet Disord. 2020;21(1):272.
5. Schupper AJ, Eskandari R, Kosnik-Infinger L, et al. A multicenter study investigating the surgeon experience with a robotic-assisted exoscope as part of the neurosurgical armamentarium. World Neurosurg. 2023;173:e571–7.
6. Matern U, Koneczny S. Safety, hazards and ergonomics in the operating room. Surg Endosc. 2007;21(11):1965–9.
7. Szeto GPY, Ho P, Ting ACW, Poon JTC, Cheng SWK, Tsang RCC. Work-related musculoskeletal symptoms in surgeons. J Occup Rehabil. 2009;19(2):175–84.
8. Auerbach JD, Weidner ZD, Milby AH, Diab M, Lonner BS. Musculoskeletal disorders among spine surgeons: results of a survey of the Scoliosis Research Society membership. Spine (Phila Pa 1976). 2011;36(26):E1715–21.
9. Raman AG, Parikh N, Gupta R, et al. Augmenting career longevity: an analysis of ergonomics training among 134 neurological surgeons. World Neurosurg. 2023;173:e11–7.
10. Gadjradj PS, Ogenio K, Voigt I, Harhangi BS. Ergonomics and related physical symptoms among neurosurgeons. World Neurosurg. 2020;134:e432–41.

11. Mamelak AN, Danielpour M, Black KL, Hagike M, Berci G. A high-definition exoscope system for neurosurgery and other microsurgical disciplines: preliminary report. Surg Innov. 2008;15(1):38–46.
12. Mamelak AN, Nobuto T, Berci G. Initial clinical experience with a high-definition exoscope system for microneurosurgery. Neurosurgery. 2010;67(2):476–83.
13. Schupper AJ, Price G, Hadjipanayis CG. Robotic-assisted digital exoscope for resection of cerebral metastases: a case series. Oper Neurosurg (Hagerstown). 2021;21(6):436–44.
14. Montemurro N, Scerrati A, Ricciardi L, Trevisi G. The exoscope in neurosurgery: an overview of the current literature of intraoperative use in brain and spine surgery. J Clin Med. 2022;11(1):223.
15. Ricciardi L, Chaichana KL, Cardia A, et al. The exoscope in neurosurgery: an innovative "point of view". A systematic review of the technical, surgical and educational aspects. World Neurosurg. 2019;124:136–44.
16. Calloni T, Roumy LG, Cinalli MA, et al. Exoscope as a teaching tool: a narrative review of the literature. Front Surg. 2022;9:878293.
17. Doglietto F, Belotti F, Panciani P, Poliani PL, Fontanella MM. High-definition 3-dimensional exoscope for 5-ALA glioma surgery: 3-dimensional operative video. Oper Neurosurg (Hagerstown). 2020;18(3):E82.
18. Murai Y, Sato S, Yui K, et al. Preliminary clinical microneurosurgical experience with the 4K3-dimensional microvideoscope (ORBEYE) system for microneurological surgery: observation study. Oper Neurosurg (Hagerstown). 2019;16(6):707–16.
19. Takahashi S, Toda M, Nishimoto M, et al. Pros and cons of using ORBEYE™ for microneurosurgery. Clin Neurol Neurosurg. 2018;174:57–62.
20. Nossek E, Schneider JR, Kwan K, et al. Technical aspects and operative nuances using a high-definition 3-dimensional exoscope for cerebral bypass surgery. Oper Neurosurg (Hagerstown). 2019;17(2):157–63.
21. Acha JL, Contreras L, Lopez K, et al. Neurovascular microsurgical experience through 3-dimensional exoscopy: case report and literature review. World Neurosurg. 2023;174:63–8.
22. Rossmann T, Veldeman M, Nurminen V, Huhtakangas J, Niemelä M, Lehecka M. 3D exoscopes are noninferior to operating microscopes in aneurysm surgery: comparative single-surgeon series of 52 consecutive cases. World Neurosurg. 2023;170:e200–13.
23. Oertel JM, Burkhardt BW. Vitom-3D for exoscopic neurosurgery: initial experience in cranial and spinal procedures. World Neurosurg. 2017;105:153–62.
24. Rechav Ben-Natan A, Agarwal N, Shabani S, et al. Use of an exoscope for enhanced visualization of a Schwab grade 5 osteotomy to correct kyphotic deformity. Neurosurg Focus Video. 2022;6(1):V19.
25. Encarnacion Ramirez M, Peralta Baez I, Nurmukhametov R, et al. Comparative survey study of the use of a low cost exoscope vs. microscope for anterior cervical discectomy and fusion (ACDF). Front Med Technol. 2022;4:1055189.
26. Yao Y-W, Yao Z-P, Jiang M, et al. Three-dimensional high-definition exoscope in minimally invasive transforaminal lumbar interbody fusion: a retrospective cohort study. Orthop Surg. 2023;15(1):187–96.
27. Srinivasan ES, Crutcher CL 2nd, Shaffrey CI, Gottfried ON, Than KD. Posterior cervical decompression and fusion with exoscope: 2-dimensional operative video. Oper Neurosurg (Hagerstown). 2022;22(2):e83.
28. Siller S, Zoellner C, Fuetsch M, Trabold R, Tonn J-C, Zausinger S. A high-definition 3D exoscope as an alternative to the operating microscope in spinal microsurgery. J Neurosurg Spine. 2020;33(5):705–14.
29. Vetrano IG, Acerbi F, Falco J, D'Ammando A, Devigili G, Nazzi V. High-definition 4K 3D exoscope (ORBEYETM) in peripheral nerve sheath tumor surgery: a preliminary, explorative, pilot study. Oper Neurosurg (Hagerstown). 2020;19(4):480–8.
30. Rodriguez-Unda NA, Wu DS. Exoscope for upper extremity peripheral nerve surgery: revision carpal tunnel release with epineurolysis and hypothenar fat flap. Cureus. 2022;14(2):10–4.
31. Miura I, Kohara K, Kawamata T. An experience with an exoscope system (ORBEYE) for surgery for tarsal tunnel syndrome: a case report. Cureus. 2022;14:e28045.
32. Kueckelhaus M, Kolbenschlag J, Kirschniak A, et al. New technologies in microsurgery: potential, indications and economical aspects - report of the consensus workshop of the German-Speaking Society for Microsurgery of Peripheral Nerves and Vessels (DAM). Handchir Mikrochir Plast Chir. 2022;54(6):507–15.
33. Krishnan KG, Schöller K, Uhl E. Application of a compact high-definition exoscope for illumination and magnification in high-precision surgical procedures. World Neurosurg. 2017;97:652–60.
34. Miura I, Kohara K, Kawamata T. An experience with an exoscope system (ORBEYE) for surgery for tarsal tunnel syndrome: a case report. Cureus. 2022;14(8):12–5.

14 Surgical Management of Peripheral Nerve Sheath Tumors

Maxwell T. Laws, Luis F. Castro, and Prashant Chittiboina

14.1 Introduction

Peripheral nerve sheath tumors (PNSTs) arise from one of the layers that surround peripheral nerves. These layers include Schwann cells, endoneurium, or the perineurium. They can be divided into benign PNST and malignant PNST (MPNST). Benign PNSTs include schwannomas, neurofibromas, granular cell tumors, and hybrid tumors [1]. The most common type of benign PNST is schwannomas, and they typically present in the head and neck region [2]. Peripheral nerve schwannomas most commonly occur as sporadic, solitary lesions but can also occur in association with inherited conditions such as neurofibromatosis type 2 (NF2) or schwannomatosis (together, termed NF2/SWN) [3]. Neurofibromas, are the second most common type of benign PNST, and they can be classified into localized, diffuse, plexiform, or atypical subtypes [4]. They can occur as solitary lesions or as part of an inherited disorder such as neurofibromatosis type 1 (NF1). Neurofibromatosis type 1 and 2 are autosomal dominant disorders with full allelic penetrance and variable expressivity [5]. NF1-related neurofibromas have a protean presentation and are challenging to manage surgically. The principles of management of NF1 neurofibromas can inform the surgical management of all PNSTs. The rest of this chapter will focus primarily on neurofibromas and NF1.

M. T. Laws
Surgical Neurology Branch, National Institute of Neurological Disorders and Stroke, National Institutes of Health, Bethesda, MD, USA

Department of Neurological Surgery, University of Virginia Health System, Charlottesville, VA, USA

Neurosurgery Unit for Pituitary and Inheritable Diseases, National Institute of Neurological Diseases and Stroke, National Institutes of Health, Bethesda, MD, USA
e-mail: maxwell.laws@nih.gov

L. F. Castro
Texas Tech University Medical Sciences Center, Lubbock, TX, USA
e-mail: luis.f.castro@ttuhsc.edu

P. Chittiboina (✉)
Surgical Neurology Branch, National Institute of Neurological Disorders and Stroke, National Institutes of Health, Bethesda, MD, USA

Neurosurgery Unit for Pituitary and Inheritable Diseases, National Institute of Neurological Diseases and Stroke, National Institutes of Health, Bethesda, MD, USA
e-mail: prashant.chittiboina@nih.gov

14.2 Histology

Neurofibromas and schwannomas are both classified by the World Health Organization (WHO) as grade I lesions. The histological defining features for schwannomas include alternating Antoni A hypercellular areas and Antoni B hypocellular areas [4, 6]. Neurofibromas are made up of

I. G. Vetrano, V. Nazzi (eds.), *Peripheral Nerve Sheath Tumors*,
https://doi.org/10.1007/978-3-031-76382-3_14

Schwann cells with thin, comma-shaped nuclei and scant cytoplasm, interspersed collagen fibers, endothelial cells, and mast cells [4]. Neurofibromas will stain positively for S-100 and SOX-10 proteins as well as EMA and CD34 staining [4, 7]. Plexiform neurofibromas have a multinodular appearance with interspersed nerve-like structures [4, 6]. Atypical neurofibromatous neoplasm with uncertain biologic potential (ANNUBP) is a type of neurofibromatous tumor that may exhibit high mitotic rate, cellular atypia, nuclear atypia, loss of typical architecture, or hypercellularity [7]. These tumor types are generally considered to be premalignant lesions [8]. MPNSTs are graded using the Fédération Nationale des Centres de Lutte Contre le Cancer (FNCLCC) grading system [9]. MPNST tumors are classified as either low-grade (grade I) or high-grade (grades II and III) based on mitotic count, tumor necrosis, and differentiation scores. Low-grade MPNSTs are the least common and account for approximately 15% of MPNSTs, while high-grade MPNSTs account for 85% of tumors [1]. High histological grade has been found to serve as an independent risk factor for poor prognosis [10].

It is unclear whether the malignant progression of benign PNSTs requires an intermediary transformation into ANNUBP; however, it is known that most MPNSTs in the setting of NF1 arise from plexiform neurofibromas [11]. The development of a distinct nodular lesion >3 cm in diameter within an existing plexiform neurofibroma is particularly suggestive of progression to an atypical neurofibroma and in some cases to MPNST. Apart from the NF1 gene mutation, the malignant progression sequence likely involves mutations to TP53, CDKN2A, TERT, EGFR, and SUZ12 genes [11].

14.3 Clinical Presentation

Neurofibromas are most commonly asymptomatic unless there is compression of surrounding nerves or structures. Oftentimes, cosmesis is the main concern but it can also present with pain, tingling, and numbness to the area. If the neurofibroma is compressing a nerve or a nerve plexus, it can present with radicular pain along the affected dermatome [12].

Macroscopically, neurofibromas can present as a localized mass, diffuse-type neurofibroma, or plexiform neurofibroma. Localized neurofibromas can occur in the epidermis or subcutaneous layer. They present as well-defined solitary masses. On palpation of the lesion, they can display a characteristic buttonhole or dimple sign [13]. Diffuse neurofibromas present as poorly demarcated dermal lesions with thickened skin that may invade underlying adipose tissue and entrap surrounding nerves. Plexiform neurofibromas are also slow-growing lesions that can display hyperpigmentation and hypertrichosis [14]. Rarely, plexiform neurofibromas of the leg can occur with lymphangiomatosis and lead to clinically striking elephantiasis neuromatosa [15]. On magnetic resonance imaging, neurofibromas present as fusiform masses that typically encase the parent nerve, whereas schwannomas tend to appear round and tend to displace the nerve [16]. As mentioned before, neurofibromas can occur in the setting of an inherited genetic disorder such as NF1 or NF2. In this case, neurofibromatosis patients may also present with café-au-lait spots, axillary or inguinal freckling, cutaneous or plexiform neurofibromas, iris hamartomas, optic pathway gliomas, and osseous abnormalities [17].

14.4 Epidemiology, Frequency of Surgeries, and Tumor Growth

Neurofibromatosis type 1 affects approximately 1 in 3000 live births [13]. In one study by Plotkin and colleagues (2012), 145 patients with NF1, NF2, or Schwannomatosis underwent whole-body MRI to study their tumor burden [18]. Among patients with NF1, the median number tumor count for all types of tumors was 4.5 with a median tumor volume of 107.9 mL. Similarly, patients with NF2 had a median tumor count of 2 with a median tumor volume of 69.5 mL. Hsu et al. retrospectively analyzed the number of surgical interventions among NF1 patients over a

25-year period [19]. Patients were divided into two groups, those whose disease process had craniofacial involvement and those without craniofacial involvement. Most patients (62.3%) underwent two or more surgical interventions with the average around 3 for patients in both groups. Despite the wide variety of anatomical presentations, there was no statistically significant difference in the average number of surgical interventions over the 25-year period between the craniofacial and non-craniofacial groups. Plexiform neurofibromas experience their fastest growth during childhood and usually slow down by adolescence [20]. During this stage of development, plexiform neurofibromas typically outpace the body growth rate. Nodular neurofibromas typically come later in life, and both nodular neurofibromas and plexiform neurofibromas will experience a small reduction in tumor volume over time [21].

14.5 Surgical Indications

Surgeons of all specialties encounter neurofibromas. It is important to first understand the regional anatomy and surgical considerations. A facial neurofibroma, common among children, will be approached differently than a retroperitoneal mass in an adult. Neurofibromas have a propensity for growth along the brachial plexus, upper extremity, lower extremity, and lumbosacral plexus [22]. While schwannomas demonstrate eccentric growth related to the parent nerve, neurofibromas typically grow as a dominant nodule in ring-like, concentric patterns and can subsume multiple nearby nerve fascicles. This dominant nodule is usually asymmetrically oriented along the course of the parent nerve, as it displaces the normal nerve fascicles. Through this asymmetric growth, the tumor capsule usually resides right below the epineurium of the parent nerve, giving surgeons a plane to remove the tumor without damaging the parent nerve. It is generally understood that the parent fascicle for a neurofibroma is nonfunctioning. Advancements in neuromonitoring and tumor mapping have made it possible to identify the nonfunctioning tumor fascicle from the functioning nerve roots.

Common indications for surgical intervention of neurofibromas include pain, paresthesia, functional compromise, and potential for malignancy (Table 14.1). Localized and diffuse-type neurofibromas rarely experience malignant transformation, but plexiform neurofibromas carry a 10% lifetime risk of malignancy [14]. Therefore, surgical intervention can help reduce the risk of malignant transformation. Since plexiform neurofibromas experience their fastest growth during childhood, surgery might be beneficial for prepubertal NF1 patients. Unfortunately, despite plexiform neurofibromas being the most surgically indicated lesions to remove, given their potential malignant transformation, they are also comparatively more challenging to resect than other lesion subtypes.

Some studies have looked at the utility of whole-body positron emission tomography (PET) scans to determine differences in PET uptake between benign PNSTs and MPNSTs. It is generally understood that [^{18}F]-FDG avidity on PET increases as the lesion transforms to a more cancerous phenotype (Fig. 14.1). A systematic review of 12 studies by Treglia and colleagues suggested using a maximum standardized uptake value (SUV_{max}) value of 3.5 as the cutoff to distinguish

Table 14.1 Indications for neurofibroma resection

Pain
Neurologic symptoms
• Weakness
• Paresthesias
• Autonomic effects
• Numbness
Cosmesis, disfigurement
Concerns for malignancy
• MRI findings:
– Increased neurofibroma growth ≥20% in tumor volume over 1-year
– ADC value <1.85 × 10^{-3} mm^2/s
– Peripheral enhancement
– Perilesional edema
– Intratumoral cystic change
• [^{18}F]-FDG PET/CT findings:
– Preliminary SUV_{max} >3.5

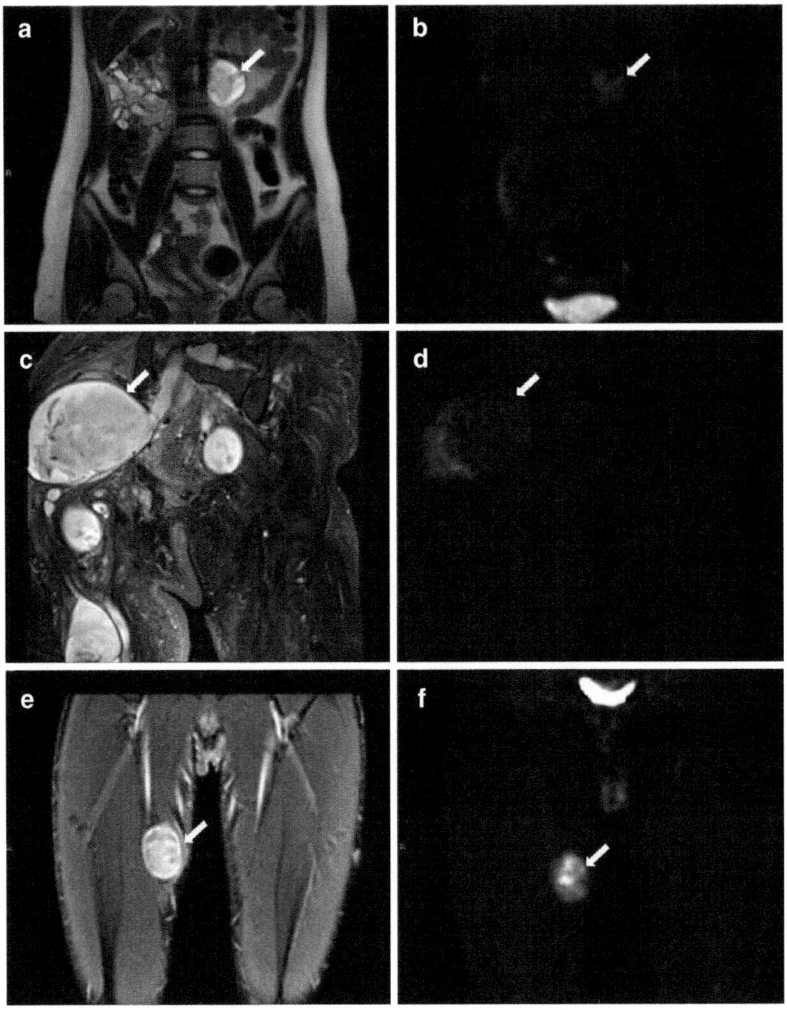

Fig. 14.1 Glucose metabolism increases in neurofibromas as they advance toward malignancy. Representative coronal non-contrast MR and PET images of representative neurofibromas included in this chapter. (**a**) T2-weighted MR of benign isolated nodular left retroperitoneal neurofibroma. (**b**) PET scan of the same patient shows a subtle focal area of increased radiotracer uptake (standard uptake valve, SUV = 1.2) in the benign nodular neurofibroma. (**c**) STIR MR of atypical nodular neurofibroma along the right sciatic nerve. (**d**) PET scan of the same patient shows an area of moderate radiotracer uptake (SUV = 6.2) limited to the lateral aspect of the lesion. (**e**) STIR MR of low-grade MPNST lesion arising from the right femoral nerve. (**f**) PET scan shows an area of hypermetabolic activity and increased radiotracer uptake (SUV = 13.9) in the same lesion

between benign PNSTs and MPNSTs [23]. The sensitivity associated with this cutoff ranged between 89% and 100% with a specificity that ranged between 72% and 100%. Since then, additional studies have validated this SUV_{max} cutoff value of 3.5 [8, 24, 25]. However, it is important to note that multicenter clinical trials are needed to establish cutoff values more definitively as part of a whole-body surveillance plan in NF1 patients.

14.6 General Principles for Resection

14.6.1 Preoperative Planning

Prior to surgery, patients undergo electromyographic and nerve conduction studies to establish baseline nerve function. MRI studies are obtained and used for preoperative decision-making.

Under some circumstances, a percutaneous biopsy may be considered. These circumstances involve patient factors, such as patients who are medically comorbid, those with challenging surgical locations, and those with borderline surgical indications, and therefore, verification of pathology could be necessary for surgical or patient decision-making.

Understanding of local blood supply is valuable, especially in the case of plexiform neurofibromas, which can be especially well-vascularized. Preoperative angiography and embolization have been reported by others [26]. There is no consensus currently on when to utilize preoperative embolization. We do not utilize tumor embolization for neurofibromas in our practice as it is challenging to predict whether the embolic agents will cause unintended ischemia of the normal vascular territory. Furthermore, most intraoperative bleeding from a neurofibroma is caused by the vasa vasorum and venous flow, and therefore, it is unclear whether embolization would reduce intraoperatively blood loss significantly in these surgeries. Depending on the anatomical presentation of the tumor, realistic expectations, and discussion of potential complications should be established before the procedure. These complications include risks of anesthesia, wound complications, and neurological injury.

14.6.2 Intraoperative Planning

Anesthesia choice should be determined based off surgeon preferences and the scope of surgery. The patient should be positioned to allow for extensive exposure to visualize the distal and proximal aspects of the originating nerve. Intraoperative electromyographic (EMG) monitoring should be established, which usually consists of motor evoked potentials and sensory evoked responses. A bipolar stimulating probe for manual stimulation is set up in the neuromonitoring system. Incision planning is guided by a combination of palpation, ultrasound, and interpretation of preoperative imaging. For most lesions within the extremities, a linear incision along the course of the nerve is suitable. Small or transverse incisions should be avoided.

14.6.3 Initial Approach and Tumor Mapping

An initial skin incision is made sharply and carried through the subcutaneous tissue layers. Judicious use of monopolar electrocautery set on a low setting and sharp dissection with Metzenbaum scissors are used to dissect further and open the fascia. Depending on surgeon preference and lesion location, either loupe or operating microscope magnification can be used to assist with visualization. The exposure is extended in the distal and proximal directions to allow for the identification of the proximal and distal ends of the tumor and the associated parent nerve root. The tumor should then be mobilized circumferentially from adjacent structures. Coursing nerves and blood vessels are tagged with vessel loops. The tumor and the parent nerve should be inspected visually and verified with stimulation mapping if possible (Fig. 14.2d) [27]. Bipolar stimulation is used to map the tumor surface. Tumor surface regions with minimal EMG stimulation or absent nerve fascicles are identified. Importantly, traversing, functioning nerve fascicles are identified. When possible, functioning nerve fascicles are assigned to known, named nerves.

14.6.4 Tumoral Incision and Circumferential Dissection

Once the tumor and adjacent nerve fascicles have been carefully identified, an initial incision through the perineurium is usually made near the equator of the tumor parallel to the parent nerve. The tumor capsule is identified within the perineurium. Extracapsular dissection is performed, with a microdissector used to tease the transversing fascicles gently off the tumor capsule. The dissection is carried from the equator to the poles of the tumor. Stimulation is used throughout the

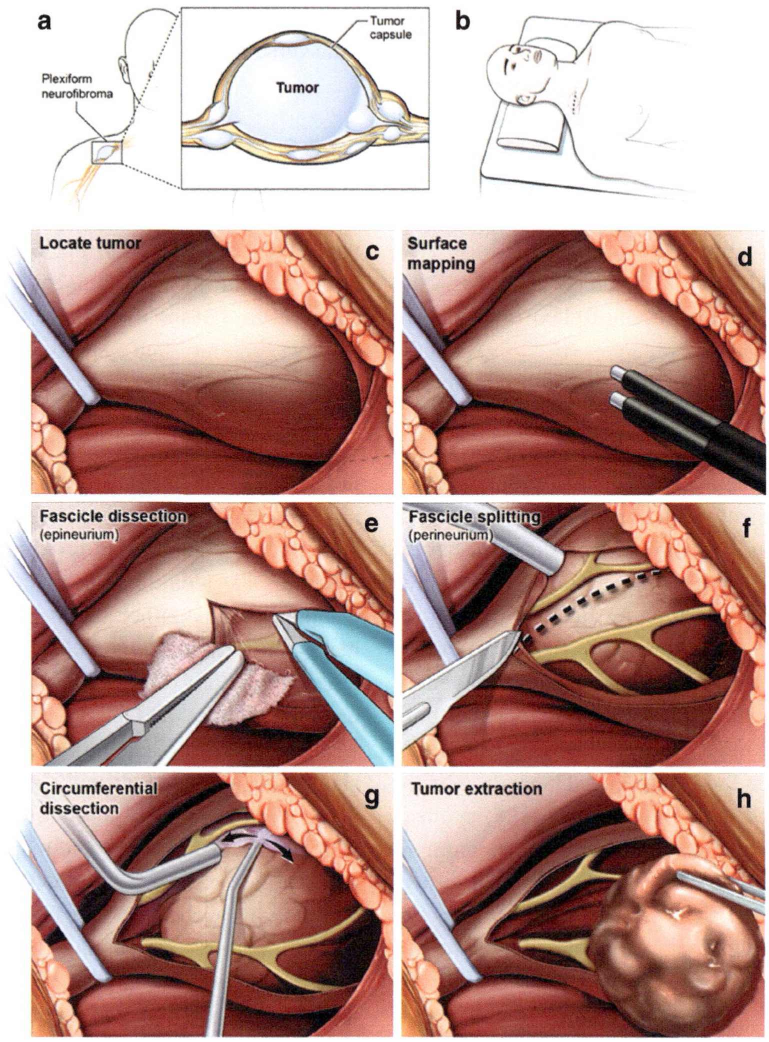

Fig. 14.2 Surgical illustration outlining a stepwise approach to safe dissection of neurofibroma dominant nodule. (**a**) Plexiform neurofibroma illustrated arising from the right brachial plexus showing an eccentric growth pattern. (**b**) The patient is positioned supine and initial skin incision is made. (**c**) Dissection is performed through the subcutaneous tissue layers, fascia, and muscle until the parent nerve is identified and isolated. (**d**) The tumor surface is mapped through bipolar stimulation and neuromonitoring and the lesion is visually inspected. (**e**) Epineurium overlying the tumor is incised and careful dissection is performed to expose the nerve fascicles. (**f**) The tumor capsule and adjacent nerve fascicles are identified, and an incision is made through the perineurium. (**g**) Extracapsular dissection is performed in a circumferential manner to tease the transversing fascicles off the tumor capsule. (**h**) Once the proximal and distal poles of the tumor are defined, the adjoining parent fascicles are incised, and the tumor is removed. Artists: Alan Hoofring and Erina He. Medical Arts, Office of Research Services, National Institutes of Health

microdissection. Once the tumor capsule is circumferentially dissected and the capsular margins are defined (Fig. 14.2g), the proximal and distal aspects of the tumor are marked with sutures, and then the parent fascicle is incised. This technique allows for the preservation of the capsular margin. Capsular margin is confirmed by inking (with India Ink or other suitable dyes) and histopathological examination for capsular breach. The tumor bed is examined for any residual tumor, meticulous hemostasis is obtained, and the wound is closed. Subfascial drains may be considered in some circumstances if there are concerns for postoperative hematoma. It is important to document intraoperative capsular breaches or if the tumor capsule or tissue invasion by the underlying tumor.

14.6.5 Postoperative Course

The patients are admitted to the general floor for monitoring and pain management. Depending on their surgery, they are mobilized on postoperative day #0. Postoperative drains are removed on postoperative day #1. We do not routinely order postoperative CT or MRI scans in these cases unless there is concern for tumor residual or an untoward complication. Patients are usually discharged home on postoperative day #1 or #2.

14.7 Neurofibroma Management at the National Institutes of Health

The National Institutes of Health (NIH) has a robust multidisciplinary team with active clinical trials aimed at the surveillance and treatment of NF1 and its tumor manifestations. Patients enrolled in the NF1 natural history or interventional trials typically undergo surveillance whole-body MRI and PET/CT every 1–3 years (often more frequently) [28]. This longitudinal imaging approach allows us to trend tumor growth, diffusion-weighted imaging (DWI)/apparent diffusion coefficient (ADC), and [^{18}F]-FDG-PET uptake. Sizable tumor burdens carried by patients make prophylactic resection of lesions prohibitive. With routine imaging, we are more readily able to identify dominant nodules likely to transform to ANF or pre-malignancy. Patients referred to our neurosurgical practice who demonstrated clinical symptoms, volumetric growth on MRI >20% per year, and/or increased metabolic avidity on [^{18}F]-FDG-PET imaging were considered candidates for surgery.

14.7.1 Our Practice

We reviewed our patient records at the NIH from July 2013 to July 2023 and identified 29 NF1 patients who collectively underwent 41 surgeries with 62 lesions resected and analyzed pathologically. Please see Fig. 14.3, which outlines the lesions resected, their location, size, and pathology result. A preponderance of lesions was resected within the upper torso, neck, and shoulders (31/62, 50%). Depending on the complexity, we may involve general surgery for brachial plexus lesions, or thoracic surgery for chest wall lesions. There were 18 (29%) lesions removed from the lower torso, abdomen, and lumbar areas. We have worked our urology colleagues to robotically approach retroperitoneal lesions, which we have published elsewhere [29]. We had fewer lesions removed from the upper and lower extremities, 5 (8%) and 7 (11.3%), respectively. Our complicate rate was similar to others, the most common being postoperative hematoma (3/41, 7.3%) and transient weakness or numbness (6/41, 15%) [22].

Owing to our frequent screening of patients, we had few instances of MPNSTs (3/62, 5%). The overwhelming majority of tumors were benign or premalignant subtypes (59/62, 95%). In our cases of MPNSTs, the malignant tissue was noted to have transformed deep within preexisting neurofibromas. Thus, we were able to use the benign tumor parenchyma as a plane to achieve safe en bloc resection of the malignant tissue. There were no amputations required as a part of MPNST treatment within this timeframe.

Many of the dominant nodules that are resected at the NIH are within a plexiform neurofibroma (34/62, 55%). Given the likelihood of nerve transection associated with a gross total resection of a plexiform neurofibroma (34/62, 55%), we take a more targeted approach and only remove lesions that show concerning growth or increased metabolic activity. Fisher et al. (2022) provide a comprehensive multidisciplinary approach to plexiform neurofibroma management [30]. Resection of large and atypical plexiform neurofibromas is best performed with input

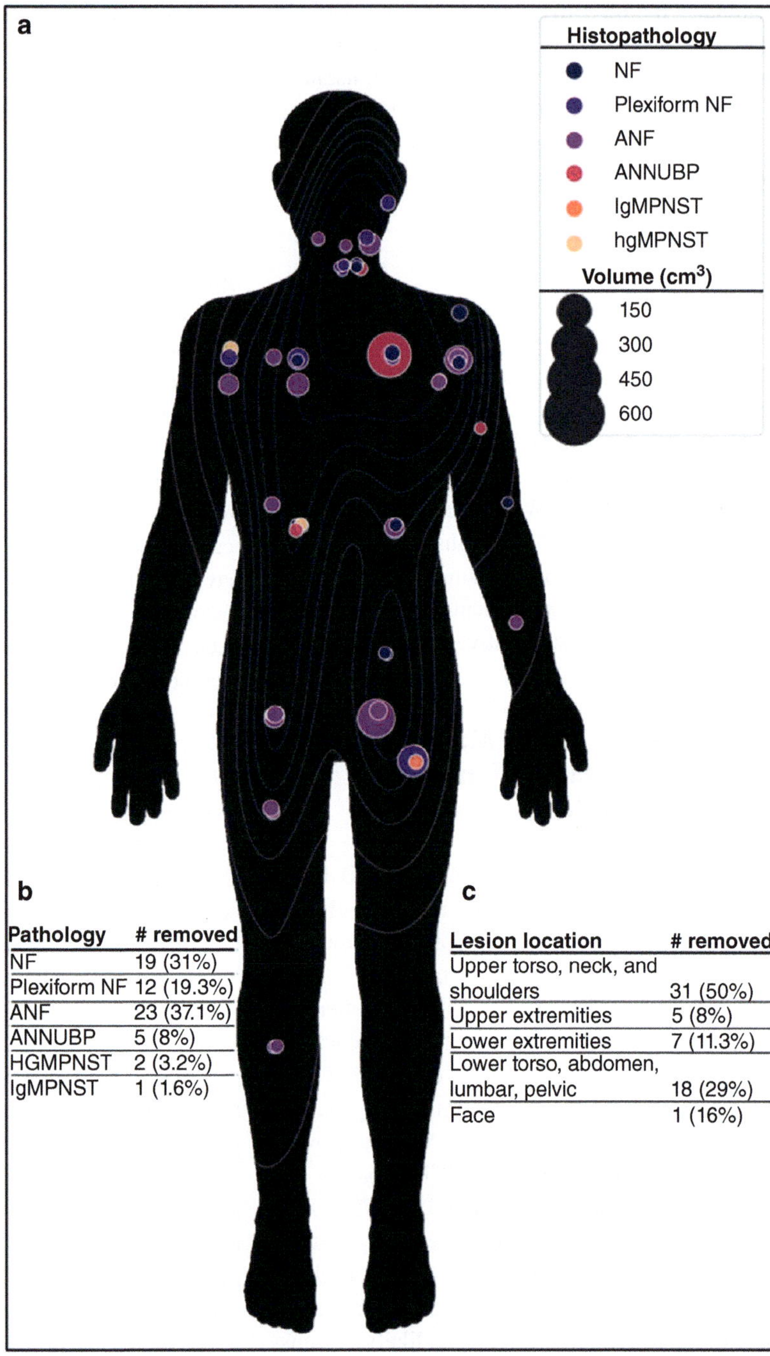

Pathology	# removed
NF	19 (31%)
Plexiform NF	12 (19.3%)
ANF	23 (37.1%)
ANNUBP	5 (8%)
HGMPNST	2 (3.2%)
lgMPNST	1 (1.6%)

Lesion location	# removed
Upper torso, neck, and shoulders	31 (50%)
Upper extremities	5 (8%)
Lower extremities	7 (11.3%)
Lower torso, abdomen, lumbar, pelvic	18 (29%)
Face	1 (16%)

Fig. 14.3 Neurofibroma surgical cases at NIH. (**a**) Tumors resected at the NIH mapped over a human silhouette. Colors and sizes of the circles represent the pathologic grade and tumor volume respectively. (**b**) Table of final pathology grade for the tumors removed. (**c**) Tumor location summary

from the appropriate multidisciplinary team, oftentimes a combination of surgeons (general, plastic, orthopedic, and neurologic, etc.) and neurofibroma medical specialists (pediatric oncologists, neuro-oncologists, neurology, etc.). Hopefully, with the introduction of selumetinib, a MEK-1/2 inhibitor, and other promising drugs, medical therapy will become a more viable option for the management of neurofibromas [30].

14.8 Conclusions

Surgery remains the first-line choice for the management of neurofibromas and other PNSTs. Our experience has demonstrated that, with regular surveillance imaging, it is possible to identify dominant nodules displaying atypical features prior to their malignant transformation. By utilizing tumor mapping with neurostimulation coupled with meticulous microdissection, we can safely remove these dominant nodules while minimizing injury to the parent nerve. Continued advancements in medical therapy and surgical techniques are poised to further enhance outcomes for patients with neurofibromas and, consequently, improve their quality of life.

Funding This project was supported by the Intramural Research Program of the National Institute for Neurological Diseases and Stroke.

References

1. Rodriguez FJ, Folpe AL, Giannini C, Perry A. Pathology of peripheral nerve sheath tumors: diagnostic overview and update on selected diagnostic problems. Acta Neuropathol. 2012;123:295–319.
2. Hilton DA, Hanemann CO. Schwannomas and their pathogenesis. Brain Pathol. 2014;24(3):205–20.
3. Burns R, Niendorf K, Steinberg K, et al. Genetic testing to gain diagnostic clarity in neurofibromatosis type 2 and schwannomatosis. Am J Med Genet A. 2022;188(8):2413–20.
4. Magro G, Broggi G, Angelico G, et al. Practical approach to histological diagnosis of peripheral nerve sheath tumors: an update. Diagnostics. 2022;12(6):1463.
5. Crowe FW. A clinical, pathological and genetic study of multiple neurofibromatosis. Neurology. 1974;24:1144–51.
6. Farma JM, Porpiglia AS, Vo ET. Benign neurogenic tumors. Surg Clin North Am. 2022;102(4):679–93.
7. Belakhoua SM, Rodriguez FJ. Diagnostic pathology of tumors of peripheral nerve. Neurosurgery. 2021;88(3):443–56.
8. Nelson CN, Dombi E, Rosenblum JS, et al. Safe marginal resection of atypical neurofibromas in neurofibromatosis type 1. J Neurosurg. 2019;133(5):1516–26.
9. Trojani M, Contesso G, Coindre JM, et al. Soft-tissue sarcomas of adults; study of pathological prognostic variables and definition of a histopathological grading system. Int J Cancer. 1984;33(1):37–42.
10. Cai W, Kassarjian A, Bredella MA, et al. Tumor burden in patients with neurofibromatosis types 1 and 2 and schwannomatosis: determination on whole-body MR images. Radiology. 2009;250(3):665–73.
11. Mattox AK, Douville C, Silliman N, et al. Detection of malignant peripheral nerve sheath tumors in patients with neurofibromatosis using aneuploidy and mutation identification in plasma. eLife. 2022;11:e74238.
12. Strike SA, Puhaindran ME. Nerve tumors of the upper extremity. Clin Plast Surg. 2019;46(3):347–50.
13. Messersmith L, Krauland K. Neurofibroma. Treasure Island (FL): StatPearls Publishing; 2019.
14. Parsons CM, Canter RJ, Khatri VP. Surgical management of neurofibromatosis. Surg Oncol Clin N Am. 2009;18(1):175–96.
15. Ponti G, Pellacani G, Martorana D, et al. Giant elephantiasis neuromatosa in the setting of neurofibromatosis type 1: a case report. Oncol Lett. 2016;11(6):3709–14.
16. Kakkar C, Shetty CM, Koteshwara P, Bajpai S. Telltale signs of peripheral neurogenic tumors on magnetic resonance imaging. Indian J Radiol Imaging. 2015;25(04):453–8.
17. Neurofibromatosis. Conference statement. National Institutes of Health consensus development conference. Arch Neurol. 1988;45(5):575–8.
18. Plotkin SR, Bredella MA, Cai W, et al. Quantitative assessment of whole-body tumor burden in adult patients with neurofibromatosis. PLoS One. 2012;7(4):e35711.
19. Hsu C-K, Denadai R, Chang C-S, et al. The number of surgical interventions and specialists involved in the management of patients with neurofibromatosis type I: a 25-year analysis. J Pers Med. 2022;12(4):558.
20. Dombi E, Solomon J, Gillespie AJ, et al. NF1 plexiform neurofibroma growth rate by volumetric MRI: relationship to age and body weight. Neurology. 2007;68(9):643–7.
21. Ly KI, Merker VL, Cai W, et al. Ten-year follow-up of internal neurofibroma growth behavior in adult patients with neurofibromatosis type 1 using whole-body MRI. Neurology. 2023;100(7):e661–70.
22. Kim DH, Murovic JA, Tiel RL, Moes G, Kline DG. A series of 397 peripheral neural sheath tumors: 30-year experience at Louisiana State University Health Sciences Center. J Neurosurg. 2005;102(2):246–55.
23. Treglia G, Taralli S, Bertagna F, et al. Usefulness of whole-body fluorine-18-fluorodeoxyglucose positron emission tomography in patients with neurofibromatosis type 1: a systematic review. Radiol Res Pract. 2012;2012:431029.
24. Derlin T, Tornquist K, Münster S, et al. Comparative effectiveness of 18F-FDG PET/CT versus whole-body MRI for detection of malignant peripheral nerve sheath tumors in neurofibromatosis type 1. Clin Nucl Med. 2013;38(1):e19–25.
25. Salamon J, Veldhoen S, Apostolova I, et al. 18 F-FDG PET/CT for detection of malignant peripheral nerve sheath tumours in neurofibromatosis type 1: tumour-

to-liver ratio is superior to an SUV max cut-off. Eur Radiol. 2014;24:405–12.
26. Tovo Filho R, Carnevale FC, Curi TZ, et al. Surgery combined with embolization in the treatment of plexiform neurofibroma: case report and literature review. JAAD Case Rep. 2020;6(5):462–4.
27. Ball JR, Biggs MT. Operative steps in management of benign nerve sheath tumors. Neurosurg Focus. 2007;22(6):E7.
28. Akshintala S, Baldwin A, Liewehr DJ, et al. Longitudinal evaluation of peripheral nerve sheath tumors in neurofibromatosis type 1: growth analysis of plexiform neurofibromas and distinct nodular lesions. Neuro-Oncology. 2020;22(9):1368–78.
29. Curry BP, Alvarez R, Widemann BC, et al. Robotic nerve sheath tumor resection with intraoperative neuromonitoring: case series and systematic review. Oper Neurosurg (Hagerstown). 2022;22(2):44–50.
30. Fisher MJ, Blakeley JO, Weiss BD, et al. Management of neurofibromatosis type 1-associated plexiform neurofibromas. Neuro-Oncology. 2022;24(11):1827–44.

Surgical Management of Pediatric PNST

15

Julian Zipfel and Martin U. Schuhmann

15.1 General Principles

Peripheral nerve sheath tumors (PNST) in children are a heterogenous group of mostly benign tumors, such as neurofibromas and schwannomas as well as premalignant tumors (atypical neurofibromatous neoplasia of uncertain biological potential, ANNUBP) and malignant neoplasms (malignant peripheral nerve sheath tumor, MPNST). They can be associated with significant neurological impairment, including sensorimotor deficits, pain, and other disability, depending on anatomical localization [1]. Tumors warranting surgical therapy in the pediatric population are often associated with neurofibromatosis Type 1 (NF1) and less often NF2-associated schwannomatosis (NF2). Sporadic PNST in children represents about 10% of cases [2]. Sometimes, diagnosis of a tumor predisposition syndrome may be missed due to diagnostic criteria developing at a later age.

The results of pediatric PNST surgery have not been published abundantly in scientific literature. Data on children and adolescents is scarce with one case series from our department [2] and a very small cohort of seven pediatric patients with eight peripheral nerve tumors [3]. Further, larger purely pediatric series does not exist; in mixed series, pediatric cases usually have not been evaluated separately [4–10]. Limited data is mostly due to the rare nature of peripheral nerve sheath tumors in children and their association with also rare genetic tumor predisposition syndromes (Table 15.1).

The distribution of histopathological findings depends greatly on the underlying disease: in NF1 basically, neurofibromas, whereas in NF2 and SMARCB1-/LZTR1-associated schwannomatosis (SWNT), schwannomas are found.

A special form of PNST is cutaneous plexiform neurofibromas (PNF) in children with NF1. Purely cutaneous PNF are, as opposed to most other PNF manifestations (see below), generally unproblematic concerning symptoms. Surgical resection is mostly indicated due to growth and cosmetic aspects. A proper diagnostic workup to evaluate the whole extent of the PNF concerning the underlying subcutaneous tissue is needed. Early surgery of cutaneous PNF manifestations in children should be considered as all tumors grow over time, sometimes overproportionally, and the cosmetic result is clearly depending on the size of the scar. Furthermore, too large tumors might not be totally resectable, and incomplete excisions will lead to a recurrence.

MPNST are associated with significant preoperative pain. Rarely, perineuriomas or ganglio-

J. Zipfel (✉) · M. U. Schuhmann
Department of Neurosurgery, Section of Pediatric Neurosurgery, University Hospital Tübingen, Tübingen, Germany
e-mail: julian.zipfel@med.uni-tuebingen.de; martin.schuhmann@med.uni-tuebingen.de

I. G. Vetrano, V. Nazzi (eds.), *Peripheral Nerve Sheath Tumors*,
https://doi.org/10.1007/978-3-031-76382-3_15

Table 15.1 Overview of case series

	No. of patients	% NF1	% NF2	% SWNT	% sporadic	Other
Zipfel et al. [2]	82	77%	10%	2%	11%	
Guedes et al. [3]	7	57% ($n = 4$)	–	–	43% ($n = 3$)	3 MPNST, also Ewing Sarcoma and desmoid tumor
Chick et al. [5]	6	–	–	100%?	–	6 patients with "sporadic schwannomatosis"
Carvi y Nievas et al. [7]	9	–	–	–	–	Only 1 child with Lipofibroma
Collins-Sawaragi et al. [11]	36	100%	–	–	–	Most tumors craniofacial, 28% surgical treatment
An et al. [12]	11	73% ($n = 8$)	–	–	27% ($n = 3$)	Only MPNST

neuromas are encountered in the pediatric population.

Generally, preoperative neurological deficits in children with PNST are rather rare but if present may severely affect young patients. The major indication for surgery is pain, followed by significant tumor growth. Surgical principles concerning tumor resection should be adhered to just as in the adult population. In experienced hands, surgical outcome concerning the extent of tumor resection, the relief of preoperative symptoms, and a low rate of new permanent deficits is good [2].

15.1.1 Indication for Surgery

In an overwhelming proportion of cases, especially in the setting of phacomatosis, PNST are primarily asymptomatic lesions. Furthermore, in phacomatosis, they mostly appear non-solitary. The mere existence of an asymptomatic, non or slowly growing non-cutaneous peripheral nerve tumor is not an indication for surgery per se. Surgical resection of peripheral nerve tumors in the pediatric population is usually indicated in one of the following situations:

- Local or radiating pain
- Motor or sensory deficit associated with the PNST
- Mass effect leading to impairment of movement or cosmetic deformity
- Cutaneous PNF with documented growth
- Overproportionally fast growth of a circumscribed tumor–indicating the possibility of ANNUBP or MPNST in case of NF1
- Increasing infiltration of adjacent tissue in NF1-associated PNFs

In clinical practice, most interventions in children are indicated by significant tumor growth, followed by pain, mass effect, and suspicion of malignancy [2].

15.1.2 Preoperative Workup

A proper preoperative workup includes a thorough clinical and neurological examination—often already identifying the affected nerve via associated pain (e.g., Hofmann-Tinel sign radiating in a typical dermatome representation), sensory, or motor deficits. Furthermore, palpation and eliciting a Hofmann-Tinel sign help in identifying the problematic PNST in the setting of multiple tumors.

Another preoperative necessity is good-quality imaging, using high-resolution MRI [13–15] and ultrasound as modalities of choice. The sonographic appearance of PNST is usually hypoechoic with hyperechogenic enhancement behind the tumor, Neurofibromas cannot be distinguished from schwannomas by ultrasound or MRI [16–19]. The tumor-nerve anatomy and relation should be clear before surgery including the association and topical relation to vascular structures and adjacent tissues (e.g., infiltration of muscle, bones, tendons, etc., especially in PNF). In cases of suspected malignancy (fast growth and/or marked sensory/motor deficits), FDG-PET scanning, preferably by PET MRI, for metabolic characterization should be considered (when not significantly delaying therapy for months) [20]. A tumor with high metabolic activity should be biopsied in a separate procedure. Alternatively, an intraoperative frozen section needs to be done, the consequences of a malignant intraoperative biopsy need to be communicated, and a radical surgical approach according to sarcoma surgery criteria has to be consented beforehand by parents.

Being familiar and experienced with peripheral nerve ultrasound is useful for neurosurgeons, as this technique is rather easy to learn and adaptive and can be used intraoperatively as well.

15.1.3 Special Considerations for Surgery in Children

Several aspects of surgical treatment of children have to be recognized when treating pediatric patients. Most cases are associated with NF1. Other pathologies associated with NF1, developmental deficits, and problematic social aspects also have to be considered. Especially in smaller children, not all clinical signs of NF1 or NF2 may be present yet, thus diagnosis may be missed. As mentioned above, in cutaneous PNF an early resection is cosmetically favorable and should be preferred over a yearlong follow-up with finally large manifestations.

Furthermore, an effect of neurological deficits, either due to the tumor itself or as the result of surgical treatment, will persist much longer than in adults. On the other hand, neurological rehabilitation and nerve regrowth may be significantly better as compared to the adult age group. Lastly, surgery under regional anesthesia is usually not feasible in children, especially not below 10 years. Therefore, indication for resection has to be made thoughtful, as often multiple tumors have to be considered, and repeated surgery for other phacomatosis-associated pathologies may be necessary. In children with phacomatosis, the relationship between neurosurgeon and patient may be life-long with multiple surgeries necessary over time.

15.2 PNST in Neurofibromatosis/Schwannomatosis

In our own institutional pediatric series with a total of 82 patients and 206 tumors, 78.8% had NF1, 9.8% NF2, 2.4% schwannomatosis, and 11% no hereditary disease [2].

Neurofibromas are mostly associated with NF1, and schwannomas occur mostly in patients with NF2, SWNT, or as sporadic occurrence [21].

Solitary benign peripheral nerve sheath tumors in children without association to an underlying genetically determined disorder are rarely described, and most cases published in larger institutional series were neurofibromas [4, 22]. Our own abovementioned series contained nine sporadic cases: three schwannomas, two ganglioneuromas, two MPNST, one peripheral neurofibroma, and one perineurioma.

The correct interpretation of the clinical significance of a certain pathology like neurofibroma or schwannoma is key to further management and diagnosis. The existence of a peripheral nerve tumor together with the other diagnostic criteria of schwannomatosis [23], NF1 [20], or NF2 [24] can pave the road to final diagnosis. Especially in children, diagnosis may be overlooked primarily [25].

15.2.1 NF1

NF1 is a neurocutaneous, autosomal-dominant genetic disease with a prevalence of about 1/3000 in the pediatric population [26]. Apart from the central nervous system, any peripheral nerve and the skin as well as other organs can be affected. The interindividual phenotype can vary immensely [27]. Peripheral nerve sheath tumors are significant factors in the morbidity of the pathology [28]. Plexiform neurofibromas are a

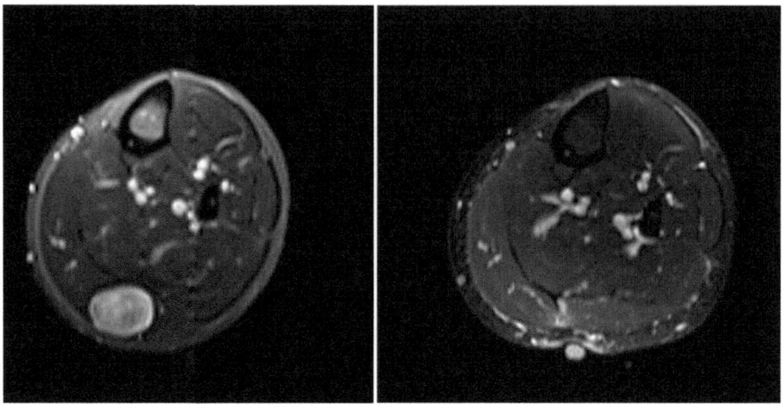

Fig. 15.1 Example of radiologically unsuspicious PNST of the sural nerve in a patient with NF1; left preoperative MRI—intracapsular resection was performed—histology revealed high-grade MPNST; radical re-resection including the scar and a skin-muscle was performed—no tumor rest was found; right postoperative MRI

specific manifestation typical for and confined to NF1, and 50% of patients are affected [26]. NF1 is associated with impaired quality of life and repeated hospitalization [29].

Benign nerve sheath tumors in NF1 can either grow diffusely as plexiform neurofibromas (PNF) with local infiltration of surrounding tissue and organs or as neurofibromas clearly related to a defined peripheral nerve, either as a solitary manifestation in an elsewise mostly normal nerve or as a circumscribed local manifestation in a diffusely changed "neurofibromatous" nerve (see Fig. 15.1). Purely cutaneous PNF can already be found in infants with NF1 and differ distinctly from the typical NF1-associated "adult" cutaneous neurofibromas that mostly occur after puberty. Cutaneous PNF infiltrate the skin and the subcutaneous tissue diffusely and continue to grow mostly significantly during the first decades of life [26]. As mentioned above, these tumors should be addressed surgically at an early age.

After the removal of a neurofibroma, the existence of NF1 needs to be ruled out or established according to clinical criteria or genetic testing if not known already.

15.2.2 NF2-Associated Schwannomatosis (NF2)

NF2 is also an autosomal-dominant genetic pathology, defined by bilateral vestibular schwannomas. Its incidence is about 1/10 as compared to NF1 [30]. In children with NF2, extradural peripheral nerve schwannomas with the need for surgical intervention occur much less frequently than in NF1 patients. Furthermore, plexiform schwannomas or cutaneous schwannomas, although existent, are as well rarer than in NF1. In rare cases, plexiform transformation of whole nerves, mostly confined to a certain quadrant or limb, may be observed (see Fig. 15.2).

If a peripheral schwannoma is removed in a child or adolescent, the existence of an underlying NF2 has to be ruled out actively by performing MRIs of the head and the spine. If those surveillance scans are negative for lesions compatible with vestibular schwannomas, meningiomas, or ependymomas, SWNT has to be considered in case more than one schwannoma is present.

15.2.3 SMARCB1-/LZTR1-Associated Schwannomatosis (SWNT)

SWNT is the third entity of genetic peripheral nerve sheath tumor predisposition. It is characterized by multiple peripheral schwannomas and differentiated from NF2 by the lack of bilateral vestibular schwannomas or other intradural NF2 manifestations like meningiomas or ependymomas [31, 32]. It is associated with SMARCB1 or LZTR1 mutations and exists in a sporadic and a familial pattern.

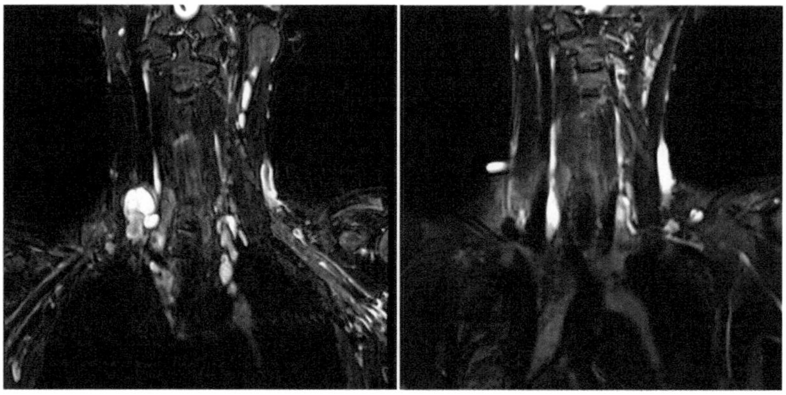

Fig. 15.2 Left pre- and right postoperative MRI of a patient with NF1 and a painful neurofibroma of the right cervical region

SWNT-associated schwannomas however mostly affect adults but can be found in adolescents. Since an age <30 years is considered an exclusion criterion for the sporadic form [6], SWNT should be considered in any adolescent with a schwannoma, and a whole-body MRI should be performed to identify or rule out further manifestations.

15.3 Interdisciplinarity

In surgery of PNST in children, a vast spectrum of localizations and clinical appearances are encountered. NF1 is the predominant underlying disorder where peripheral nerve surgery is necessary in the pediatric age group. As described earlier in this book, interdisciplinary teams are necessary to assess and manage the known multiple aspects of NF1. Furthermore, surgically relevant PNST can appear in any anatomical localization whilst being present mostly in the extremities and torso. In a large cohort of pediatric NF1 patients, the predominant location of PNST was craniofacial with a high rate of disfigurement. Most patients were managed conservatively though [11].

Expertise in peripheral nerve tumor surgery and function-preserving resection techniques is needed to manage pediatric PNST cases well from a surgical point of view. This expertise can be provided by specifically trained pediatric neurosurgeons, pediatric surgeons, plastic surgeons, and—concerning manifestations in the face—maxillofacial surgeons. The usual pediatric surgical approach to “remove it all” will lead invariably to neurological deficits and is inadequate for benign PNST.

On the other hand, in cases with suspicion or confirmation of malignancy after initial workup as described above, intraoperative adherence to the criteria of sarcoma surgery is key. This means a “switch” to a radical approach with ideally clean resection margins with the sacrifice of function is mandatory for cure.

As shown in our abovementioned series, a significant number of PNST were found intra-abdominally, in the retroperitoneum, or in the thoracic cavity. Therefore, interdisciplinary planning and a joint approach are necessary in these situations, including endoscopic approaches, where feasible or open laparotomy and thoracotomy.

Neurosurgeons or plastic surgeons therefore need interdisciplinary operations together with pediatric surgeons to adapt their techniques of (mini) thoracotomy, thoracoscopic, or laparoscopic surgery to perform surgery on smaller tumors as minimally invasive as possible but still applying the classical function-preserving techniques of microneurosurgery of the peripheral nerve [33].

When encountering PNST of the facial region, cosmetic aspects have to be considered, and experienced plastic and/or maxillofacial surgeons have to be included.

Managing pediatric PNST includes developmental and neuropsychological aspects, especially in the setting of NF1. Therefore, neuropediatricians must be involved, as discussed in other chapters of this book. Furthermore, specialists in electrophysiology in children and peripheral nerve ultrasound as well as MR neurography and PET may be needed for proper preoperative workup. This allows for better planning concerning surgical approaches and indications for resection.

15.4 Microsurgical Technique and Outcome

As the majority of PNST are benign lesions, preservation of function is one of the most important aspects of surgical intervention. The standard resection technique is intracapsular enucleation. After carefully choosing an ideal approach with consideration of neurovascular structures and cosmetic aspects, the tumor is identified, and a safe entry zone is defined. PNST, apart from cutaneous and plexiform tumors, usually possess a capsule, bearing the axons of the tumor-carrying nerve. Nerve preservation thus includes identification of the capsule and choosing an entry zone in an area without sensory or motor nerve fascicles. Microscopic visualization and intraoperative monitoring including direct stimulation are necessary adjuncts. After opening the capsule, a dissection plane is prepared circumferentially around the tumor preserving the capsule, which contains the functional fascicles, preferentially in the bimanual technique. Finally, the PSNT can often be bluntly dissected and removed from the capsule. Often the small tumor-bearing fascicle—usually not neurologically intact at the moment of surgery—needs to be transected on both ends of the tumor.

Cutaneous neurofibromas, being clearly limited in their extension to the skin, are excised in sano. Cutaneous PNF may, if skin or deeper tissue extension and involvement allows for it, also be resected in sano. In larger tumors, plastic reconstruction of the skin defect by flap techniques may be necessary. In situations, where tumors are too large or involve non-resectable tissue, subtotal resection may be considered and no loss of function should occur. It is important to remember that these tumors are benign albeit invasive growth. Careful preparation and preoperative planning are needed, as dissection through the PNF can lead to extensive bleeding, and affected remaining skin edges may be prone to impairment of wound closure and infection.

In non-cutaneous PNF, subtotal resection may also be necessary in cases of invasion of neurovascular structures or organs. The principles of intracapsular preparation can often not be employed in diffusely infiltrating tumors.

With intracapsular resection, the outcome is generally good, concerning neurological function. Pain can often be reduced. Improvement of symptoms has been reported in about 75% of patients, and deterioration in about 2.4%. Complete resection is feasible in about 90% of cases [2].

Of course, functional outcome is the most important measure in benign lesions. Especially in children, lifelong disability may have severe psychological and socioeconomic effects and should be avoided by all means.

15.5 Other Treatment Options

Alternative therapeutic options are usually reserved for cases of non-operable or subtotal resectable tumors; most of them will be PNF.

Radiation of benign peripheral nerve tumors is no option for peripheral nerve tumors in children, especially in the setting of an underlying phacomatosis. The alteration of a tumor suppressor gene product predisposes to secondary malignancies in the radiation field or malignant transformation of irradiated tumors [34]. Furthermore, a surgical or observational approach does not carry the long-term effects of radiotherapy. In many cases, residual tumor masses can be followed, and—if indicated—repeated surgical reduction is feasible. In many cases, tumors will not necessarily keep growing, especially after reaching adulthood.

In the setting of NF1, inhibition of the disease-specific alteration of the MAP-kinase pathway with so-called MEK inhibitors is a promising option of pharmacological intervention in cases where a tumor reduction should be performed; however it is surgically not possible or associated with too many risks [8, 9, 35].

In MPNST, radiotherapy can enhance local control in selected cases, but there is a significant risk of secondary malignancies [36]. Other chapters of this book will discuss nonsurgical therapeutic options in more detail.

15.6 Surgical Treatment of MPNST in Children

The most critical pathology of PNST, MPNST, still poses a diagnostic and therapeutic challenge with high morbidity and mortality [1, 21]. Most MPNST arise from previously benign neurofibromas transforming directly to MPNST or via a transitional state of enhanced proliferative activity defined as ANNUBP.

In NF1, MPNST usually develop from PNF, though not from cutaneous PNF. Most of these tumors appear in the torso, proximal extremities, and cervical region. Symptoms are usually pain, overproportional tumor growth, or rather fast developing neurological deficits. Up to 5–10% of patients with NF1 develop an MPNST in their lifetime as compared to 0.001% of the general population. The 2-year overall survival is reported at about 73%, and the 2-year event-free survival is 59% in children [12].

When suspecting MPNST, PET-MR or PET-CT and MR-spectroscopy are nowadays the methods of choice—if available in the specific health system—for identifying high-metabolic lesions. Most importantly, the surgical approach to tumors with the suspicion of malignancy has to be adapted. Some MPNST may present unsuspiciously looking like benign tumors (see Fig. 15.3). Intracapsular resection may lead to a breach of the tumor capsule resulting in the dissemination of tumor cells locally or even systemically. Therefore, known MPNST should be surgically removed according to sarcoma criteria en bloc with tumor-free margins in order to reduce the risk of recurrence. This includes sacrificing the associated nerve or nerve root. In the treatment of MPNST, complete excision is the most important factor concerning cure and survival. Of course, postoperative neurological deficits are therefore very common and occur in 30–40% of patients. Because of the dichotomy of surgical approaches in benign or malignant cases, a reliable diagnosis is key to planning an operation and informing patients and parents. Obtaining histological tissue samples via biopsy is the most reliable measure.

In our cohort, MPNST were associated with significant preoperative motor and sensory impairment. Since radical surgical excision is the only curative therapeutic option, these deficits did persist or worsen. Early surgical resection of suspicious lesions is of great importance, as these tumors mostly develop from previously benign neurofibromas. In our experience, resection while still biologically benign or as a precancerous ANNUBP has a favorable outcome. Data sug-

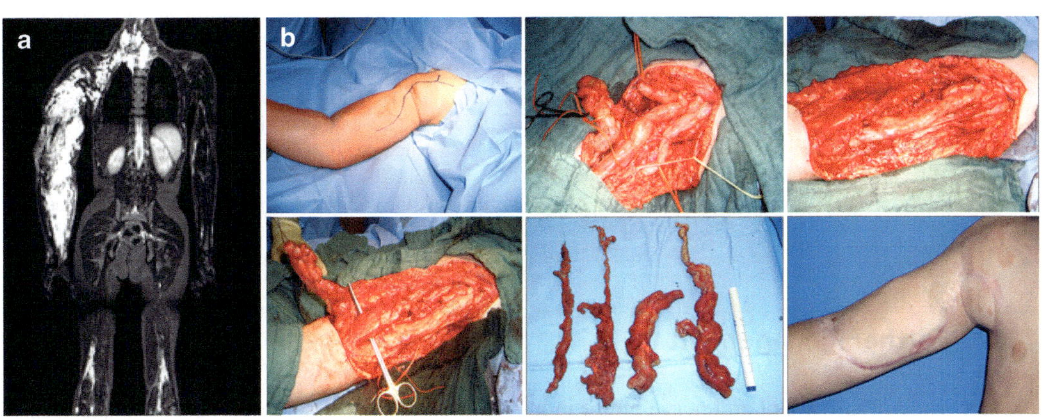

Fig. 15.3 (**a**) Preoperative MRI of a patient with segmental NF1, massive plexiform tumor manifestations of the right arm (**b**) pre-/intra- and postoperative photo documentation

gests that strict surveillance of patients with NF1 with a low threshold for surgery in suspicious lesions leads to diagnosis of MPNST in only about 20% of tumors, which means that 80% of tumors were removed in time prior to malignant transformation [2]. Prospective data on this concept of resection before malignant transformation occurs is still missing, though.

In the surgical treatment of MPNST, interdisciplinarity is just as or even more important as in other PNST. Pediatric surgeons are trained in sarcoma surgery, utilizing excisions with safe margins and en bloc resection without direct manipulation of tumor tissue. Most (pediatric) neurosurgeons might not be accustomed to these principles, which are important in MPNST treatment. The attempt of intracapsular resection in MPNST might be a severe mistake, which can lead to contamination of the surgical field with malignant cells, thus resulting in a worse prognosis for the patient. Therefore, the correct preoperative interpretation of imaging characteristics, addition of PET data, and the visual aspect of surgery combined with the experience of surgeons in an interdisciplinary setting are most important to choosing the correct "mode of surgery." When inconclusive, a two-stage approach with upfront or intraoperative biopsy is mandatory. Choosing a trajectory or approach, which on the one hand considers functional integrity and on the other hand allows for secondary surgical excision of a malignant tumor, can be challenging.

In cases of significant neurological deficit after surgery, especially paresis, nerve reconstruction by nerve transplants can be considered but should be discussed upfront in light of the overall treatment plan since subsequent chemo- and radiotherapy might greatly diminish the chances of reinnervation. Alternatively, tendon or muscle transfers and cutaneous flaps can be utilized [36].

15.7 Further Management

After removal of any peripheral nerve tumor in a child, a surveillance ultrasound screening of all extremities and brachial plexus and neck should be performed as a first diagnostic step to identify other, smaller and not palpable tumors of the peripheral nerves, which would be characteristic for an underlying phacomatosis. Depending on the tumor entity and the existence or absence of a phacomatosis, further workup might include full-body MR-imaging and genetic testing from blood and if inconclusive from tumor tissue.

In experienced hands, surgical resection of PNST in children is safe and effective. Young patients benefit from early interventions in tumors with significant growth, pain, motor deficit, or suspected malignancy. Interdisciplinary care of patients is needed, due to the complex nature of underlying diseases. Special requirements of surveillance, diagnostics, and interdisciplinary surgery in localizations not commonly accessed by pediatric neurosurgeons alone are needed.

References

1. Wolters PL, Martin S, Merker VL, et al. Patient-reported outcomes of pain and physical functioning in neurofibromatosis clinical trials. Neurology. 2016;87(7 Suppl 1):S4–S12.
2. Zipfel J, Al-Hariri M, Gugel I, et al. Correction to: surgical management of peripheral nerve sheath tumours in children, with special consideration of neurofibromatoses. Childs Nerv Syst. 2021;37(9):2953.
3. Guedes F, Brown RS, Torrao-Junior FJL, Barbosa DAN, Ravanini GAG, Amorim RMP. Pediatric peripheral nerve tumors: clinical and surgical aspects. Childs Nerv Syst. 2019;35(12):2289–97.
4. Chick G, Alnot JY, Silbermann-Hoffman O. Benign solitary tumors of the peripheral nerves. Rev Chir Orthop Reparatrice Appar Mot. 2000;86(8):825–34.
5. Chick G, Victor J, Hollevoet N. Six cases of sporadic schwannomatosis: topographic distribution and outcomes of peripheral nerve tumors. Hand Surg Rehabil. 2017;36(5):378–83.
6. Chick G, Victor J, Poujade T, Hollevoet N. Sporadic schwannomatosis: a systematic review following the 2005 consensus statement. J Neurol Surg A Cent Eur Neurosurg. 2018;79(5):408–15.
7. Carvi y Nievas MN, Archavlis E, Unkel B. Delayed outcome from surgically treated patients with benign nerve associated tumors of the extremities larger than 5 cm. Neurol Res. 2010;32(6):563–71.
8. Rosenbaum T, Wimmer K. Neurofibromatosis type 1 (NF1) and associated tumors. Klin Padiatr. 2014;226(6–7):309–15.

9. Karajannis MA, Ferner RE. Neurofibromatosis-related tumors: emerging biology and therapies. Curr Opin Pediatr. 2015;27(1):26–33.
10. Mautner VF, Tatagiba M, Guthoff R, Samii M, Pulst SM. Neurofibromatosis 2 in the pediatric age group. Neurosurgery. 1993;33(1):92–6.
11. Collins-Sawaragi YC, Ferner R, Vassallo G, et al. Location, symptoms, and management of plexiform neurofibromas in 127 children with neurofibromatosis 1, attending the National Complex Neurofibromatosis 1 service, 2018-2019. Am J Med Genet A. 2022;188(6):1723–7.
12. An HY, Hong KT, Kang HJ, et al. Malignant peripheral nerve sheath tumor in children: a single-institute retrospective analysis. Pediatr Hematol Oncol. 2017;34(8):468–77.
13. Chee DW, Peh WC, Shek TW. Pictorial essay: imaging of peripheral nerve sheath tumours. Can Assoc Radiol J. 2011;62(3):176–82.
14. Fayad LM, Wang X, Blakeley JO, et al. Characterization of peripheral nerve sheath tumors with 3T proton MR spectroscopy. AJNR Am J Neuroradiol. 2014;35(5):1035–41.
15. Soldatos T, Fisher S, Karri S, Ramzi A, Sharma R, Chhabra A. Advanced MR imaging of peripheral nerve sheath tumors including diffusion imaging. Semin Musculoskelet Radiol. 2015;19(2):179–90.
16. Reynolds DL Jr, Jacobson JA, Inampudi P, Jamadar DA, Ebrahim FS, Hayes CW. Sonographic characteristics of peripheral nerve sheath tumors. AJR Am J Roentgenol. 2004;182(3):741–4.
17. Winter N, Dohrn MF, Wittlinger J, Loizides A, Gruber H, Grimm A. Role of high-resolution ultrasound in detection and monitoring of peripheral nerve tumor burden in neurofibromatosis in children. Childs Nerv Syst. 2020;36(10):2427–32.
18. Widmann G, Riedl A, Schoepf D, Glodny B, Peer S, Gruber H. State-of-the-art HR-US imaging findings of the most frequent musculoskeletal soft-tissue tumors. Skeletal Radiol. 2009;38(7):637–49.
19. Telleman JA, Stellingwerff MD, Brekelmans GJ, Visser LH. Nerve ultrasound in neurofibromatosis type 1: a follow-up study. Clin Neurophysiol. 2018;129(2):354–9.
20. Bergqvist C, Servy A, Valeyrie-Allanore L, et al. Neurofibromatosis 1 French national guidelines based on an extensive literature review since 1966. Orphanet J Rare Dis. 2020;15(1):37.
21. Harder A, Wesemann M, Hagel C, et al. Hybrid neurofibroma/schwannoma is overrepresented among schwannomatosis and neurofibromatosis patients. Am J Surg Pathol. 2012;36(5):702–9.
22. Kehoe NJ, Reid RP, Semple JC. Solitary benign peripheral-nerve tumours. Review of 32 years' experience. J Bone Joint Surg Br. 1995;77(3):497–500.
23. Plotkin SR, Blakeley JO, Evans DG, et al. Update from the 2011 international schwannomatosis workshop: from genetics to diagnostic criteria. Am J Med Genet A. 2013;161A(3):405–16.
24. Mehta GU, Huynh H, Lekovic GP. Peripheral nerve sheath tumors in neurofibromatosis type 2: surgical and histopathologic features. Clin Neurol Neurosurg. 2019;190:105649.
25. Angelova-Toshkina D, Holzapfel J, Huber S, et al. Neurofibromatosis type 1: a comparison of the 1997 NIH and the 2021 revised diagnostic criteria in 75 children and adolescents. Genet Med. 2022;24(9):1978–85.
26. Dunning-Davies BM, Parker AP. Annual review of children with neurofibromatosis type 1. Arch Dis Child Educ Pract Ed. 2016;101(2):102–11.
27. Garcia-Romero MT, Parkin P, Lara-Corrales I. Mosaic neurofibromatosis type 1: a systematic review. Pediatr Dermatol. 2016;33(1):9–17.
28. Kiuru M, Busam KJ. The NF1 gene in tumor syndromes and melanoma. Lab Investig. 2017;97(2):146–57.
29. Doser K, Hove H, Ostergaard JR, et al. Cohort profile: life with neurofibromatosis 1 - the Danish NF1 cohort. BMJ Open. 2022;12(9):e065340.
30. Campian J, Gutmann DH. CNS tumors in neurofibromatosis. J Clin Oncol. 2017;35(21):2378–85.
31. Blakeley JO, Plotkin SR. Therapeutic advances for the tumors associated with neurofibromatosis type 1, type 2, and schwannomatosis. Neuro-Oncology. 2016;18(5):624–38.
32. Thomas AK, Egelhoff JC, Curran JG, Thomas B. Pediatric schwannomatosis, a rare but distinct form of neurofibromatosis. Pediatr Radiol. 2016;46(3):430–5.
33. Urla C, Fuchs J, Grimm A, et al. Interdisciplinary surgical approach enables complete tumor resection with preservation of neurological function in specific conditions of pediatric solid malignancies. J Cancer Res Clin Oncol. 2023;149:4497–507.
34. Grill J, Dhermain F, Habrand JL. Risks of radiation therapy in patients with neurofibromatosis. Int J Radiat Oncol Biol Phys. 2009;75(2):632.
35. Gross AM, Wolters PL, Dombi E, et al. Selumetinib in children with inoperable plexiform neurofibromas. N Engl J Med. 2020;382(15):1430–42.
36. Martin E, Flucke UE, Coert JH, van Noesel MM. Treatment of malignant peripheral nerve sheath tumors in pediatric NF1 disease. Childs Nerv Syst. 2020;36(10):2453–62.

16 Functional Status in Malignant Peripheral Nerve Sheath Tumors

Enrico Martin and J. Henk Coert

16.1 Introduction

Malignant peripheral nerve sheath tumors (MPNST) are rare in the population. Because of its aggressive behavior, an MPNST is considered to be part of the group of soft tissue sarcomas (STS). They generally harbor poor prognosis with high rates of local recurrence and metastases. Median survival varies between 5 and 8 years [1–3]. In STS literature, the focus of research has generally remained on improving oncological outcomes and diminishing postoperative complications. However, MPNSTs are generally large tumors requiring major resection of soft tissues, in turn resulting in significant morbidity. To date, STS arising in extremities are primarily treated by limb-sparing surgery. Nevertheless, amputations are not uncommon and although their indications differ, major neurovascular involvement is still often seen as one [4, 5]. Resection of any type of nerve has been shown to affect function and quality of life after resection of an STS [6, 7]. As MPNST have their origin in nerve tissue, they are among the STS with the highest risk for postoperative function loss. Yet functional outcomes still have earned little attention in literature.

What makes MPNST a unique subtype of STS is that 35–40% of cases are related to neurofibromatosis type 1 (NF1) [1, 2]. Many NF1 patients will develop (plexiform) neurofibromas that have the potential to develop into an MPNST. These benign counterparts may cause high levels of functional disability already because of pain and in some cases motor and sensory deficits. This subdues many NF1 patients to frequent controls, scans, and biopsies. Correctly managing their functional disabilities in the case of MPNST is therefore important to improve quality of life.

Over the past decades, oncological outcomes in any STS, including MPNSTs, have not improved. Regardless, long-term survivors are common, especially among children [8]. Quality of life and functional status are therefore important factors to keep in mind when treating MPNST patients. Prevention of function loss may be obtained by reducing the need for amputations or the need for resection of adjacent neurovascular structures. So-called limb-sparing surgery and close-margin surgery go hand in hand with radiotherapy in order to reduce local recurrence rates and benefit from equal survival outcomes. Its use may nonetheless be seen as

E. Martin (✉) · J. H. Coert
Department of Plastic and Reconstructive Surgery, University Medical Center Utrecht, Utrecht, The Netherlands
e-mail: e.martin-2@umcutrecht.nl; j.h.coert@umcutrecht.nl

I. G. Vetrano, V. Nazzi (eds.), *Peripheral Nerve Sheath Tumors*,
https://doi.org/10.1007/978-3-031-76382-3_16

controversial in pediatric populations, as it has not been shown to improve overall survival yet could induce secondary malignancies and growth defects. Similarly, it has yet to be determined what the role of radiotherapy should be in the case of NF1-related MPNSTs, especially in those encased by plexiform neurofibromas.

Despite attempts to manage the extent of surgical margins, many cases may result in significant morbidity due to the resection of functional structures. In such cases, functional reconstructions may become more important. The reconstruction of resected muscle groups, tendons, and motor and sensory nerves is achievable even in oncologic settings. Their consideration is therefore an important tool in the surgical management options of MPNST.

In this chapter, we discuss postoperative function in MPNST and how their outcomes may be improved. We do so by revisiting the need for wide resection margins in MPNST, the use of radiotherapy, and the role of functional reconstructions.

16.2 The Role of Surgical Margins

Although the options for restoration of function and prevention of nerve pain through reconstruction are plentiful, the prevention of function loss is preferred. While benign peripheral nerve sheath tumors are treated by nerve-sparing, intracapsular resections, MPNST are generally treated by wide resections [2, 9]. So-called macroscopically complete resection margins have thus far been the only proven therapy to increase overall survival in MPNST [2, 9, 10]. Neither radiotherapy nor chemotherapy has been proven to improve survival in localized MPNST. Currently, there are studies being performed to investigate whether certain subgroups might benefit from additional chemotherapy in a non-metastatic setting [11, 12]. Because of the aggressive nature of MPNSTs, many surgeons advocate for wide resection margins [13]. For nerve surgeons, this stands in contrast to the treatment of many other benign peripheral nerve sheath tumors that can often undergo intracapsular resection. Nevertheless, there is a difference between macroscopically complete tumor resections without the resection of non-invaded structures (close margin surgery) and wide resection margins. The latter may result in higher rates of microscopically free resection margins (R0), while the first may result in both R0 and microscopically positive margins (R1). However, obtaining wide resection margins increases the indication for amputation [14].

Current guidelines recommend administering radiotherapy in any case of R1 resections to reduce local recurrences [15, 16]. However, they already recommend doing so in any high-grade, deep-seated tumor larger than 5 cm, which in practice makes up the largest proportion of MPNSTs. This is similar to pediatric MPNST treatment guidelines [17]. The sequence of radiotherapy to surgery is slowly shifting toward administering preoperatively. Studies have shown that both adjuvant and neoadjuvant administration have similar oncological outcomes, but in the case of neoadjuvant administration, total dosage and radiation fields can be reduced [18, 19]. It is speculated that late radiation-induced morbidity is reduced because of this; however, this has not been proven in literature to date. Especially, as neoadjuvant administration does increase the risk of major postoperative complications by almost three-fold [18, 20]. Overall, the role of any radiotherapy in limb-sparing surgery is widely accepted. The effect of R1 resection margins compared to R0 on overall survival remains a subject of debate. Several large retrospective studies have however not shown that R1 resections were independently associated with worse survival in a multivariate model [1, 2, 9, 10]. While ideally R0 margins should be strived for as it may obviate the need for radiotherapy, this needs to be evaluated case-by-case in relation to possible morbidity. MPNSTs may both arise from major nerve bundles and smaller axons. In the first case, the resection of the originating nerve is generally inevitable. Nevertheless, due to their expansive growth, in some cases, the tumor may come close to or even partially encase adjacent nerves. Studies have shown that in such circumstances, epineural dissection of the encased nerve is technically feasible and oncologically safe whenever it is being combined with radiotherapy [21]. The nerve should however not be encased by more than 50–75% of the tumor. This may directly have significant implications for brachial plexus tumors for which forequarter amputations are currently not uncommon.

16.3 Current Functional Outcomes in MPNST

A recent international survey among surgeons involved in the surgical care of MPNST patients sought to investigate current practices regarding the prevalence of postoperative functional deficits and how they anticipate or treat them [22]. With at least 35% of MPNSTs originating from major nerves, on top of all tumors that undergo resection of adjacent neurovascular structures, morbidity is thought to be high [23, 24]. The survey reported average postoperative motor deficits in almost 40% of patients as well as neuropathic pain in over 40% [22]. Most commonly surgeons found a combination of neuropathic pain, motor disability, and sensory loss (70%). Despite this high number of postoperative deficits, only 53% of respondents always considered functional status preoperatively. Over 40% of respondents would also never consider performing less extensive resections in order to preserve certain function.

The MONACO study was one of the first to objectify current practices and outcomes regarding postoperative function in MPNST [23]. This international collaboration between the Mayo Clinic and Dutch sarcoma-treating hospitals was set up, in part, to investigate perioperative functional status among MPNST patients using retrospective data. The study included 756 patients (33.4% NF1 related), of which approximately 16% presented with motor deficits and 13% with sensory deficits. In total, 658 patients were treated surgically. They found that postoperative motor deficits were present in 35% of surgically treated patients. In those with known Medical Research Council muscle grade (MRC), most patients developed deficits with less than MRC grade 3 muscle power, which may have been an indication for reconstruction. Whenever patients initially presented with motor deficits, almost 60% of deficits will remain when they are not reconstructed. Postoperative sensory deficits were even more common. In 55% of cases, any form of sensory deficits was documented, and overall, around 25% of all sensory deficits included at least partial sensory loss of the hand or feet. Again, sensory deficits at presentation were likely to persist postoperatively in most cases. Tumors arising in or around the brachial plexus or at a pelvic site have the highest risk of developing both motor and sensory deficits (>50% of cases). Risk factors for developing significant (less than MRC grade 3) postoperative motor deficits were evaluated in a multivariate model. This showed that NF1, extremity/brachial plexus/pelvic tumor sites, a major nerve of origin, deep-seated tumors, and a motor deficit at presentation were all independent predictors. For the development of sensory deficits of the hand or feet, the same tumor sites, a major nerve of origin, and sensory deficit at presentation were independent predictors. Amputations make up a significant proportion of anticipated deficits. A total of 61 patients underwent amputation at any point in time, most of which were carried out during initial treatment. This indicates that amputation is carried out in over 20% of extremity MPNSTs. Despite rates of significant morbidity being this high, functional reconstructions are rarely performed [22, 23, 25]. In the MONACO study, 26 patients underwent any form of functional reconstruction and in literature, <150 patients have been reported undergoing functional reconstruction after any type of soft tissue sarcoma of the extremities in 2019 [25].

Besides motor and sensory deficits, chronic neuropathic pain after resection of MPNST remains a significant burden in physical functioning. The extent of this burden has not been studied in MPNST patients specifically. Among adult amputees, including those with oncologic amputations, residual limb pain is seen in over 50% commonly related to neuroma formation [26]. In traumatic cases of amputation, studies are showing that the use of techniques such as targeted muscle reinnervation and regenerative peripheral nerve interface can heavily reduce neuropathic pains [27]. These techniques can be used in both a direct fashion by rewiring transected nerves at the time of amputation, as well as in a secondary fashion whenever neuropathic pain arises. However, these applications are currently rarely performed in oncologic cases but can also be performed in symptomatic neuroma treatment outside of an amputation setting.

16.4 Incorporating Functional Reconstructions

Functional reconstructions are regarded as reconstructions that have the potential to restore lost function. These include all types of nerve reconstructions, such as grafting or nerve transfers, free-functioning muscle transfers (FFMT), and tendon grafting or transfers. Also, free vascularized soft tissue flaps with accompanying sensory nerve coaptation can be seen as a form of functional reconstruction. Nevertheless, they are still rarely used in the setting of MPNST surgery. The problem is likely twofold. Firstly, most patients are operated on by surgical oncologists of whom many do not incorporate a reconstructive or nerve surgeon in their multidisciplinary consultations [22]. This seems partially caused by a lack of knowledge of the possibilities and outcomes of functional reconstructions. However, the use of reconstructions has been shown not to affect oncological outcomes [20]. Secondly, some reconstructive surgeons are hesitant toward the expected outcomes of functional reconstructions in the oncologic setting [22]. Not only because of the use of radiotherapy and possibly chemotherapy but also because of unclear expectations on survival. In the same survey however, regardless of surgical subspecialty, surgeons agreed that patients with an expected median survival of more than 3 years could be suitable candidates for functional reconstruction. This indicates that most localized MPNST patients should be seen as suitable patients. While radiotherapy does increase the risk of postoperative wound complications, there is currently no evidence in literature that it significantly impacts the success rates of functional reconstructions [23, 25]. The same goes for chemotherapy. Nerve regeneration is of course even more precarious. However, both preclinical and clinical evidence do not suggest adverse effects of multimodal treatment on the regeneration of nerves [23, 25, 28–31]. Performing nerve reconstructions in irradiated fields may require the removal of poorly vascularized and fibrous tissue [32].

Although the preservation of functional anatomy is preferred; in some MPNST cases, their resection is inevitable. Using functional reconstructions in case of (anticipated) lost function can be a powerful tool. Each case requires a personalized approach as to what type of functional reconstruction best suits its needs. In some cases, for example, the addition of vascularized soft tissue or skin is necessary to close the surgical site. There is increasing evidence that, especially in cases who undergo preoperative radiotherapy and are prone to develop postoperative wound infections, vascularized tissue transfers should be used more often in order to reduce major wound complications [33, 34]. Further considerations are in regard to type of function loss. While motor deficits can be reconstructed in numerous ways, the loss of sensation requires a nerve reconstruction by definition. With a shift toward limb salvage surgery in the past decades, in essence, a fully functional limb is what should be strived for, both in motor function and in sensory function. In general, the reconstruction of motor deficits of the proximal part of the arms and legs is most commonly reconstructed with the use of a latissimus dorsi flap [35–37]. More distally, a gracilis flap, which is smaller in size, can be used. For reconstructions of defects in the hand and feet tendon reconstructions are often needed, either by transfers or grafts like in Hunter's procedure. More proximal defects of, for instance, brachial plexus, tumors may require the use of distal nerve transfers known from traumatic plexus lesion reconstruction [23, 38]. Peripheral nerve lesions specifically can be reconstructed by both nerve grafting techniques and nerve transfers. The latter has gained popularity over the past decades due to their superior outcomes after traumatic lesions [39]. In many injuries, nerve transfers show faster recovery due to their proximity to end target reinnervation, and in turn better muscle power compared to nerve grafting. In MPNST surgery specifically, it could mean that the reconstruction can also be performed outside of any irradiated field. The ideal type and combination of reconstruction of loss of specific peripheral nerve motor function are debated and still studied on, comparing tendon transfers to nerve transfers or even a combination in some cases [40, 41].

16.5 Future Perspectives in MPNST Surgery

In order to optimize both oncological and functional outcomes, close collaboration between ablative and reconstructive (nerve) surgeons is important. Ideally, preoperative planning is combined and discussed within a multidisciplinary team (MDT), including at least oncologists and radiotherapists. All options regarding the timing and use of radiotherapy, close margin surgery, and (functional) reconstructions should be discussed as each option affects the other. The options should be tailored to tumor—and patient characteristics and of course the expectations and wishes of patients themselves. Oncologic calculators can be helpful both in MDT and patient consultation. Their use can not only enlighten expected survival and odds of cancer recurrence but may also help in identifying what patients should be eligible for functional reconstruction and less extensive resections. Currently, two calculators for STS are available, Sarculator and PERSARC which have been shown to aid in identifying high-risk patients [42, 43]. Their drawback, however, is that they are not designed for pediatric cases, non-extremity sites (Sarculator does include trunk wall tumors), and do not account for tumor-specific variables of interest. NF1 has, for instance, been shown to affect overall survival in MPNST [3, 23, 44]. Because of the complexity of MPNST treatment, their care should ideally be centralized, possibly even more than other STS. In future, studies should be encouraged to further explore functional status and quality of life, including pain. This information can give further insights into how we could prevent morbidity as well as what patients we should treat for functional defects.

References

1. Martin E, Coert JH, Flucke UE, et al. A nationwide cohort study on treatment and survival in patients with malignant peripheral nerve sheath tumours. Eur J Cancer. 2019;124:77–87. https://doi.org/10.1016/j.ejca.2019.10.014.
2. Valentin T, Le Cesne A, Ray-Coquard I, et al. Management and prognosis of malignant peripheral nerve sheath tumors: the experience of the French Sarcoma Group (GSF-GETO). Eur J Cancer. 2016;56:77–84. https://doi.org/10.1016/j.ejca.2015.12.015.
3. Miao R, Wang H, Jacobson A, et al. Radiation-induced and neurofibromatosis-associated malignant peripheral nerve sheath tumors (MPNST) have worse outcomes than sporadic MPNST. Radiother Oncol. 2019;137:61–70. https://doi.org/10.1016/j.radonc.2019.03.015.
4. Alamanda VK, Crosby SN, Archer KR, Song Y, Schwartz HS, Holt GE. Amputation for extremity soft tissue sarcoma does not increase overall survival: a retrospective cohort study. Eur J Surg Oncol. 2012;38(12):1178–83. https://doi.org/10.1016/j.ejso.2012.08.024.
5. Ghert MA, Davis AM, Griffin AM, et al. The surgical and functional outcome of limb-salvage surgery with vascular reconstruction for soft tissue sarcoma of the extremity. Ann Surg Oncol. 2005;12(12):1102–10. https://doi.org/10.1245/ASO.2005.06.036.
6. Watson KL, Al Sannaa GA, Kivlin CM, et al. Patterns of recurrence and survival in sporadic, neurofibromatosis type 1-associated, and radiation-associated malignant peripheral nerve sheath tumors. J Neurosurg. 2017;126(1):319–29. https://doi.org/10.3171/2015.12.JNS152443.
7. Davidge KM, Wunder J, Tomlinson G, Wong R, Lipa J, Davis AM. Function and health status outcomes following soft tissue reconstruction for limb preservation in extremity soft tissue sarcoma. Ann Surg Oncol. 2010;17(4):1052–62. https://doi.org/10.1245/s10434-010-0915-5.
8. Martin E, Coert JH, Flucke UE, et al. Neurofibromatosis-associated malignant peripheral nerve sheath tumors in children have a worse prognosis: a nationwide cohort study. Pediatr Blood Cancer. 2020;67(4):e28138. https://doi.org/10.1002/pbc.28138.
9. Stucky C-CH, Johnson KN, Gray RJ, et al. Malignant peripheral nerve sheath tumors (MPNST): the Mayo Clinic experience. Ann Surg Oncol.

2012;19(3):878–85. https://doi.org/10.1245/s10434-011-1978-7.

10. Anghileri M, Miceli R, Fiore M, et al. Malignant peripheral nerve sheath tumors: prognostic factors and survival in a series of patients treated at a single institution. Cancer. 2006;107(5):1065–74. https://doi.org/10.1002/cncr.22098.
11. Gronchi A, Ferrari S, Quagliuolo V, et al. Histotype-tailored neoadjuvant chemotherapy versus standard chemotherapy in patients with high-risk soft-tissue sarcomas (ISG-STS 1001): an international, open-label, randomised, controlled, phase 3, multicentre trial. Lancet Oncol. 2017;18(6):812–22. https://doi.org/10.1016/S1470-2045(17)30334-0.
12. Acem I, van Houdt WJ, Grünhagen DJ, et al. The role of perioperative chemotherapy in primary high-grade extremity soft tissue sarcoma: a risk-stratified analysis using PERSARC. Eur J Cancer. 2022;165:71–80. https://doi.org/10.1016/j.ejca.2022.01.013.
13. Martin E, Slooff W-BM, van Houdt WJ, van Dalen T, Verhoef C, Coert JH. Oncological treatment considerations differ across surgical subspecialties treating malignant peripheral nerve sheath tumors: an international survey. Sarcoma. 2020;2020:6406439. https://doi.org/10.1155/2020/6406439.
14. Rosenberg SA, Tepper J, Glatstein E, et al. The treatment of soft-tissue sarcomas of the extremities: prospective randomized evaluations of (1) limb-sparing surgery plus radiation therapy compared with amputation and (2) the role of adjuvant chemotherapy. Ann Surg. 1982;196(3):305–15.
15. von Mehren M, Randall RL, Benjamin RS, et al. Soft tissue sarcoma, version 2.2016, NCCN clinical practice guidelines in oncology. J Natl Compr Cancer Netw. 2016;14(6):758–86.
16. Casali PG, Abecassis N, Aro HT, et al. Soft tissue and visceral sarcomas: ESMO-EURACAN Clinical Practice Guidelines for diagnosis, treatment and follow-up. Ann Oncol. 2018;29(Suppl 4):iv51–67. https://doi.org/10.1093/annonc/mdy096.
17. van Noesel MM, Orbach D, Brennan B, et al. Outcome and prognostic factors in pediatric malignant peripheral nerve sheath tumors: an analysis of the European Pediatric Soft Tissue Sarcoma Group (EpSSG) NRSTS-2005 prospective study. Pediatr Blood Cancer. 2019;66(10):e27833. https://doi.org/10.1002/pbc.27833.
18. O'Sullivan B, Davis AM, Turcotte R, et al. Preoperative versus postoperative radiotherapy in soft-tissue sarcoma of the limbs: a randomised trial. Lancet. 2002;359(9325):2235–41. https://doi.org/10.1016/S0140-6736(02)09292-9.
19. Haas RL, Gronchi A, van de Sande MAJ, et al. Perioperative management of extremity soft tissue sarcomas. J Clin Oncol. 2018;36(2):118–24. https://doi.org/10.1200/JCO.2017.74.7527.
20. Slump J, Bastiaannet E, Halka A, et al. Risk factors for postoperative wound complications after extremity soft tissue sarcoma resection: a systematic review and meta-analyses. J Plast Reconstr Aesthet Surg. 2019;72(9):1449–64. https://doi.org/10.1016/j.bjps.2019.05.041.
21. Clarkson PW, Griffin AM, Catton CN, et al. Epineural dissection is a safe technique that facilitates limb salvage surgery. Clin Orthop Relat Res. 2005;438:92–6.
22. Martin E, Slooff W-BM, van Houdt WJ, van Dalen T, Verhoef C, Coert JH. Surgical strategies and the use of functional reconstructions after resection of MPNST: an international survey on surgeons' perspective. Orthoplastic Surg. 2021;4:12–9. https://doi.org/10.1016/j.orthop.2021.03.001.
23. Martin E, Pendleton C, Verhoef C, Spinner RJ, Coert JH. Morbidity and function loss after resection of malignant peripheral nerve sheath tumors. Neurosurgery. 2022;90(3):354–64. https://doi.org/10.1093/neuros/nyab342.
24. Vauthey JN, Woodruff JM, Brennan MF. Extremity malignant peripheral nerve sheath tumors (neurogenic sarcomas): a 10-year experience. Ann Surg Oncol. 1995;2(2):126–31.
25. Martin E, Dullaart MJ, van de Sande MAJ, van Houdt WJ, Schellekens PPA, Coert JH. Resuscitating extremities after soft tissue sarcoma resections: are functional reconstructions an overlooked option in limb salvage? A systematic review. Eur J Surg Oncol. 2019;45(10):1762–9. https://doi.org/10.1016/j.ejso.2019.05.024.
26. List EB, Krijgh DD, Martin E, Coert JH. Prevalence of residual limb pain and symptomatic neuromas after lower extremity amputation: a systematic review and meta-analysis. Pain. 2021;162(7):1906–13. https://doi.org/10.1097/j.pain.0000000000002202.
27. McNamara CT, Iorio ML. Targeted muscle reinnervation: outcomes in treating chronic pain secondary to extremity amputation and phantom limb syndrome. J Reconstr Microsurg. 2020;36(4):235–40. https://doi.org/10.1055/s-0039-1700559.
28. Brandt K, Evans GR, Ang KK, et al. Postoperative irradiation: are there long-term effects on nerve regeneration? J Reconstr Microsurg. 1999;15(6):421–5. https://doi.org/10.1055/s-2007-1000121.
29. Evans GR, Brandt K, Ang KK, et al. Peripheral nerve regeneration: the effects of postoperative irradiation. Plast Reconstr Surg. 1997;100(2):375–80.
30. Brandt K, Evans GR, Gurlek A, et al. The effects of preoperative irradiation on peripheral nerve regeneration. Ann Plast Surg. 1998;40(3):277–82.
31. Zhu Y, Zhou S, Xu W, Liu L, Lu H, Yang W. Effects of postoperative radiotherapy on vascularized nerve graft for facial nerve repair in a rabbit model. J Oral Maxillofac Surg. 2019;77(11):2339–46. https://doi.org/10.1016/j.joms.2019.04.012.
32. Evans GRD, Brandt K. Peripheral nerve regeneration: the effects of postoperative irradiation. Plast Reconstr Surg. 2003;111(6):2023–4. https://doi.org/10.1097/01.PRS.0000056837.37545.58.
33. Slump J, Hofer SOP, Ferguson PC, et al. Flap reconstruction does not increase complication rates follow-

ing surgical resection of extremity soft tissue sarcoma. Eur J Surg Oncol. 2018;44(2):251–9. https://doi.org/10.1016/j.ejso.2017.11.015.
34. Kapalschinski N, Goertz O, Harati K, et al. Plastic surgery in the multimodal treatment concept of soft tissue sarcoma: influence of radiation, chemotherapy, and isolated limb perfusion on plastic surgery techniques. Front Oncol. 2015;5:268. https://doi.org/10.3389/fonc.2015.00268.
35. Doi K, Kuwata N, Kawakami F, Hattori Y, Otsuka K, Ihara K. Limb-sparing surgery with reinnervated free-muscle transfer following radical excision of soft-tissue sarcoma in the extremity. Plast Reconstr Surg. 1999;104(6):1679–87.
36. Grinsell D, Di Bella C, Choong PFM. Functional reconstruction of sarcoma defects utilising innervated free flaps. Sarcoma. 2012;2012:315190. https://doi.org/10.1155/2012/315190.
37. Mundinger GS, Prucz RB, Frassica FJ, Deune EG. Concomitant upper extremity soft tissue sarcoma limb-sparing resection and functional reconstruction: assessment of outcomes and costs of surgery. Hand (N Y). 2014;9(2):196–204. https://doi.org/10.1007/s11552-013-9567-9.
38. Spiliopoulos K, Williams Z. Brachial plexus reconstruction following resection of a malignant peripheral nerve sheath tumor: case report. Neurosurgery. 2011;69(1):E245–50. https://doi.org/10.1227/NEU.0b013e31821867de.
39. Maldonado AA, Bishop AT, Spinner RJ, Shin AY. Five operations that give the best results after brachial plexus injury. Plast Reconstr Surg. 2017;140(3):545–56. https://doi.org/10.1097/PRS.0000000000003620.
40. Meals CG, Meals RA. Tendon versus nerve transfers in elbow, wrist, and hand reconstruction: a literature review. Hand Clin. 2013;29(3):393–400. https://doi.org/10.1016/j.hcl.2013.04.010.
41. Patterson JMM, Russo SA, El-Haj M, Novak CB, Mackinnon SE. Radial nerve palsy: nerve transfer versus tendon transfer to restore function. Hand (N Y). 2022;17(6):1082–9. https://doi.org/10.1177/1558944720988126.
42. Pasquali S, Colombo C, Pizzamiglio S, et al. High-risk soft tissue sarcomas treated with perioperative chemotherapy: improving prognostic classification in a randomised clinical trial. Eur J Cancer. 2018;93:28–36. https://doi.org/10.1016/j.ejca.2018.01.071.
43. van Praag VM, Rueten-Budde AJ, Jeys LM, et al. A prediction model for treatment decisions in high-grade extremity soft-tissue sarcomas: personalised sarcoma care (PERSARC). Eur J Cancer. 2017;83:313–23. https://doi.org/10.1016/j.ejca.2017.06.032.
44. Carli M, Ferrari A, Mattke A, et al. Pediatric malignant peripheral nerve sheath tumor: the Italian and German soft tissue sarcoma cooperative group. J Clin Oncol. 2005;23(33):8422–30. https://doi.org/10.1200/JCO.2005.01.4886.

17 MEK Inhibitors and Other Medical Therapies for Benign Peripheral Nerve Sheath Tumors

Brigitte C. Widemann, Eva Dombi, and Andrea M. Gross

Neurofibromatosis type 1 (NF1) is a common autosomal dominant genetic tumor predisposition condition [1]. It is caused by the haploinsufficiency of the *NF1* gene, which results in overactivation of the Ras pathway. As a consequence of this, people with NF1 develop multiple tumor and non-tumor manifestations, including histologically benign, borderline, and malignant peripheral nerve sheath tumors (MPNST). Plexiform neurofibromas (PNF) are histologically benign complex tumors, which can extend across the length of a nerve and involve multiple fascicles and branches. These tumors can result in substantial clinical morbidities, including pain, disfigurement, neurologic and motor dysfunction, as well as airway compromise and vision loss [2, 3]. Surgical management of PNF is challenging, and indications for surgery depend on several considerations. Importantly, complete surgical resection of PNF is frequently not feasible, and regrowth of tumors after surgery has been observed [4]. Therefore, there has been an unmet need for the development of effective medical treatments for PNF. Multiple agents targeting Ras pathway activation in adult cancers including malignant melanoma have been in development, some of which have received regulatory approval [5]. Capitalizing on the availability of Ras pathway targeting agents for other conditions, a series of clinical trials have been performed over the past 20 years with the goal of identifying agents that slow the progression of PNF or cause tumor shrinkage [6]. Recently, selumetinib, an inhibitor of the mitogen-activated protein kinase kinase (MEK) downstream of Ras, has demonstrated clinical benefit for inoperable PNF in children and received regulatory approval in many countries [7, 8]. In this chapter, we review the development of medical treatments for PNF and the current experience with MEK inhibitors (MEKi) for histologically benign NF1 peripheral nerve sheath tumors, and describe considerations for future clinical trials.

B. C. Widemann (✉) · E. Dombi · A. M. Gross
Pediatric Oncology Branch, Center for Cancer Research, National Cancer Institute, Bethesda, MD, USA
e-mail: widemanb@mail.nih.gov

17.1 Challenges in the Development of Medical Therapies for PNF

Plexiform neurofibromas have posed substantial challenges to the development of medical therapies for several reasons. Many PNF arise in early childhood, demonstrate the most rapid growth rate in young children, and result in the development of clinical problems at early age [2, 6, 9]. Of note, the clinical manifestations are highly variable, depending on the location of the PNF. For these reasons, most clinical trials with

I. G. Vetrano, V. Nazzi (eds.), *Peripheral Nerve Sheath Tumors*,
https://doi.org/10.1007/978-3-031-76382-3_17

targeted agents have focused on children with PNF in contrast to adults, where PNF typically do not show substantial growth [9, 10]. However, the ability of young children to take oral medications such as tablets or capsules, which may have to be swallowed whole, is limited. In addition, assessing the risks and potential benefits of a new medical treatment for young developing children is of great importance, especially given the anticipated need for prolonged duration of treatment. This combined with the fact that PNF are histologically benign and frequently not life-threatening made it difficult to receive support from pharmaceutical companies to conduct PNF trials with investigational agents in children.

Another challenge was the development of trial designs and endpoints taking the unique aspects of PNF into consideration. Anticipating the need for prolonged administration of medical therapies in individuals with a non-cancer condition, tolerability of the medical treatment was critical with the goal of not impacting daily activities of living [11]. This is in contrast to the development of treatments targeting aggressive cancers where treatment doses frequently are escalated to the maximum tolerated dose (MTD). Measurement of treatment response was another challenge. In clinical trials for solid cancers, typically one-dimensional RECIST and two-dimensional WHO criteria are used to measure treatment response or progression. Given the much slower growth rate of PNF compared to cancers, applying these criteria would result in extended treatment durations with a potentially inactive investigational agent. The development of volumetric MRI analysis to assess PNF was critical to the establishment of meaningful response assessment in clinical trials allowing the detection of smaller changes in tumor size more sensitively and reproducibly [12]. In contrast to one-dimensional and two-dimensional criteria for PNF volumetric response, criteria are defined by smaller changes: A partial response is defined by a PNF volume decrease of ≥20% and progressive disease by a ≥20% PNF volume increase compared to baseline [13]. Utilization of these response criteria in clinical trials allowed for the detection of time to progression within an acceptable time frame. In an analysis comparing one-dimensional, two-dimensional, and volumetric response criteria in a placebo-controlled double-blind crossover trial of the farnesyltransferase inhibitor R115777 (tipifarnib) versus placebo, the median time to progression for 60 participants enrolled in this trial was 14.3 months for volumetric analysis, 52.5 months for two-dimensional analysis, and the median time to progression was not reached for one-dimensional analysis [14]. Volumetric response criteria therefore have become the standard for most clinical trials conducted for PNF [13].

17.2 Medical Treatment Trials for PNF

All published prior medical treatment trials directed at PNF in NF1 were recently described in a review discussing the management of PNF [6]. Agents evaluated in phase I and phase II trials included antiangiogenic agents (peginterferon alpha-2b), a farnesyltransferase inhibitor (R115777-tipifarnib), receptor tyrosine kinase inhibitors (sorafenib, cabozantinib, imatinib), mTOR inhibitors (sirolimus, everolimus), tumor microenvironment targeting agents (pexidartinib, imatinib), and most recently MEKi (selumetinib, mirdametinib, trametinib, and binimetinib). Primary trial endpoints for phase II studies were progression-free survival or PNF shrinkage. Prior to the evaluation of MEKi, only a few agents demonstrated clinical activity measured by either resulting in tumor shrinkage or prolongation of progression-free survival compared to the placebo arm of the tipifarnib trial. The phase II trial of peginterferon alpha-2b was the only study to result in more than doubling in progression-free survival of progressively growing PNF with a median time to progression of 29.4 months compared to 10.6 months on the placebo arm of the previously conducted double-blind crossover phase II trial of R115777 (tipifarnib). However, partial responses were seen in only 2 of the 82 patients on this trial [15]. On a phase II trial of the C-KIT and PDGFR inhibitor imatinib, partial responses were described in 6 of 36 participants.

However, partial responses were only recorded in patients with very small PNF volume (<25 mL) [16] and not in bulky tumors.

17.3 Phase I and II Trials of the MEKi Selumetinib (SPRINT) for Children and Young Adults

A phase I trial of the oral MEKi selumetinib (NCT01362803) was the first to demonstrate partial responses in the majority of patients (17/24 patients), many of whom had large PNF (median baseline tumor volume 1205 mL, range 29 to 8744 mL) [17]. This phase I trial used a standard 3 + 3 dose escalation design with the goal of determining the MTD based on the tolerability of selumetinib during the first three treatment cycles (1 cycle = 28 days). The MTD in this study was 25 mg/m^2 body surface area per dose administered as capsules, which had to be swallowed whole, approximately every 12 h on a continuous dosing schedule. This dose corresponds to approximately 60% of the recommended dose in adults with cancers. Overall, selumetinib was well tolerated in this study. Dose-limiting toxicities included asymptomatic elevation of creatine phosphokinase, cellulitis, urticaria, decreased left ventricular ejection fraction, mucositis, and rash. Importantly, all dose-limiting toxicities were completely reversible upon holding the study drug. In addition to tumor shrinkage, anecdotal clinical improvement was observed in this trial including improvement in pain, function, and decrease in disfigurement. However, this trial did not include patient-reported outcome (PRO) measures or functional assessments to prospectively and objectively assess clinical changes.

In order to evaluate if the modest PNF volume decreases observed in the phase I trial (median decrease in PNF volume, −31%, range, −5.8 to −47) are associated with clinical benefit, the study team along with the NCI Cancer Therapy Evaluation Program as a trial sponsor and AstraZeneca designed a phase II registration study (SPRINT) that included patient and observer reported outcome measures in all patients as well as functional outcome measures, depending on PNF location and existing or potential for PNF related morbidities [8]. Stratum 1 of the trial required that patients with inoperable PNF had at least one PNF-related morbidity such as disfigurement, pain, motor dysfunction, or vision loss at the time of enrollment. Stratum 2 allowed enrollment of patients at risk for the development of clinically significant morbidity but not yet having clinically significant morbidity at the time of enrollment. Stratum 1 was the key stratum for regulatory submission, and response rate by volumetric MRI analysis was the primary endpoint. In addition, key secondary endpoints were PROs, including pain interference and pain intensity as well as quality of life measures, standardized photography of visible PNF, and functional outcome measures, depending on the location of the PNF. For example, a patient with orbital PNF would undergo serial detailed vision evaluations, whereas a patient with an airway PNF would undergo sleep studies, pulmonary function, and walk tests. This trial enrolled 50 children (median age 10.2 years) from four participating sites within 1 year. The confirmed objective response rate was 68% and 56% of the participants had a durable partial response defined as a response lasting for at least 12 months after being first established. This trial demonstrated an improvement in pain intensity and pain interference, other PROs, as well as disfigurement in some patients and an improvement in functional outcome measures. Similar to the prior phase I trial, selumetinib was overall well tolerated, and no new safety signals were observed. In comparison to prior medical trials and to the placebo arm of the previously conducted phase 2 trial of R115777 (tipifarnib) for children and young adults with progressive PNF, selumetinib improved progression-free survival, and the median time to progression was not reached at the time of submission of the new drug application to the USA Food and Drug Administration (FDA). An additional comparison of patients treated with selumetinib on the SPRINT trial to age-matched control patients enrolled in the NIH Natural History study not receiving selumetinib demonstrated that selumetinib clearly altered the natural history of PNF growth. The median progression-free survival for children on selumetinib at 3 years was 84% compared to 15% for children not receiving selumetinib in the natural history study. Based on these findings, the FDA approved selumetinib for children 2 years and older with symptomatic inoperable NF1 PNF in 2020 [7, 8]. Subsequent regulatory approvals have been received in many countries, including the European Union. An updated analysis of the long-term safety and efficacy of selumetinib, which included all participants from the phase I selumetinib study and stratum 1 of the phase II study and provided up to 5 years of additional data, was recently published [18]. For the 74 children (median age 10.3 years; range 3–18.5) enrolled, the overall confirmed partial response rate was 70% (52/74), and the median duration of treatment was 57.5 cycles (range 1–100). Responses were generally sustained with 59% (n = 44) lasting ≥12 cycles. Tumor pain intensity (n = 19, P = 0.015) and pain interference (n = 18, P = 0.0059) showed durable improvement from baseline to 48 cycles. While no new safety signals were identified, some participants developed known selumetinib-related adverse events, such as decreased left ventricular ejection fraction, for the first time after several years of treatment highlighting the need for safety evaluations over prolonged time periods. Participants on SPRINT are continued to be monitored for possible long-term selumetinib-related toxicities. This analysis allowed for the first time calculating a median progression-free survival for the combined phase I and II selumetinib study of approximately 6.7 years (88 cycles of treatment), significantly longer than the previously published age-matched natural history cohort with a median PFS of 1.3 years [8]. The results for SPRINT stratum 2 for children with inoperable PN without clinically significant morbidity were also recently reported [19]. Selumetinib resulted in a confirmed partial response in 72% of participants. No new PNF-related symptoms developed while on selumetinib, and PRO measures indicated declines in tumor-related pain intensity. The stratum 2 data supported that selumetinib treatment may prevent the development of PNF-related morbidities and the development of a secondary prevention trial of selumetinib in young children.

17.4 Secondary Prevention Trial of Selumetinib for Children with NF1

While selumetinib provides clinical benefit in children with symptomatic PNF, many PNF morbidities, such as disfigurement, are not fully reversible despite achieving some PNF shrinkage with treatment [8]. Based on the hypothesis that treatment with selumetinib in children with asymptomatic PNF will be tolerable and prevent the development of PNF-related morbidities, a secondary prevention trial of selumetinib was developed and will soon begin enrollment. Given that PNF morbidities develop early in life, this trial will require treatment of very young children [2], and a clinical trial assessing a selumetinib granule formulation for children 1–7 years old is currently ongoing (NCT05309668). This multicenter trial will be coordinated by the US NF Clinical Trials Consortium and assess the ability of selumetinib to prevent the development of PNF growth and morbidity in children with asymptomatic NF1 PNF in high-risk locations. The incidence of asymptomatic PNF in any location in children with NF1 and no known PNF will be determined through whole-body MRI evaluation. Children ≤8 years old with a diagnosis of NF1 and an asymptomatic PNF in a high-risk location will then be randomized to either treatment with selumetinib or observation to determine the impact on PNF growth and PNF-related symptoms. Participants who develop progressive disease (PNF volume increase ≥20% from baseline) on the observation arm will be eligible to start treatment with selumetinib.

17.5 Phase II Trial of Selumetinib for Adults with NF1 and Inoperable PNF

While PNF grow most rapidly in young children, they also cause substantial morbidity in adults with NF1. An NCI-sponsored single-institution phase II trial of selumetinib (NCT02407405) therefore assessed the imaging response and clinical benefit of selumetinib in adults with NF1 and inoperable and symptomatic or growing PNF. Response assessment by MRI and patient-reported and functional outcome measures were nearly identical to the pediatric SPRINT trial. In addition, the adult trial incorporated mandatory serial biopsies of PNF for analysis of tumor phospho ERK, MEK, and AKT and other pharmacodynamic correlates. This study enrolled 33 participants (median age 35.6, range 18.3–60.2, 11 female) and has reported an objective response rate of 63.6% (21/33 participants), similar to the pediatric trials [20]. In addition, from baseline to cycle 12, tumor pain intensity and pain interference improved. Tumor biopsy analyses confirmed decreased phosphorylation ratios of ERK1/2 on treatment with selumetinib comparing baseline to 1–3 months on treatment with selumetinib, and no compensatory phosphorylation of AKT1/2/3. All adverse events were reversible and similar to those in the pediatric trial, with acneiform rash, asymptomatic increase in creatine kinase and alanine aminotransferase, dry skin, pruritis, and limb edema being the most common. An Alexion-sponsored phase 3 trial of selumetinib for adults with NF1-PNF (KOMET, NCT04924608) is ongoing.

17.6 Trials with Other MEKi for PNF

In addition to selumetinib, clinical trials with other MEK inhibitors for PNF have been conducted or are ongoing as reviewed by Fisher et al. [6] These include trametinib, binimetinib, and mirdametinib. Published results to date demonstrate a similar adverse event profile and confirm the clinical activity of the different MEK inhibitors. A phase I/II trial of trametinib for children with NF1 PNF described responses in 12 or 26 participants. A phase II trial of mirdametinib for patients 16 years and older reported 8 partial responses out of 19 participants' total. A phase II trial of binimetinib reported 13 partial responses out of 25 adult participants enrolled with 20 being evaluable for response and 14 partial responses out of 20 pediatric participants enrolled. Finally, results of a pivotal phase 2b trial of mirdametinib in children and adults with inoperable symptomatic NF1 PNF were recently reported [21]. Mirdametinib was administered as a capsule or dispersible tablet without regard to food in 3 weeks on, 1 week off 28-day cycles. The objective response rate for 58 adult patients was 41% and for children 52%. This trial also reported statistically significant improvement in pain and quality of life measures. A head-to-head comparison of the efficacy and tolerability of the various MEKi in NF1 PNF has not been performed to date, and as of August 2024, selumetinib is the only MEKi with regulatory approval for NF1 PNF. However, with the possibility of additional regulatory approvals, specific drug characteristics such as half-life (5–7 h for selumetinib versus 4–5 days for trametinib) [5], central nervous system penetration and formulation may help with treatment considerations for specific patient populations.

17.7 Trials with MEKi for Other Forms of Benign Neurofibromas

Cutaneous neurofibromas (CNF) in NF1 can be debilitating due to disfigurement and emotional psychological burden and have an unmet need for the development of effective treatments [22–25]. Just like PNF, these tumors are characterized by biallelic loss of *NF1* as the only genomic alteration. It is thus possible that MEKi may demonstrate clinical activity in CNF. In addition, the superficial location of CNF may make these tumors amenable to topical treatments. An NCI CTEP-sponsored pilot trial assessing the ability of the MEKi selumetinib [26] to reduce the volume of CNF measured by calipers recently reported initial results [23]. Adults with NF1 and CNF received selumetinib 50 mg BID for up to 24 cycles (1 cycle = 28 days) with restaging visits after every 4 cycles. Photography and volumetric measurements of CNFs were used to track the number and volume of tumors. The Skindex-29 quality of life assessment, a patient-reported outcome measure, was used to quantify the effect of treatment on symptoms, functioning, emotions, itching, and pain. Eleven participants enrolled, and the median best response across all participants and tumors was a 28.5% decrease in volume from baseline, similar to the volume reduction observed in selumetinib trials for PNF. The study reported some visible improvement in CNF burden on treatment in standardized photographs; however, only four participants completed the full 24 cycles of treatment. Reversible selumetinib-related adverse events such as dry skin and rash were frequent and highlighted the challenges of treating CNF with a drug that often results in dermatologic adverse events. Results of a phase IIa trial with NFX-179, a topical gel MEKi, were recently reported [27]. This trial randomized participants to the treatment of five target CNF with three different concentrations of NFX-179 or vehicle for 28 days. NFX-179 was well tolerated and did not result in local or systemic toxicities. In participants treated with the highest concentration (0.5%), 20% of CNF showed a ≥50% volume reduction compared to 6% in the vehicle group. This study demonstrated the potential therapeutic benefit of a topical treatment of CNF.

Spinal neurofibromas (SNF) in NF1 can cause severe morbidity through spinal cord compression, which can simultaneously occur at several levels of the spine. The spinal portion of these tumors is small and frequently not amenable to volumetric MRI analysis. A retrospective qualitative analysis assessed the effect of selumetinib on SNF in children and adults enrolled in phase 2 studies of selumetinib for PN at the National Cancer Institute [28]. The study identified 24 patients with SNF and serial spine MRI imaging. Eighteen patients showed subtle to marked improvement in SNF burden as evidenced by decreased cord deformity, five remained stable, and no worsening was observed during treatment. While this was a retrospective study, it suggested that selumetinib may decrease and prevent the worsening of cord compression and could possibly benefit select patients with spinal cord compression who are not candidates for surgical intervention. Consideration has also been given to MEKi therapy prior to performing surgery when possible [29].

17.8 Safety Evaluations and Considerations for MEKi Treatment in PNF

Selumetinib and other MEKi administered at the recommended doses have demonstrated overall safety and tolerability in children and adults with NF1 and PNF. Importantly, treatment-limiting adverse events, such as an asymptomatic decrease in left ventricular ejection fraction, have been reversible upon holding drug [7, 8]. However, new adverse events can occur after a prolonged duration of treatment, and thus safety evaluations should be performed throughout the duration of therapy with a MEKi [18]. On the phase II registration trial of selumetinib SPRINT for children with NF1 PNF, participants will be monitored for 5–7 years after completion of therapy [8]. In addition to recommendations for evaluation based on the product label, supportive care guidelines have been developed for the monitoring and management of the most frequent MEKi toxicities, including cutaneous toxicities (acneiform rash, eczema, paronychia, and mucositis), gastrointestinal toxicities, as well as monitoring for rare and potentially serious toxicities such as ocular or cardiac toxicity [5, 30]. These include periodic physical and skin examination, laboratory studies, echocardiogram, ophthalmology evaluation, and MRI evaluation of the target PNF prior to treatment and at regular intervals after the start of therapy [30]. In addition, consensus recommendations for the management of PNF and initiation of medical treatments have been developed [6, 31].

The required duration of treatment with MEKi has not been defined yet, but current experience suggests that prolonged treatment will be required for most patients [18]. This is based on observed regrowth of PNF after discontinuation of MEKi treatment [8, 17], particularly in younger patients and in patients with actively growing PNF prior to initiation of MEKi treatment [32]. In patients with PNF in critical locations, such as airway or SNF, close clinical observation after discontinuing MEKi treatment is recommended, as even small tumor regrowth can result in clinical compromise [28]. Longitudinal analyses of the growth behavior of PNF after stopping MEKi are ongoing.

17.9 Other Active Medical Therapies for NF1 PNF

PNF are composed of multiple cell types, and interactions between tumorigenic Schwann cells and cells in the tumor microenvironment, such as mast cells, macrophages, and endothelial cells, contribute to tumor initiation and maintenance in preclinical models providing a rationale for the evaluation of agents targeting the TME [33]. Based on the preclinical activity of cabozantinib, a receptor tyrosine kinase inhibitor of multiple kinases including c-Kit, VEGFR2, and AXL in a transgenic mouse model of NF1 neurofibroma, the NF Clinical Trials Consortium developed and conducted a phase II trial of cabozantinib (NCT02101736) for patients ≥16 y/o with NF1 PNF [34]. This trial is based on a completed phase I trial of cabozantinib for children with refractory solid cancers coordinated by the Children's Oncology Group [35]. Of 19 participants, 8 (42%) had a partial response associated with a significant reduction in tumor pain intensity and pain interference. Similar to prior preclinical trials with MEKi in genetically engineered mouse models of NF1, these studies accurately predicted the clinical activity of cabozantinib in patients with NF1 PNF. Cabozantinib is thus the first drug with a different mechanism of action than MEKi to demonstrate activity in clinical trials enrolling patients with large PNF.

17.10 Conclusion

After many years of work, meaningful clinical trial designs and endpoints have been developed, which allow for timely and accurate assessment of the activity of novel agents for benign peripheral nerve sheath tumors in NF1. Clinical trials with several MEKi have documented tumor shrinkage in the majority of participants, and this was associated with clinical benefit. The MEKi selumetinib is currently approved in many countries for the treatment of children with inoperable, symptomatic PNF, and trials with several MEKi are ongoing or have been completed. In addition, the receptor tyrosine kinase inhibitor cabozantinib has demonstrated activity in NF1 PNF. None of the trials to date have resulted in complete responses, and regrowth of PNF after discontinuation of treatment has been observed. Thus, there is a continued need for the development of safe and more effective agents, combinations of agents, and prevention strategies.

References

1. Hirbe AC, Gutmann DH. Neurofibromatosis type 1: a multidisciplinary approach to care. Lancet Neurol. 2014;13:834–43.
2. Gross AM, Singh G, Akshintala S, et al. Association of plexiform neurofibroma volume changes and development of clinical morbidities in neurofibromatosis 1. Neuro-Oncology. 2018;20:1643–51.
3. Prada CE, Rangwala FA, Martin LJ, et al. Pediatric plexiform neurofibromas: impact on morbidity and mortality in neurofibromatosis type 1. J Pediatr. 2012;160:461–7.
4. Needle MN, Cnaan A, Dattilo J, et al. Prognostic signs in the surgical management of plexiform neurofibroma: the Children's Hospital of Philadelphia experience, 1974–1994. J Pediatr. 1997;131:678–82.
5. de Blank PMK, Gross AM, Akshintala S, et al. MEK inhibitors for neurofibromatosis type 1 manifestations: clinical evidence and consensus. Neuro-Oncology. 2022;24:1845–56.
6. Fisher MJ, Blakeley JO, Weiss BD, et al. Management of neurofibromatosis type 1-associated plexiform neurofibromas. Neuro-Oncology. 2022;24:1827–44.
7. Casey D, Demko S, Sinha A, et al. FDA approval summary: selumetinib for plexiform neurofibroma. Clin Cancer Res. 2021;27:4142–6.
8. Gross AM, Wolters PL, Dombi E, et al. Selumetinib in children with inoperable plexiform neurofibromas. N Engl J Med. 2020;382:1430–42.
9. Akshintala S, Baldwin A, Liewehr DJ, et al. Longitudinal evaluation of peripheral nerve sheath tumors in neurofibromatosis type 1: growth analysis of plexiform neurofibromas and distinct nodular lesions. Neuro-Oncology. 2020;22:1368–78.
10. Ly KI, Merker VL, Cai W, et al. Ten-year follow-up of internal neurofibroma growth behavior in adult patients with neurofibromatosis type 1 using whole-body MRI. Neurology. 2023;100:e661–70.

11. Gross AM, Widemann BC. Clinical trial design in neurofibromatosis type 1 as a model for other tumor predisposition syndromes. Neurooncol Adv. 2020;2:i134–40.
12. Solomon J, Warren K, Dombi E, Patronas N, Widemann B. Automated detection and volume measurement of plexiform neurofibromas in neurofibromatosis 1 using magnetic resonance imaging. Comput Med Imaging Graph. 2004;28:257–65.
13. Dombi E, Ardern-Holmes SL, Babovic-Vuksanovic D, et al. Recommendations for imaging tumor response in neurofibromatosis clinical trials. Neurology. 2013;81:S33–40.
14. Widemann BC, Dombi E, Gillespie A, et al. Phase 2 randomized, flexible crossover, double-blinded, placebo-controlled trial of the farnesyltransferase inhibitor tipifarnib in children and young adults with neurofibromatosis type 1 and progressive plexiform neurofibromas. Neuro-Oncology. 2014;16:707–18.
15. Jakacki RI, Dombi E, Steinberg SM, et al. Phase II trial of pegylated interferon alfa-2b in young patients with neurofibromatosis type 1 and unresectable plexiform neurofibromas. Neuro-Oncology. 2017;19:289–97.
16. Robertson KA, Nalepa G, Yang FC, et al. Imatinib mesylate for plexiform neurofibromas in patients with neurofibromatosis type 1: a phase 2 trial. Lancet Oncol. 2012;13:1218–24.
17. Dombi E, Baldwin A, Marcus LJ, et al. Activity of selumetinib in neurofibromatosis type 1-related plexiform neurofibromas. N Engl J Med. 2016;375:2550–60.
18. Gross AM, Dombi E, Wolters PL, et al. Long-term safety and efficacy of selumetinib in children with neurofibromatosis type 1 on a phase 1/2 trial for inoperable plexiform neurofibromas. Neuro-Oncology. 2023;25:1883–94.
19. Gross AM, Glassberg B, Wolters PL, et al. Selumetinib in children with neurofibromatosis type 1 and asymptomatic inoperable plexiform neurofibroma at risk for developing tumor-related morbidity. Neuro-Oncology. 2022;24:1978–88.
20. Gross A, O'Sullivan Coyne G, Dombi E, et al. Phase II trial of the MEK 1/2 inhibitor selumetinib in adults with neurofibromatosis type 1 (NF1) and inoperable, symptomatic plexiform neurofibromas (PN) – a look at clinical responses and pharmacodynamics. Brussels: Children's Tumor Foundation; 2024.
21. Moertel C, Hirbe A, Shuhaiber H, et al. ReNeu: a pivotal phase 2b trial of mirdametinib in children and adults with neurofibromatosis type 1 (NF1)-associated symptomatic inoperable plexiform neurofibroma (PN). J Clin Oncol. 2024;42:3016.
22. Blakeley JO, Wolkenstein P, Widemann BC, et al. Creating a comprehensive research strategy for cutaneous neurofibromas. Neurology. 2018;91:S1–4.
23. Cannon A, Jarnagin K, Korf B, et al. Clinical trial design for cutaneous neurofibromas. Neurology. 2018;91:S31–7.
24. Cannon A, Pichard DC, Wolters PL, et al. Perspective of adults with neurofibromatosis 1 and cutaneous neurofibromas: implications for clinical trials. Neurology. 2021;97:S15–24.
25. Cannon A, Sarin KY, Petersen AK, et al. Perspectives of adolescents with neurofibromatosis 1 and cutaneous neurofibromas: implications for clinical trials. Clin Trials. 2024;21(1):67–72.
26. Reid OH, Baldwin LA, Gross AM, et al: Treatment of cutaneous neurofibromas (cNF) in neurofibromatosis type 1 (NF1) with the MEK inhibitor seluemtinib, Children's Tumor Foundation, Scottsdale, AZ, 2023.
27. Sarin KY, Bradshaw M, O'Mara C, et al. Effect of NFX-179 MEK inhibitor on cutaneous neurofibromas in persons with neurofibromatosis type 1. Sci Adv. 2024;10:eadk4946.
28. Jackson S, Baker EH, Gross AM, et al. The MEK inhibitor selumetinib reduces spinal neurofibroma burden in patients with NF1 and plexiform neurofibromas. Neurooncol Adv. 2020;2:vdaa095.
29. Xu AL, Suresh KV, Gomez JA, et al. Consensus-based best practice guidelines for the management of spinal deformity and associated tumors in pediatric neurofibromatosis type 1: screening and surveillance, surgical intervention, and medical therapy. J Pediatr Orthop. 2023;43:e531–7.
30. Klesse LJ, Jordan JT, Radtke HB, et al. The use of MEK inhibitors in neurofibromatosis type 1-associated tumors and management of toxicities. Oncologist. 2020;25:e1109–16.
31. Armstrong AE, Belzberg AJ, Crawford JR, Hirbe AC, Wang ZJ. Treatment decisions and the use of MEK inhibitors for children with neurofibromatosis type 1-related plexiform neurofibromas. BMC Cancer. 2023;23:553.
32. Roman O, Gross A, Whitcomb P, et al. The effect of selumetinib treatment interruptions on plexiform neurofibroma growth. Brussels: Children's Tumor Foundation; 2024.
33. White EE, Rhodes SD. The *NF1+/−* immune microenvironment: dueling roles in neurofibroma development and malignant transformation. Cancers (Basel). 2024;16(5):994.
34. Fisher MJ, Shih CS, Rhodes SD, et al. Cabozantinib for neurofibromatosis type 1-related plexiform neurofibromas: a phase 2 trial. Nat Med. 2021;27:165–73.
35. Chuk MK, Widemann BC, Minard CG, et al. A phase 1 study of cabozantinib in children and adolescents with recurrent or refractory solid tumors, including CNS tumors: trial ADVL1211, a report from the Children's Oncology Group. Pediatr Blood Cancer. 2018;65:e27077.

18 Systemic Treatment in Malignant Peripheral Nerve Sheath Tumors: Current Practice and Future Possibilities

Enrico Martin and Ingrid M. E. Desar

18.1 Introduction

Malignant peripheral nerve sheath tumors (MPNSTs) are a relatively rare subtype of soft tissue sarcoma (STS) originating from nervous tissue. They generally tend to be aggressive tumors with high rates of local and distant recurrence. In contrast to benign peripheral nerve sheath tumors, they require complete resection with a certain margin often combined with radiotherapy and in some cases chemotherapy. In approximately 35–40% of cases, MPNSTs are related to neurofibromatosis type 1 (NF1) [1, 2]. In other cases, MPNSTs are sporadic or radiation-induced (~10%). MPNSTs harbor complex genetics, but the *NF1* gene is commonly affected. *NF1* loss results in loss of neurofibromin, a Ras-inhibiting enzyme [3]. Ras activation results in the upregulation of several downward pathways, including phosphoinositide 3-kinase (PI3K) and mitogen-activated protein kinase (MAPK) [4]. However, *NF1* deactivation alone is not enough for MPNSTs to develop. Numerous pathways have been shown to affect MPNST pathogenesis, including loss of function in *CDKN2A, TP53, SUZ12,* and *PTEN* genes, as well as amplification of several receptors tyrosine kinases including epidermal growth factor receptor (EGFR), MET, and platelet-derived growth factor receptor (PDGFR) [5–9]. Also, increasing evidence suggests an important role in tumor microenvironment interacting with dysregulated Schwann cells promoting MPNST formation, such as initiating neo-angiogenesis [10].

Despite our increasing understanding of the complex biology MPNSTs harbor, prognosis has stagnated in the past decades. Median survival for localized disease is 5–8 years. However, 10–20% of all patients present with metastatic or unresectable disease, and 40% will develop metastases over time [1, 2, 11–17]. NF1 patients harbor a twofold higher risk of developing metastases [11–13]. Median survival of metastatic MPNST, both synchronous and metachronous, is 8–12 months [11]. Most metastases occur in lung sites, similar to other STS [11, 18]. More rarely, metastases occur in the liver, lymph node, or bone sites. In 80% of cases, metastases occur only at one anatomical site. In selected patients with oligometastatic disease limited to one anatomical site, metastasectomy may result in a disease-free interval of more than 1 year [19]. Nevertheless, systemic treatment remains the

E. Martin
Department of Plastic and Reconstructive Surgery, University Medical Center Utrecht, Utrecht, The Netherlands

Department of Surgical Oncology, Erasmus Medical Center Cancer Institute, Rotterdam, The Netherlands
e-mail: e.martin-2@umcutrecht.nl

I. M. E. Desar (✉)
Department of Medical Oncology, Radboud University Medical Center, Nijmegen, The Netherlands
e-mail: ingrid.desar@radboudumc.nl

I. G. Vetrano, V. Nazzi (eds.), *Peripheral Nerve Sheath Tumors*,
https://doi.org/10.1007/978-3-031-76382-3_18

mainstay treatment in unresectable and metastatic STS.

The chemoresistant nature of MPNST combined with their aggressive clinical behavior calls for new treatment regimens targeting specific pathways. Targeted therapies, immunotherapy, and oncolytic viruses have shown promising results in many other cancer types, but their use has yet to be proven in STS and MPNST specifically. These regimens have the potential to explicitly target cancer cells limiting systemic side effects and maximizing their action. One such drug, tyrosine kinase inhibitor pazopanib, has been proven effective as a second-line regimen in any STS, which has shown significant delay in disease progression, but not yet in overall survival [20].

This chapter will address current practice and indications for chemotherapy in metastatic MPNST. While MPNST-specific noncytotoxic treatment regimens are not yet implemented, results of clinical trials are provided for targeted therapies and immunotherapy.

18.2 Chemotherapy

First-line palliative chemotherapy consists of an anthracycline such as doxorubicin or epirubicin and may be combined with ifosfamide [19, 21]. MPNSTs are unfortunately part of the chemoresistant STS group, which is even more so in NF1 patients [22–25]. The clinical response rate of chemotherapy in MPNST is only 20% when doxorubicin is combined with ifosfamide [24]. While the addition of ifosfamide may improve progression-free survival (PFS), overall survival (OS) does not seem to improve in all types of STS, and its addition comes at the cost of increased toxicity [26]. Some studies suggest a partial response rate of 8–18% in NF1 compared to 55–60% in sporadic cases [22–24]. To date, no rationale has been proven to treat MPNSTs differently to other STS in this setting. In a large study investigating histotype-tailored neoadjuvant chemotherapy in four STS subtypes including etoposide plus ifosfamide for MPNST, no benefit was shown compared to standard anthracycline-based chemotherapy [27]. In a large retrospective study, 53% of patients with synchronous metastases were treated with chemotherapy, while metachronous metastases only received chemotherapy in 38% of cases [11]. The use of chemotherapy, albeit differences in regimens, was independently associated with better survival [11]. Guidelines do support using chemotherapy in metastatic and unresectable disease to patients that are physically fit enough, ideally using doxorubicin with or without the addition of ifosfamide [19, 21]. After failure of first-line treatment, other cytotoxic treatments can be considered following general STS algorithms and all with limited success rates.

In adults, trials investigating the use of chemotherapy in localized resectable disease of any STS have shown conflicting results of effect on OS [28–31]. For instance, the 2008 Sarcoma Meta-Analysis Collaboration did suggest an advantage of adjuvant chemotherapy on OS in localized STS [29]. The subsequent European Organization for Research and Treatment of Cancer Soft Tissue and Bone Sarcoma Group (EORTC-STSBG) 62,931 trial did not find any difference in PFS and OS [31]. However, they argued that this may have been caused by a relatively large proportion of low-risk patients, and post hoc analysis of this trial selecting high-risk patients based on the Sarculator nomogram (<60% probability of 10-year survival) showed that these patients did benefit from adjuvant chemotherapy on both PFS and OS [32]. Because of this current guidelines do state that in localized disease, chemotherapy can be offered to physically fit patients that harbor high-risk STS [19]. European and American guidelines report conflicting preferences for neoadjuvant or adjuvant timing in these patients [19]. Doxorubicin combined with ifosfamide remains the standard regimen for neoadjuvant chemotherapy since more histotype-tailored schedules—etoposide combined with ifosfamide in the case of MPNST—failed to succeed [27].

A recent international survey among surgeons treating MPNST showed heterogeneity in opinion regarding indications for chemotherapy use in localized disease [33]. Only tumor sizes

>10 cm and macroscopically incomplete resection margins were viewed as indications for chemotherapy by more than 50% of respondents. Nevertheless, some surgeons thought chemotherapy should be considered in any localized MPNST, more so among neurosurgeons. There was no consensus on the timing of systemic treatment. The Dutch national cancer registry reports that chemotherapy was only used in 6.7% of resectable disease and in 24% of unresectable disease [2].

For MPNSTs arising from retroperitoneal or pelvic sites, special considerations can be taken into account. Although these tumor sites are rare, their overall survival in localized disease is significantly worse compared to other sites, 13 vs. 72 months, respectively [2]. Surgical resection in these tumors still remains the mainstay treatment but carries high morbidity due to their common need for resection of adjacent organs. Unlike extremity and trunk wall STS, the use of radiotherapy for local control has been proven less effective in these types of STS [34]. The effect of chemotherapy in localized disease is still being investigated, yet it has been used more commonly in this setting compared to other tumor sites in MPNST [2].

While, biologically, MPNSTs in children behave similarly to their adult counterparts, the use of multimodal treatment differs. This is partially because tumors tend to be large in children as well, which means they can be larger proportionally to body size compared to adult MPNST. Secondly, childhood cancers, in general, are more aggressively treated by pediatric oncologists. In pediatric MPNST, the use of doxorubicin combined with ifosfamide has become the standard of care in many cases of even localized disease [21, 22]. Current guidelines for pediatric MPNST are referred to in nonrhabdomyosarcoma soft tissue sarcoma (NRSTS). In pediatric MPNST, chemotherapy may be given in unresectable localized disease, with or without radiotherapy, possibly facilitating resectability after neoadjuvant treatment [35]. In localized disease, so-called intermediate-risk patients (high-grade and >5 cm or unresectable) chemoradiotherapy is advised either postoperatively or in a neoadjuvant setting [21, 36]. In the EpSSG trial, 27 pediatric MPNST patients were included in the neoadjuvant chemotherapy arm in which one patient experienced a complete response, four a partial response, seven a minor partial response (20–50% reduction of tumor volume), and nine stable diseases [21]. Interestingly, in this trial, no difference was seen in response rate between NF1 and sporadic MPNST, in contrast to other studies [25, 37].A retrospective nationwide study in the Netherlands showed that over the last few decades, <35% of children received chemotherapy in localized disease [38]. Its use has however been increasing over the past decade as the treatment of pediatric oncology has been centralized to one Dutch dedicated pediatric oncology center. OS has also seemed to be improved for pediatric MPNST when comparing cases before 2005 and after, in contrast to their adult counterparts [2, 38]. This may be due to centralization or can be the effect of an increasing use of chemotherapy in localized disease as well as the implementation of doxorubicin and ifosfamide as the standard regimen [22].

Overall, the use of chemotherapy in nonmetastatic MPNST is currently still disputed, especially among adults. Several authors as well as the most recent guidelines suggest that survival calculators may aid in choosing for adjuvant chemotherapy in localized STS. In adult extremity STS, the PERSARC and Sarculator are two of such instruments, of which the latter may also be used in trunk wall and retroperitoneal tumors. Both calculators have been shown to distinguish high-risk STS patients who benefit from additional chemotherapy in nonmetastatic disease [32, 39]. However, while these instruments have the ability to discriminate prognosis for different subtypes of STS including MPNST, they lack certain generalizability and do not include possible subtype-specific prognosticators. A recent validation study also showed that while the Sarculator has a good overall performance, MPNSTs had the lowest concordance index, overall overestimating survival [40]. An MPNST-specific calculator is currently underway based on the MONACO database, including all tumor sites, ages, and tumor-specific variables [2, 11].

For instance, in a multivariable model predicting metastases in MPNST, NF1, tumor size larger than 5 cm, high grade, rhabdomyoblastic differentiation (Triton tumors), and macroscopically positive margins (R2) were all independent predictors of metastatic disease [11].

As the role of chemotherapy is still evolving in nonmetastatic MPNST, its use should ideally be discussed early on in a multidisciplinary team in a case-by-case setting with special attention to the anticipated merit in NF1-associated MPNST.

18.3 Targeted Therapies

Targeted therapies often function as inhibitors of certain proteins that may be upregulated in specific cancers. They may be targeting membrane receptors such as receptor tyrosine kinases (RTKs), cytoplasmic proteins, and even nuclear proteins. Numerous studies have investigated potential actionable pathways in vitro and in vivo. Multiple cytoplasmic, nuclear, and membrane targets have shown to be potential therapeutic targets [41]. Receptor tyrosine kinase (RTK) inhibition, especially those including the inhibition of vascular endothelial growth factor receptor (VEGFR), has shown promising results in vivo [41]. There is increasing evidence that angiogenesis plays a crucial role in the progression of benign peripheral nerve sheath tumors to MPNST [10]. VEGF levels are, for instance, increased in MPNST compared to their benign counterparts, and higher VEGF levels were also correlated to a worse prognosis [42, 43]. The exact mechanisms of neovascularization in MPNST are however still not completely understood as numerous factors play their role. Currently, the only accredited targeted therapy in STS is pazopanib, a multikinase inhibitor including VEGFR inhibition [20]. The PALETTE trial showed that pazopanib induces modest prolongation of PFS but did not significantly affect overall survival [20]. Nishida et al. studied 12 advanced MPNST patients and partial response and stable disease in 50% of patients after 3 months (Table 18.1) [51]. A retrospective study of pazopanib in chemotherapy-treated metastatic STS showed reasonable effects in which MPNSTs ($n = 5$) were among the best responders at a partial response rate of 20% and median PFS and OS of 6.5 and 8.9 months, respectively [53]. In a curative setting, one trial showed that pazopanib as an adjunct to doxorubicin and ifosfamide prior to surgery of any STS provided higher rates of pathological response, especially in MPNST [54]. Nevertheless, these studies have not yet proven a beneficial effect of pazopanib on OS, and most have not stratified for MPNST specifically. Other RTK inhibitors have not been proven effective in MPNST patients to date [44–47].

The inhibition of mTOR, a downstream actionable target in the Ras-PI3K pathway, has shown the most promising results on tumor growth and survival in in vivo studies. MEK inhibition, another downstream Ras target, seems logical as well and has already proven its use in plexiform neurofibromas [55]. Expression of phosphorylated-MEK has been detected in more than 90% of MPNST samples, compared to phosphorylated-mTOR, which is expressed in only 50% of samples [4]. In vivo studies show promising results for MEK inhibition in MPNST but more so when combined with other agents [41]. Although the combination of mTOR inhibition with MEK inhibition has shown better anti-tumorigenic effects, their combination may come at a cost of increased toxicity [56, 57]. Clinical trials have yet only studied the effect of mTOR inhibition in MPNST [48, 50]. Widemann et al. showed that the combination of everolimus (mTOR inhibitor) and bevacizumab only led to stable disease in three patients and progressive disease in the remaining 22 [48]. Kim et al. combined mTOR inhibitor sirolimus with Hsp90 inhibitor ganetespib and also found no clinical benefit of these regimens in advanced MPNST patients [50]. The combination of pexidartinib (multikinase inhibitor including c-Kit) and sirolimus did show promising results in a phase I trial, including six MPNST patients [52]. Stable disease was achieved in 3/6 patients for at least 18 weeks. Phase II of this trial is still ongoing and will likely include more MPNST patients (Table 18.2).

Table 18.1 Published trials on targeted therapies in MPNST (as per 01-04-2023)

Author, year	Phase	Tumor inclusion	N	Drug	Pathway	Outcome RR	PFS	OS
Albritton, 2006 [44]	II	MPNST	20	Erlotinib	EGFR	1 SD, 19 PD	2 months	4 months
Chugh, 2009 [45]	II	Advanced STS (MPNST)	5	Imatinib	Multikinase (incl. c-Kit)	1 SD, 4 PD	*NA*	*NA*
Maki, 2009 [46]	II	Advanced STS (MPNST)	12	Sorafenib	Multikinase (incl. VEGFR)	3 SD, 9 PD	1.7 months	4.9 months
Schuetze, 2016 [47]	II	Advanced STS (MPNST)	14	Dasatinib	Multikinase (incl. BCR/ABL)	14 PD	2-month: 14% 4-month: 7%	*NA*
Widemann, 2013 [48]	II	Advanced STS (MPNST)	25	Everolimus + bevacizumab	mTOR, VEGF	3 SD, 22 PD	*NA*	*NA*
Dickson, 2016 [49]	II	Advanced STS (MPNST)	10	Alisertib	AURKA	No response (SD and PD)	13 weeks 12-week: 60%	69 weeks
Kim, 2020 [50]	II	MPNST	10	Sirolimus + ganetespib	mTOR, Hsp90	1 SD, 9 PD	*NA*	*NA*
Nishida, 2021 [51]	II	MPNST	12	Pazopanib	Multikinase (incl. VEGFR)	1 PR, 5 SD, 6 PD	5.4 months	10.6 months
Manji, 2021 [52]	I	Advance STS (MPNST)	6	Sirolimus + pexidartinib	mTOR, CSF-1R	3SD, 3PD	19 weeks	145 weeks

AURKA aurora kinase A, *CR* complete response, *CSF1-R* colony-stimulating factor-1 receptor, *EGFR* endothelial growth factor receptor, *MPNST* malignant peripheral nerve sheath tumor, *mTOR* mammalian target of rapamycin, *N* number of MPNST patients included, *NA* not available, *OS* overall survival, *PD* progressive disease, *PFS* progression-free survival, *PR* partial response, *RR* response rate according to RECIST criteria, *SD* stable disease, *VEGFR* vascular endothelial growth factor receptor

Table 18.2 Ongoing trials of targeted therapies in MPNST (as per 01-04-2023)

Study	Phase	Tumor inclusion	Treatment	Target	N
NCT04222413	I	Advanced solid tumors (MPNST)	Metarrestin	RNA polymerase	64
NCT05245500	I/II	MTAP deleted tumors (MPNST)	MRTX1719	PRMT5	339
NCT05275478	I/II	MTAP deleted tumors (MPNST)	TNG908	PRMT5	170
NCT03433183	II	MPNST	Sirolimus + selumetinib	mTOR, MEK	21
NCT02601209	II	Advanced STS (MPNST)	Sapanisertib +/− pazopanib	mTOR, multikinase	137
NCT02584647	II	STS (MPNST)	Pexidartinib + sirolimus	Multikinase, mTOR	49
NCT03009201	I	Advanced STS (MPNST)	Doxorubicin + ribociclib	CDK4/6	36

CDK cyclin-dependent kinase, *MPNST* malignant peripheral nerve sheath tumor, *MTAP* methylthioadenosine phosphorylase, *mTOR* mammalian target of rapamycin, *N* number, *PRMT5* protein arginine methyltransferase 5, *STS* soft tissue sarcoma

Cellular targets other than Ras pathway proteins and RTKs have been studied as well. The nuclear target of Aurora kinase A (AURKA), an epigenetic regulator, showed promising results in vivo but failed to show efficacy in humans. No responses were observed and median PFS was 13 weeks in STS [49]. One study evaluating the effect of selinexor combined data from two trials and several patients receiving selinexor as single-patient use, resulting in outcomes of nine MPNST patients, two of which received additional doxorubicin [58]. Stable disease was found in four patients and partial responses in three. One NF1-associated MPNST patient receiving selinexor monotherapy showed stable disease and remained stable for over than 13 months. Recently, deletion of methylthioadenosine phosphorylase (*MTAP*) has been proven to exist in 25–50% of MPNSTs [59, 60]. *MTAP* is frequently lost due to its close proximity to the commonly lost *CDKN2A* gene. In vitro studies show that loss of *MTAP* function leads to increased PRMT5 activity which is an actionable target. This has led to two independent trials currently ongoing that study the effect of PRMT5 inhibition on solid tumors including MPNST. Several other targeted drugs, including CDK4/6 and HDAC inhibitors, show promising preclinical results but still have to be tested in clinical trials [41].

18.4 Immunotherapy and Oncolytic Viruses

Immunotherapy has increasingly been popularized since its success in tumors like melanoma, renal cell cancer, and lung cancer. Unfortunately, several trials among pooled sets of STS patients have not shown efficacy overall [61–63]. Different reasons may underlie these results, including variability in immune tumor microenvironment (TME) between different STS subtypes as well as in STS subtypes themselves. Data on the TME of MPNST is scarce and conflicting. Inactivation of polycomb-repressive complex 2 (PRC2) is a prevalent genetic event in MPNST and is associated with an immune desert TME [64]. While PD-L1/PD1 expression is low or even absent in many MPNSTs, they have been found to show increased infiltrating cytotoxic T cells and low regulatory T cells, which could make them responsive to immune checkpoint inhibition [65, 66]. However, like other targeted therapies, the biological heterogeneity of MPNSTs makes it challenging [65]. Clinical data is still scarce, but some case reports have been published showing promising effects of immunotherapy in selected cases. A case report of a 22-year-old male with metachronous metastatic disease was treated successfully with pembrolizumab of a PD-L1 positive primary tumor [67]. In a phase I trial investigating the effect of pembrolizumab including patients with advanced solid tumors, one patient with an MPNST showed stable disease [68]. Further data is still lacking, and the recent SARC028 trial evaluating the effect of pembrolizumab in advanced sarcoma did eventually not include any MPNST patients. Currently, several trials are still being done investigating the effect of immunotherapy in MPNST (Table 18.3). One study investigating the effect of pembrolizumab was unfortunately terminated because of slow enrolment during the COVID pandemic (NCT02691026).

Oncolytic viruses may also be used as vectors for treating cancer, affecting the immune system in several ways. In MPNST, several in vivo studies have investigated the effect of oncolytic viruses showing promising results. Most studies used oncolytic herpes viruses and showed significant effect on tumor growth and survival [69–75]. The addition of other targeted therapies was studied for erlotinib, an EGFR

Table 18.3 Ongoing trials of immunotherapy and oncolytic virus therapy in MPNST (as per 01-04-2023)

Study	Phase	Tumor inclusion	Treatment	Target	N
NCT03611868	I/II	Advanced melanoma, solid tumors, and MPNST	Pembrolizumab + APG-115	PD-1, MDM2	224
NCT02834013	II	Rare tumors (MPNST)	Nivolumab + ipilimumab	PD-1, CTLA4	707
NCT04465643	II	MPNST	Nivolumab + ipilimumab	PD-1, CTLA4	18
NCT04551430	II	Metastatic STS (MPNST)	Cabozantinib + nivolumab + ipilimumab	Multikinase, PD-1, CTLA4	105
NCT04897321	I	Advanced solid tumors (MPNST)	CAR-T cells	B7H3	32
NCT04483778	I	Advanced solid tumors (MPNST)	CAR-T cells	B7H3	68
NCT03618381	I	Advanced solid tumors (MPNST)	CAR-T cells	EGFR	36
NCT05642455	I/II	Advanced solid tumors (MPNST)	Afamitresgene autoleucel	MAGE A4	20
NCT02452554	II	CD56 expressing tumors (MPNST)	Lorvotuzumab mertansine	CD56	65
NCT03719430	II	Advanced STS (MPNST)	Doxorubicin + APX005M	CD40	32
NCT02700230	I	MPNST	MV-NIS	oMV	30
NCT00931931	I	Advanced solid tumors (MPNST)	HSV1716	oHSV	18

EGFR endothelial growth factor receptor, *MPNST* malignant peripheral nerve sheath tumor, *mTOR* mammalian target of rapamycin, *N* number, *oHSV* oncolytic herpes simplex virus, *oMV* oncolytic measles virus, *STS* soft tissue sarcoma inhibitor, which did not result in improved survival in their in vivo model [71]. The addition of an AURKA inhibitor did show a synergistic effect on survival and tumor growth in another study [75]. An ongoing phase I study is still investigating the role of intratumoral administration of an oncolytic measles virus to express NF1 in operable or recurrent MPNST patients (NCT02700230). In the absence of unclear subtypes of MPNST, a combination of either cytotoxic, target therapies, or immunotherapies will likely have the highest efficacy of treatment, albeit possible increased toxicities.

References

1. Valentin T, Le Cesne A, Ray-Coquard I, et al. Management and prognosis of malignant peripheral nerve sheath tumors: the experience of the French Sarcoma Group (GSF-GETO). Eur J Cancer. 2016;56:77–84. https://doi.org/10.1016/j.ejca.2015.12.015.
2. Martin E, Coert JH, Flucke UE, et al. A nationwide cohort study on treatment and survival in patients with malignant peripheral nerve sheath tumours. Eur J Cancer. 2019;124:77–87. https://doi.org/10.1016/j.ejca.2019.10.014.
3. Basu TN, Gutmann DH, Fletcher JA, Glover TW, Collins FS, Downward J. Aberrant regulation of ras proteins in malignant tumour cells from type 1 neurofibromatosis patients. Nature. 1992;356(6371):713–5. https://doi.org/10.1038/356713a0.
4. Endo M, Yamamoto H, Setsu N, et al. Prognostic significance of AKT/mTOR and MAPK pathways and antitumor effect of mTOR inhibitor in NF1-related and sporadic malignant peripheral nerve sheath tumors. Clin Cancer Res. 2013;19(2):450–61. https://doi.org/10.1158/1078-0432.CCR-12-1067.
5. De Raedt T, Beert E, Pasmant E, et al. PRC2 loss amplifies Ras-driven transcription and confers sensitivity to BRD4-based therapies. Nature. 2014;514(7521):247–51. https://doi.org/10.1038/nature13561.
6. Beert E, Brems H, Daniels B, et al. Atypical neurofibromas in neurofibromatosis type 1 are premalignant tumors. Genes Chromosomes Cancer. 2011;50(12):1021–32. https://doi.org/10.1002/gcc.20921.
7. Legius E, Dierick H, Wu R, et al. TP53 mutations are frequent in malignant NF1 tumors. Genes Chromosomes Cancer. 1994;10(4):250–5.
8. Upadhyaya M, Spurlock G, Thomas L, et al. Microarray-based copy number analysis of neurofibromatosis type-1 (NF1)-associated malignant peripheral nerve sheath tumors reveals a role for Rho-GTPase pathway genes in NF1 tumorigenesis. Hum Mutat. 2012;33(4):763–76. https://doi.org/10.1002/humu.22044.
9. Masliah-Planchon J, Pasmant E, Luscan A, et al. MicroRNAome profiling in benign and malignant neurofibromatosis type 1-associated nerve sheath tumors: evidences of PTEN pathway alterations in early NF1 tumorigenesis. BMC Genomics. 2013;14:473. https://doi.org/10.1186/1471-2164-14-473.
10. Katz D, Lazar A, Lev D. Malignant peripheral nerve sheath tumour (MPNST): the clinical implications of cellular signalling pathways. Expert Rev Mol Med. 2009;11:e30. https://doi.org/10.1017/S1462399409001227.
11. Acem I, Martin E, van Houdt WJ, et al. The association of metastasis pattern and management of metastatic disease with oncological outcomes in patients with malignant peripheral nerve sheath tumors: a multicenter cohort study. Cancers (Basel). 2021;13(20):5115. https://doi.org/10.3390/cancers13205115.
12. Anghileri M, Miceli R, Fiore M, et al. Malignant peripheral nerve sheath tumors: prognostic factors and survival in a series of patients treated at a single institution. Cancer. 2006;107(5):1065–74. https://doi.org/10.1002/cncr.22098.
13. Wong WW, Hirose T, Scheithauer BW, Schild SE, Gunderson LL. Malignant peripheral nerve sheath tumor: analysis of treatment outcome. Int J Radiat Oncol Biol Phys. 1998;42(2):351–60.
14. Zehou O, Bularca S, Bastuji-Garin S, et al. Neurofibromatosis 1 phenotype associated to malignant peripheral nerve sheath tumours: a case-control study. J Eur Acad Dermatol Venereol. 2013;27(8):1044–7. https://doi.org/10.1111/j.1468-3083.2012.04485.x.
15. Stucky C-CH, Johnson KN, Gray RJ, et al. Malignant peripheral nerve sheath tumors (MPNST): the Mayo Clinic experience. Ann Surg Oncol. 2012;19(3):878–85. https://doi.org/10.1245/s10434-011-1978-7.
16. Miao R, Wang H, Jacobson A, et al. Radiation-induced and neurofibromatosis-associated malignant peripheral nerve sheath tumors (MPNST) have worse outcomes than sporadic MPNST. Radiother Oncol. 2019;137:61–70. https://doi.org/10.1016/j.radonc.2019.03.015.
17. Zou C, Smith KD, Liu J, et al. Clinical, pathological, and molecular variables predictive of malignant peripheral nerve sheath tumor outcome. Ann Surg. 2009;249(6):1014–22. https://doi.org/10.1097/SLA.0b013e3181a77e9a.
18. Lochner J, Menge F, Vassos N, Hohenberger P, Kasper B. Prognosis of patients with metastatic soft tissue sarcoma: advances in recent years. Oncol Res Treat. 2020;43(11):613–9. https://doi.org/10.1159/000509519.
19. Gronchi A, Miah AB, Dei Tos AP, et al. Soft tissue and visceral sarcomas: ESMO-EURACAN-GENTURIS Clinical Practice Guidelines for diagnosis, treatment

and follow-up. Ann Oncol. 2021;32(11):1348–65. https://doi.org/10.1016/j.annonc.2021.07.006.
20. van der Graaf WTA, Blay J-Y, Chawla SP, et al. Pazopanib for metastatic soft-tissue sarcoma (PALETTE): a randomised, double-blind, placebo-controlled phase 3 trial. Lancet. 2012;379(9829):1879–86. https://doi.org/10.1016/S0140-6736(12)60651-5.
21. van Noesel MM, Orbach D, Brennan B, et al. Outcome and prognostic factors in pediatric malignant peripheral nerve sheath tumors: an analysis of the European Pediatric Soft Tissue Sarcoma Group (EpSSG) NRSTS-2005 prospective study. Pediatr Blood Cancer. 2019;66(10):e27833. https://doi.org/10.1002/pbc.27833.
22. Carli M, Ferrari A, Mattke A, et al. Pediatric malignant peripheral nerve sheath tumor: the Italian and German soft tissue sarcoma cooperative group. J Clin Oncol. 2005;23(33):8422–30. https://doi.org/10.1200/JCO.2005.01.4886.
23. Ferrari A, Miceli R, Rey A, et al. Non-metastatic unresected paediatric non-rhabdomyosarcoma soft tissue sarcomas: results of a pooled analysis from United States and European groups. Eur J Cancer. 2011;47(5):724–31. https://doi.org/10.1016/j.ejca.2010.11.013.
24. Kroep JR, Ouali M, Gelderblom H, et al. First-line chemotherapy for malignant peripheral nerve sheath tumor (MPNST) versus other histological soft tissue sarcoma subtypes and as a prognostic factor for MPNST: an EORTC soft tissue and bone sarcoma group study. Ann Oncol. 2011;22(1):207–14. https://doi.org/10.1093/annonc/mdq338.
25. Higham CS, Steinberg SM, Dombi E, et al. SARC006: phase II trial of chemotherapy in sporadic and neurofibromatosis type 1 associated chemotherapy-naive malignant peripheral nerve sheath tumors. Sarcoma. 2017;2017:8685638. https://doi.org/10.1155/2017/8685638.
26. Judson I, Verweij J, Gelderblom H, et al. Doxorubicin alone versus intensified doxorubicin plus ifosfamide for first-line treatment of advanced or metastatic soft-tissue sarcoma: a randomised controlled phase 3 trial. Lancet Oncol. 2014;15(4):415–23. https://doi.org/10.1016/S1470-2045(14)70063-4.
27. Gronchi A, Ferrari S, Quagliuolo V, et al. Histotype-tailored neoadjuvant chemotherapy versus standard chemotherapy in patients with high-risk soft-tissue sarcomas (ISG-STS 1001): an international, open-label, randomised, controlled, phase 3, multicentre trial. Lancet Oncol. 2017;18(6):812–22. https://doi.org/10.1016/S1470-2045(17)30334-0.
28. Sarcoma Meta-analysis Collaboration (SMAC). Adjuvant chemotherapy for localised resectable soft tissue sarcoma in adults. Cochrane Database Syst Rev. 2000;(4):CD001419. https://doi.org/10.1002/14651858.CD001419.
29. Pervaiz N, Colterjohn N, Farrokhyar F, Tozer R, Figueredo A, Ghert M. A systematic meta-analysis of randomized controlled trials of adjuvant chemotherapy for localized resectable soft-tissue sarcoma. Cancer. 2008;113(3):573–81. https://doi.org/10.1002/cncr.23592.
30. Gortzak E, Azzarelli A, Buesa J, et al. A randomised phase II study on neo-adjuvant chemotherapy for "high-risk" adult soft-tissue sarcoma. Eur J Cancer. 2001;37(9):1096–103.
31. Woll PJ, Reichardt P, Le Cesne A, et al. Adjuvant chemotherapy with doxorubicin, ifosfamide, and lenograstim for resected soft-tissue sarcoma (EORTC 62931): a multicentre randomised controlled trial. Lancet Oncol. 2012;13(10):1045–54. https://doi.org/10.1016/S1470-2045(12)70346-7.
32. Pasquali S, Pizzamiglio S, Touati N, et al. The impact of chemotherapy on survival of patients with extremity and trunk wall soft tissue sarcoma: revisiting the results of the EORTC-STBSG 62931 randomised trial. Eur J Cancer. 2019;109:51–60. https://doi.org/10.1016/j.ejca.2018.12.009.
33. Martin E, Slooff W-BM, van Houdt WJ, van Dalen T, Verhoef C, Coert JH. Oncological treatment considerations differ across surgical subspecialties treating malignant peripheral nerve sheath tumors: an international survey. Sarcoma. 2020;2020:6406439. https://doi.org/10.1155/2020/6406439.
34. Bonvalot S, Gronchi A, Le Péchoux C, et al. Preoperative radiotherapy plus surgery versus surgery alone for patients with primary retroperitoneal sarcoma (EORTC-62092: STRASS): a multicentre, open-label, randomised, phase 3 trial. Lancet Oncol. 2020;21(10):1366–77. https://doi.org/10.1016/S1470-2045(20)30446-0.
35. Spunt SL, Hill DA, Motosue AM, et al. Clinical features and outcome of initially unresected nonmetastatic pediatric nonrhabdomyosarcoma soft tissue sarcoma. J Clin Oncol. 2002;20(15):3225–35. https://doi.org/10.1200/JCO.2002.06.066.
36. PDQ Pediatric Treatment Editorial Board. Childhood soft tissue sarcoma treatment (PDQ®): health professional version. Bethesda (MD): National Cancer Institute (US); 2019.
37. Ferrari A, Bisogno G, Macaluso A, et al. Soft-tissue sarcomas in children and adolescents with neurofibromatosis type 1. Cancer. 2007;109(7):1406–12. https://doi.org/10.1002/cncr.22533.
38. Martin E, Coert JH, Flucke UE, et al. Neurofibromatosis-associated malignant peripheral nerve sheath tumors in children have a worse prognosis: a nationwide cohort study. Pediatr Blood Cancer. 2020;67:e28138. https://doi.org/10.1002/pbc.28138.
39. Acem I, van Houdt WJ, Grünhagen DJ, et al. The role of perioperative chemotherapy in primary high-grade extremity soft tissue sarcoma: a risk-stratified analysis using PERSARC. Eur J Cancer. 2022;165:71–80. https://doi.org/10.1016/j.ejca.2022.01.013.
40. Voss RK, Callegaro D, Chiang Y-J, et al. Sarculator is a good model to predict survival in resected extremity and trunk sarcomas in US patients. Ann Surg Oncol. 2022; https://doi.org/10.1245/s10434-022-11442-2.

41. Martin E, Lamba N, Flucke UE, et al. Non-cytotoxic systemic treatment in malignant peripheral nerve sheath tumors (MPNST): a systematic review from bench to bedside. Crit Rev Oncol Hematol. 2019;138:223–32. https://doi.org/10.1016/j.critrevonc.2019.04.007.
42. Wasa J, Nishida Y, Suzuki Y, et al. Differential expression of angiogenic factors in peripheral nerve sheath tumors. Clin Exp Metastasis. 2008;25(7):819–25. https://doi.org/10.1007/s10585-008-9197-8.
43. de Vasconcelos RAT, Guimarães Coscarelli P, Vieira TM, Noguera WS, Rapozo DCM, Acioly MA. Prognostic significance of mast cell and microvascular densities in malignant peripheral nerve sheath tumor with and without neurofibromatosis type 1. Cancer Med. 2019;8(3):972–81. https://doi.org/10.1002/cam4.1977.
44. Albritton KH, Rankin C, Coffin CM, et al. Phase II study of erlotinib in metastatic or unresectable malignant peripheral nerve sheath tumors (MPNST). J Clin Oncol. 2006;24(18_suppl):9518. https://doi.org/10.1200/jco.2006.24.18_suppl.9518.
45. Chugh R, Wathen JK, Maki RG, et al. Phase II multicenter trial of imatinib in 10 histologic subtypes of sarcoma using a bayesian hierarchical statistical model. J Clin Oncol. 2009;27(19):3148–53. https://doi.org/10.1200/JCO.2008.20.5054.
46. Maki RG, D'Adamo DR, Keohan ML, et al. Phase II study of sorafenib in patients with metastatic or recurrent sarcomas. J Clin Oncol. 2009;27(19):3133–40. https://doi.org/10.1200/JCO.2008.20.4495.
47. Schuetze SM, Wathen JK, Lucas DR, et al. SARC009: phase 2 study of dasatinib in patients with previously treated, high-grade, advanced sarcoma. Cancer. 2016;122(6):868–74. https://doi.org/10.1002/cncr.29858.
48. Widemann BC, Reinke DK, Helman LJ, et al. SARC006: phase II trial of chemotherapy in sporadic and neurofibromatosis type 1 (NF1)-associated high-grade malignant peripheral nerve sheath tumors (MPNSTs). J Clin Oncol. 2013;31(15_suppl):10522. https://doi.org/10.1200/jco.2013.31.15_suppl.10522.
49. Dickson MA, Mahoney MR, Tap WD, et al. Phase II study of MLN8237 (Alisertib) in advanced/metastatic sarcoma. Ann Oncol. 2016;27(10):1855–60. https://doi.org/10.1093/annonc/mdw281.
50. Kim A, Lu Y, Okuno SH, et al. Targeting refractory sarcomas and malignant peripheral nerve sheath tumors in a phase I/II study of sirolimus in combination with ganetespib (SARC023). Sarcoma. 2020;2020:5784876. https://doi.org/10.1155/2020/5784876.
51. Nishida Y, Urakawa H, Nakayama R, et al. Phase II clinical trial of pazopanib for patients with unresectable or metastatic malignant peripheral nerve sheath tumors. Int J Cancer. 2021;148(1):140–9. https://doi.org/10.1002/ijc.33201.
52. Manji GA, Van Tine BA, Lee SM, et al. A phase I study of the combination of pexidartinib and sirolimus to target tumor-associated macrophages in unresectable sarcoma and malignant peripheral nerve sheath tumors. Clin Cancer Res. 2021;27(20):5519–27. https://doi.org/10.1158/1078-0432.CCR-21-1779.
53. Yoo KH, Kim HS, Lee SJ, et al. Efficacy of pazopanib monotherapy in patients who had been heavily pretreated for metastatic soft tissue sarcoma: a retrospective case series. BMC Cancer. 2015;15:154. https://doi.org/10.1186/s12885-015-1160-x.
54. Weiss AR, Chen Y-L, Scharschmidt TJ, et al. Pathological response in children and adults with large unresected intermediate-grade or high-grade soft tissue sarcoma receiving preoperative chemoradiotherapy with or without pazopanib (ARST1321): a multicentre, randomised, open-label, phase 2 trial. Lancet Oncol. 2020;21(8):1110–22. https://doi.org/10.1016/S1470-2045(20)30325-9.
55. Dombi E, Baldwin A, Marcus LJ, et al. Activity of selumetinib in neurofibromatosis type 1-related plexiform neurofibromas. N Engl J Med. 2016;375(26):2550–60. https://doi.org/10.1056/NEJMoa1605943.
56. Lock R, Ingraham R, Maertens O, et al. Cotargeting MNK and MEK kinases induces the regression of NF1-mutant cancers. J Clin Invest. 2016;126(6):2181–90. https://doi.org/10.1172/JCI85183.
57. Malone CF, Fromm JA, Maertens O, DeRaedt T, Ingraham R, Cichowski K. Defining key signaling nodes and therapeutic biomarkers in NF1-mutant cancers. Cancer Discov. 2014;4(9):1062–73. https://doi.org/10.1158/2159-8290.CD-14-0159.
58. Al-Ezzi E, Gounder M, Watson G, et al. Selinexor, a first in class, nuclear export inhibitor for the treatment of advanced malignant peripheral nerve sheath tumor. Oncologist. 2021;26(4):e710–4. https://doi.org/10.1002/onco.13692.
59. Kryukov GV, Wilson FH, Ruth JR, et al. MTAP deletion confers enhanced dependency on the PRMT5 arginine methyltransferase in cancer cells. Science. 2016;351(6278):1214–8. https://doi.org/10.1126/science.aad5214.
60. Cortes-Ciriano I, Steele CD, Piculell K, et al. Genomic patterns of malignant peripheral nerve sheath tumor (MPNST) evolution correlate with clinical outcome and are detectable in cell-free DNA. Cancer Discov. 2023;13(3):654–71. https://doi.org/10.1158/2159-8290.CD-22-0786.
61. Tawbi HA, Burgess M, Bolejack V, et al. Pembrolizumab in advanced soft-tissue sarcoma and bone sarcoma (SARC028): a multicentre, two-cohort, single-arm, open-label, phase 2 trial. Lancet Oncol. 2017;18(11):1493–501. https://doi.org/10.1016/S1470-2045(17)30624-1.
62. D'Angelo SP, Mahoney MR, Van Tine BA, et al. Nivolumab with or without ipilimumab treatment for metastatic sarcoma (Alliance A091401): two

open-label, non-comparative, randomised, phase 2 trials. Lancet Oncol. 2018;19(3):416–26. https://doi.org/10.1016/S1470-2045(18)30006-8.
63. Maki RG, Jungbluth AA, Gnjatic S, et al. A pilot study of anti-CTLA4 antibody ipilimumab in patients with synovial sarcoma. Sarcoma. 2013;2013:168145. https://doi.org/10.1155/2013/168145.
64. Yan J, Chen Y, Patel AJ, et al. Tumor-intrinsic PRC2 inactivation drives a context-dependent immune-desert microenvironment and is sensitized by immunogenic viruses. J Clin Invest. 2022;132(17):e153437. https://doi.org/10.1172/JCI153437.
65. Haworth KB, Arnold MA, Pierson CR, et al. Immune profiling of NF1-associated tumors reveals histologic subtype distinctions and heterogeneity: implications for immunotherapy. Oncotarget. 2017;8(47):82037–48. https://doi.org/10.18632/oncotarget.18301.
66. Shurell E, Singh AS, Crompton JG, et al. Characterizing the immune microenvironment of malignant peripheral nerve sheath tumor by PD-L1 expression and presence of CD8+ tumor infiltrating lymphocytes. Oncotarget. 2016;7(39):64300–8. https://doi.org/10.18632/oncotarget.11734.
67. Davis LE, Nicholls LA, Babiker HM, Liau J, Mahadevan D. PD-1 inhibition achieves a complete metabolic response in a patient with malignant peripheral nerve sheath tumor. Cancer Immunol Res. 2019;7(9):1396–400. https://doi.org/10.1158/2326-6066.CIR-19-0072.
68. Patnaik A, Kang SP, Rasco D, et al. Phase I study of pembrolizumab (MK-3475; anti-PD-1 monoclonal antibody) in patients with advanced solid tumors. Clin Cancer Res. 2015;21(19):4286–93. https://doi.org/10.1158/1078-0432.CCR-14-2607.
69. Liu T-C, Zhang T, Fukuhara H, et al. Oncolytic HSV armed with platelet factor 4, an antiangiogenic agent, shows enhanced efficacy. Mol Ther. 2006;14(6):789–97. https://doi.org/10.1016/j.ymthe.2006.07.011.
70. Liu T-C, Zhang T, Fukuhara H, et al. Dominant-negative fibroblast growth factor receptor expression enhances antitumoral potency of oncolytic herpes simplex virus in neural tumors. Clin Cancer Res. 2006;12(22):6791–9. https://doi.org/10.1158/1078-0432.CCR-06-0263.
71. Mahller YY, Vaikunth SS, Currier MA, et al. Oncolytic HSV and erlotinib inhibit tumor growth and angiogenesis in a novel malignant peripheral nerve sheath tumor xenograft model. Mol Ther. 2007;15(2):279–86. https://doi.org/10.1038/sj.mt.6300038.
72. Mahller YY, Sakthivel B, Baird WH, et al. Molecular analysis of human cancer cells infected by an oncolytic HSV-1 reveals multiple upregulated cellular genes and a role for SOCS1 in virus replication. Cancer Gene Ther. 2008;15(11):733–41. https://doi.org/10.1038/cgt.2008.40.
73. Maldonado AR, Klanke C, Jegga AG, et al. Molecular engineering and validation of an oncolytic herpes simplex virus type 1 transcriptionally targeted to midkine-positive tumors. J Gene Med. 2010;12(7):613–23. https://doi.org/10.1002/jgm.1479.
74. Antoszczyk S, Spyra M, Mautner VF, et al. Treatment of orthotopic malignant peripheral nerve sheath tumors with oncolytic herpes simplex virus. Neuro-Oncology. 2014;16(8):1057–66. https://doi.org/10.1093/neuonc/not317.
75. Currier MA, Sprague L, Rizvi TA, et al. Aurora A kinase inhibition enhances oncolytic herpes virotherapy through cytotoxic synergy and innate cellular immune modulation. Oncotarget. 2017;8(11):17412–27. https://doi.org/10.18632/oncotarget.14885.

19 Proton and Carbon Ion Therapy for MPNST

Maria Rosaria Fiore, Agnieszka Chalaszczyk, and Ester Orlandi

19.1 Introduction

Malignant peripheral nerve sheath tumors (MPNSTs) are rare and aggressive tumors that arise from peripheral nerves. They account for 5–10% of all soft tissue sarcomas [1] and are often associated with the genetic disorder neurofibromatosis type 1 (NF-1) [2]. NF-1 disorder is found in approximately 50% of MPNST cases, while 10% of all MPNSTs are radiation-induced, and the remaining 40% occur sporadically [3]. Although about 10–20% of all MPNSTs are diagnosed in children, there is no difference between children and adults in terms of tumor location, size, or histological grade [4]. Common characteristics of MPNSTs include a locally aggressive growth pattern, high local recurrence rates, and poor prognosis. High rates of distant metastasis are also reported [5].

Surgical tumor resection with wide margins remains the main therapy, despite the typical proximity to structures deputed to relevant function, not always allowing complete resection. MPNSTs can develop anywhere in the body but mainly occur at the location of major nerve plexuses/roots. The extremities are most commonly involved, followed by the trunk and head and neck region. MPNSTs of the extremities have superior outcomes compared to other disease sites, most likely due to an enhanced ability to achieve complete resection and improved tumor control [6]. The most common site of origin for MPNSTs is the brachial plexus, where complete resection with negative margins may result in unacceptable morbidity. In patients with head and neck MPNSTs, achieving negative margins is extremely difficult due to important critical structures (i.e., brainstem, vascular structures). This likely contributes to the poorer outcome seen in these patients [7]. The overall poor prognosis has also been demonstrated for spinal intradural sporadic MPNSTs, which are different from the usual spinal intradural tumors even after surgical treatment and adjuvant treatment [8].

The therapeutic multimodality strategy is still a challenging multidisciplinary approach. Both chemotherapy and radiotherapy are widely used for the clinical management of MPNSTs [9, 10]. Radiotherapy is mainly considered as part of the local treatment of these tumors due to their predominant pattern of local recurrence, especially for patients with positive margins where the local recurrence rate is higher. In the case of macroscopic residual disease, voluminous tumors not suitable for surgery, or local recurrence, radiotherapy also has a role. Despite the rarity of MPNSTs, several retrospective studies recommend high doses of radiotherapy to improve local

M. R. Fiore (✉) · A. Chalaszczyk · E. Orlandi ·
Centro Nazionale di Adroterapia Oncologica,
Pavia, Italy
e-mail: mariarosaria.fiore@cnao.it;
agnieszka.chalaszczyk@cnao.it;
ester.orlandi@cnao.it;

I. G. Vetrano, V. Nazzi (eds.), *Peripheral Nerve Sheath Tumors*,
https://doi.org/10.1007/978-3-031-76382-3_19

control. The main indications are for patients with large volumes (e.g., >5 cm), high-grade tumors, and/or positive margins at resection. However, while the risk of local recurrence is reduced, the benefit on overall survival is unclear. The median cumulative radiation doses needed for therapeutic efficacy are also quite high—often well in excess of 50 Gray (Gy) [11, 12].

The possibility to deliver the high doses is often inadequate due to the low radiation dose tolerance of critical structures, particularly for the brainstem, spinal cord, nerve roots, and bowel, limiting the high dose required. More recently, advances in RT techniques have allowed for safe delivery of higher RT doses than in the past. Among the newer radiotherapy techniques, proton and carbon ion radiotherapy, also called particle therapy, have aroused major interest in the last decades due to their intrinsic physical selectivity, allowing for the delivery of a high dose to the tumor while sparing critical tissues, even if they are closely adjacent to the tumor, as typically found in MPNST. Furthermore, carbon ion has a superior radiobiological efficacy, which is theoretically more effective on radioresistant tumors.

19.2 Role of Radiation Therapy (RT): Overview

19.2.1 Photon Radiation Therapy

The common indication for RT is in the adjuvant setting to improve local control after incomplete resection. Most of the available studies have been conducted using photon RT in high-volume sarcoma centers. Many of these have recommended the use of radiation therapy in an adjuvant setting, although to date only one single-institution series from Milan has shown a statistically significant increase in local control with radiation therapy. Anghileri et al. analyzed 205 patients with localized disease, primary or recurrent, who underwent surgery; 91 patients received external beam RT with doses ranging from 45 grays (Gy) to 65 Gy. They found recurrent disease, head and neck or trunk location, large tumor size, positive margins, and lack of RT to be unfavorable factors. Margin status and radiation therapy played a role, mostly related to their effect on local outcome. They recommended adjuvant radiation therapy to improve local control, which may be beneficial for survival too [11].

The first large retrospective series reporting a significant impact on local control, by Wong in 1998, analyzed 134 patients with MPNST. Seventy-three patients underwent external beam radiation therapy, including 14 who had brachytherapy and 16 who had intraoperative electron irradiation (IOERT). Despite the limitation of this old study, comparing different RT techniques, they found the cumulative radiation dose appeared to be important. The 5-year local control rate was 73% for those whose cumulative radiation dose (external beam, IOERT, and brachytherapy) exceeded 60 Gy, compared with 50% for lesser doses [12].

In 2012, Stucky et al. published an update of this study. They reviewed tumor and treatment characteristics of 175 patients to identify prognostic variables. They found high tumor grade, tumor size >5 cm, and truncal location to be the most significant adverse disease-specific survival factors for MPNST. The local recurrence rate was better than previously reported by Wong, 22% versus 42%, probably due to an evolution of the surgical techniques achieving negative margin and the efficacy of RT techniques in improving local control. The authors suggested a multimodality treatment for MPNST consisting of surgery with the intent to reach negative margins, combined with radiation therapy for tumors with aggressive characteristics such as high grade, size >5 cm, and R1 margin status [13].

A retrospective series was published by Bishop et al. in 2018 on 71 patients treated with surgery and RT for localized MPNST. Preoperative RT was also used to treat 23 patients to a median dose of 50 Gy (range, 50–60 Gy), while 48 received postoperative RT to a median dose of 64 Gy (range, 45–70 Gy). The 5-year local control and overall survival rates were 84% and 66%, respectively. They found a positive/uncertain surgical margin status associated with worse local recurrence at 5 years (28% vs. 5% for negative margins) and high rates of local control with surgery in combination with RT [14].

One of the biggest series to come out of Massachusetts General Hospital was conducted by Miao et al. They classified 280 patients who were treated for MPNST as neurofibromatosis-associated (nfMPNST, $n = 77$), radiation-induced (rMPNST, $n = 21$), or sporadic (sMPNST, $n = 182$). Of these patients, 169 received pre- and/or postoperative RT to a median dose of 59.6 Gy, while an additional 20 patients had definitive RT alone as local treatment to a median dose of 50.4 Gy. The study found that RT alone and R2 resection were correlated with worse overall survival (OS), while surgery with RT was associated with improved overall survival compared to surgery alone. For patients with positive margins or partial resection, the addition of postoperative RT reduced local recurrence. Additionally, both radiation-induced and neurofibromatosis-associated MPNSTs have a poorer prognosis than sporadic MPNSTs, and complete resection of the tumor was a significant prognostic factor for MPNST, while RT after surgery improves local control [15]. Tumor size, tumor site, primary or recurrent disease, histological grading, margin status, metastatic disease at diagnosis, association with neurofibromatosis type 1 (NF-1), and age are widely considered prognostic factors among studies. Larger tumor size, metastatic disease at diagnosis, intermediate to high grade, and partial resection are commonly identified as significant predictors for poor survival by all authors. The largest series' authors have also considered RT as a local control prognostic factor for patients with nonmetastatic MPNST at the time of diagnosis. Literature suggests that adjuvant RT after surgery could be an effective treatment for patients, especially for those with gross mass larger than 5 cm or with residual tumor, which is critical in the prognosis of head/neck or trunk primary or spinal intradural MPNSTs [16, 17]. The anatomical site has an important impact on the outcomes, and the possibility of aggressive surgery with intent of complete resection with negative margins is strictly related to the anatomic location. In patients with head and neck, trunk, or spine location, complete resection may result in unacceptable toxicity. This is a critical point for RT as well. The margin status after surgery requires adjuvant RT with high dose, where the proximity to critical healthy tissue influences the feasibility and toxicity of RT [18].

19.2.2 Proton Beam and Carbon Ion Radiotherapy

19.2.2.1 Background

High doses of RT are recommended for the treatment of MPNST in the adjuvant setting. However, conventional RT may not be tolerable due to nearby neurological structures, such as the spinal cord and brainstem, which are critical organs with low dose tolerance, especially in the head and neck or axial region. The spinal cord tolerance with conventionally fractionated RT is about 45–54 Gy, and brainstem tolerance is approximately 54–60 Gy [19]. Postoperative RT with doses of more than 60 Gy is necessary due to the radioresistance of these tumors and the difficulty in achieving radical resection.

Among advanced radiation modalities, proton-beam therapy (PBT) and carbon ion radiotherapy (CIRT) have attracted great interest due to their intrinsic physical selectivity. They allow for the delivery of high doses to the tumor while sparing critical tissues, even if they are closely adjacent to the tumor. Proton and carbon ion beams have unique intrinsic selectivity physical properties that deposit energy in tissues differently from photons. After penetrating into the body, proton and carbon ion release a low radiation dose along their pathway, delivering the maximal dose at the end of their range at a specific depth called the Bragg peak. Beyond the Bragg peak, the radiation dose decreases sharply. This pattern of irradiation allows for selective irradiation, delivering an optimal radiation dose to the tumor while exposing critical organs surrounding the tumor to a lower dose. In contrast, photons deposit dose throughout the beam path leading to low-dose around the target volume (see Fig. 19.1) [21].

From a radiobiology standpoint, carbon ion beams have distinctive properties that distinguish them from protons, including their higher radiobiological effectiveness (RBE) and increased lin-

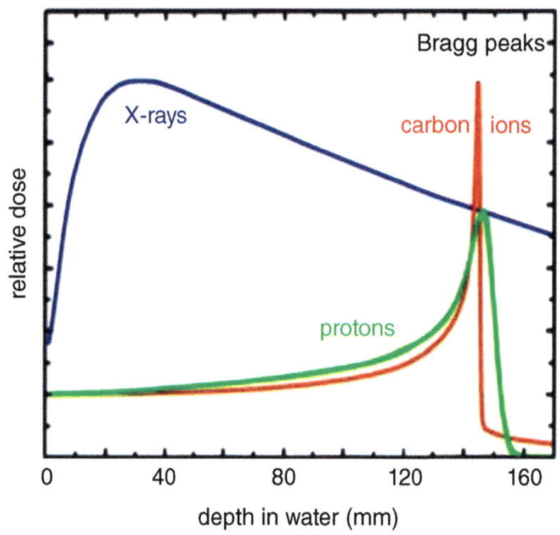

Fig. 19.1 Physical properties of protons and carbon ions in comparison to X-rays. Depth-dose distributions showing the Bragg peak for carbon ions and protons at the same range, photons deposit dose throughout the beam path leading to low dose around the target volume [20]

ear energy transfer (LET). Carbon ion beams deliver a larger mean energy per unit length of their trajectory, (i.e. greater LET), to the body tissues than photon and proton beams. Furthermore, the LET of carbon ion beams increases progressively from the initial value at the entrance point as it passes through the body, reaching its maximal value at the end of its range in the region of the Bragg peak. The high LET in the final path corresponding to a high RBE strongly reduces the repairing of radioinduced damages. Other biological effects of carbon ion, such as the reduced dependence on fractionation and cell-cycle stage, and especially the reduced oxygen enhancement factor, make it particularly indicated for the treatment of radioresistant malignancies (see Fig. 19.2) [20]. Several studies are available on radioresistant tumors treated with carbon ion, which have shown encouraging results in terms of efficacy and toxicity [22–24].

19.2.2.2 Protons and Carbon Ion Radiotherapy for MPNST

Due to the recent emerging interest in particle therapy as a therapeutic strategy, and particularly for the rarity of MPNST, the literature is lacking data about proton and carbon ion radiotherapy in selected and homogeneous series of MPNST patients. Currently, patients with MPNST treated with particle therapy can be found in cohorts of sarcoma patients [25, 26].

Bachmann et al. recently published a retrospective study reporting on 36 patients affected by MPNSTs who received PBT with curative intent between 1999 and 2020. Twenty-eight and eight patients were treated at diagnosis for tumor recurrence or progression, respectively. PBT was delivered in a postoperative (78%), preoperative (14%), and definitive (9%) setting. NF-1 disease was found in eight patients. The median prescribed total dose was 64 Gy RBE (range, 50–74) in 32 fractions. Five-year local control and overall survival were 74% and 54%, respectively. A late grade ≥ 2 toxicity rate was observed in four (11%) patients. There were no observed late grade 4 or 5 toxicities. There was no late grade >3 toxicity or radiation-induced secondary cancer [27].

RT should be used with caution in the context of MPNST. Radiation itself is a well-documented risk factor for MPNST (and other secondary malignancies) in patients with NF1, given the amplified radiation sensitivity of patients with NF1. This can lead to increased side effects from irradiation and a higher risk of secondary tumor induction. The risk of secondary MPNST should be taken into account before patients receive RT for the treatment of the primary lesion [28–31]. With PBT, the dose to organs at risk can be reduced, and hence a lower integral dose is achievable. Considering that, MPNSTs are often in a critical location (i.e., head or spine) and the young age of patients, PBT could be the optional RT treatment given the limited low doses delivered to healthy tissue.

Carbon ion radiotherapy (CIRT) is also being evaluated as a treatment option, mostly for macroscopic residual disease after surgery or unresectable tumors. Clinical experience is still limited compared to proton beam therapy. Generally, the available results of proton radiotherapy are more numerous than those with carbon ion, because more leading centers use PBT, currently operating in the United States, Europe, and Japan. According to the Particle Therapy Cooperative Group (PTCOG), by the last update of January 2023, there are 14 C-ion

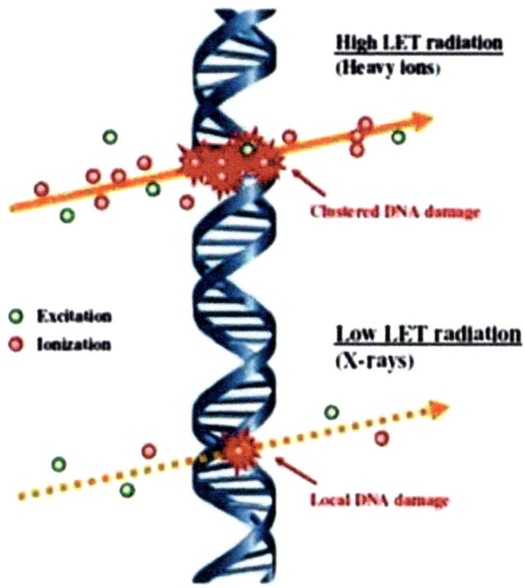

Fig. 19.2 The different quality of DNA damages between carbon ions and X-rays [20]

facilities in clinical operation and more than 100 proton facilities worldwide. The main results about radioresistant tumors come from Japan and Germany, followed by more recent experience from Italy. In a report from *the Working Group for Carbon Ion Radiotherapy for Bone and Soft Tissue Sarcomas for the National Institute of Radiological Sciences* in Japan, data on 128 inoperable soft-tissue sarcomas treated with curative intent with exclusive CIRT are described. In this series, 15 patients had MPNST, and 5-year local control was 52%. Patients were treated with a 16-fraction schedule with four fractions per week over 4 weeks. The prescribed dose ranged from 64 to 73.6 Gy (RBE). Thirty-eight patients had recurrence; in-field recurrence was observed in 25 patients, and 20% and 16% of infield recurrences were MPNST and synovial sarcoma, respectively. The authors concluded with the prospect of dose escalation in selected cases like MPNST and synovial sarcoma [32]. This proposal supports the necessity to deliver higher dose to the tumor together avoiding the lower dose close to critical organs.

Two retrospective studies are available that analyze the results of CIRT for patients with gross residual or unresectable MPNSTs. The first study was conducted by Jensen et al. in 2015 and included 11 patients who received both photons and CIRT (3 patients) or CIRT alone (8 patients). Patients who received combined photon irradiation were given 50 Gy photon intensity-modulated radiation therapy (IMRT) in 25 fractions, followed by a 24 Gy (RBE) boost with C12 irradiation in 8 fractions. Patients who received C12 irradiation were given 60–66 Gy (RBE) in 20–22 fractions. The authors reported a 2-year local control, progression-free survival, and survival rates of 65%, 56%, and 75%, respectively, with a median follow-up of 17 months. Late grade ≥2 toxicity was observed in 2 (15%) patients [33].

Similar results were observed in a retrospective series of 23 patients with unresected MPNSTs treated with definitive CIRT at the National Center for Oncological Hadrontherapy (CNAO). The median dose was 69.8 GyRBE (range, 54-76.8) delivered in a median of 16 fractions (range, 15-22). With a median follow-up of 23 months, the overall survival rates at 1 and 2 years were 82.38% and 61.51%, respectively. The 1-year and 2-year local relapse-free survival rates were 65.07% and 48.80%, respectively, and the 1-year and 2-year progression-free survival rates were 56.37% and 40.99%, respectively. Late grade ≥2 toxicity was observed in 8 patients [34]. When comparing proton therapy (PBT) versus CIRT, the local control rate appears superior in the PBT study. In both CIRT studies, the patients mainly had gross residual tumors, which usually involve healthy tissue or critical structures, forcing under dosage areas, which could explain the lower local control. On the other hand, the rates of toxicities are in line when comparing proton with CIRT. Carbon ion radiotherapy could be more effective in larger masses, achieving higher irradiation doses with fewer complications. The physical features of both particles and the advantageous radiobiological features specific to carbon ions should be taken into account in the therapeutic paradigm of these tumors.

19.3 Technical Aspects of Particle Therapy for MPNST

In 2013, the World Health Organization (WHO) classified MPNST as a soft tissue tumor [35]. Due to the rarity and complexity of these tumors, evaluation and management by a multidisciplinary team with extensive expertise and experience in the treatment of soft tissue sarcoma is particularly important. Although institutional practices may vary, adherence to evidence-based references is recommended, as significant expertise is required for efficacious radiotherapy treatment for these challenging tumors. Prior to starting RT treatment, all patients should undergo a thorough evaluation.

19.3.1 Pretreatment Evaluation

Pretreatment evaluation is the first necessary step, which includes a physical examination to evaluate the positioning and immobilization of the patient in a reproducible condition. In most cases, patients have undergone aggressive surgeries that can impair their performance status, which often affects the positioning modality. Images of the primary lesion are also required. Diagnostic imaging should include preoperative and postoperative magnetic resonance imaging (MRI) to delineate the gross tumor, adjacent areas at risk for microscopic spread, and close critical healthy organs. Additional imaging may be required at the discretion of the radiation oncologist.

19.3.2 Patient Immobilization and Computed Tomography (CT) Simulation

The goal of immobilization is to create a correct and reproducible position of the patient during treatment in order to avoid uncertainties related to inter- and intra-fraction motion. Immobilization is an important step for precise and effective radiotherapy, particularly critical for particle therapy due to the high selectivity of delivering doses. Principles of immobilization to consider are low intra-fraction motion (i.e., rigidity), low inter-fraction motion (i.e., reproducibility), and patient tolerability. Although institutional practices vary, a common approach is to use a customized thermoplastic mask and a shaped pillow in the supine or prone position. The patient should be positioned on the CT table with the shaped pillow under the region of interest, then immobilized with the mask molded to the patient's body shape, preventing movements during treatment (see Fig. 19.3). Custom vacuum bag immobilization can also be considered. The same immobilization device is utilized for the acquisition of CT images and contrast-enhanced MRI for rigid registration. A 2-mm thick slice CT is suggested. The fused images will be utilized for contouring of the volumes. For lesions located in the anterior part of the body, patients are usually positioned in a supine position. For lesions located posteriorly, the patient is usually positioned in a prone position. Oblique supine or prone positions are sometimes considered for lesions requiring a different angle of irradiation (see Fig. 19.3). These particular positions are considered in facilities where the rotating gantry is not available. Most facilities having carbon ion beams are only equipped with fixed beams, vertical, horizontal, or 45-degree oblique. So far, gantries for carbon ion radiotherapy have only been developed by three institutions, such as the University of Heidelberg in Germany, operative since 2012, the National Institute of Radiological

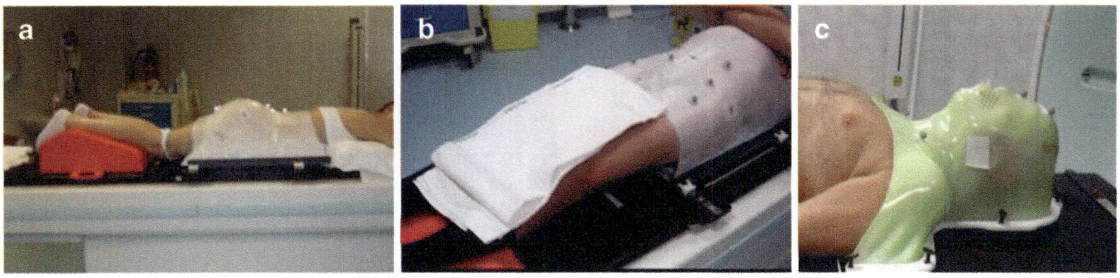

Fig. 19.3 Immobilization on CT table using the pillow and mask: (**a**) prone position, (**b**) supine position, (**c**) head and neck region. Contributed by the National Center for Oncological Hadrontherapy (CNAO)

Sciences in Japan, operative since 2016, and the East Japan Heavy Ion Cancer Therapy in Yamagata University Hospital in Japan, operative since 2022. Using highly rigid and reproducible immobilization requires daily precise setup imaging. Both orthogonal kilovoltage imaging and cone beam CT provide adequate daily setup imaging [36]. A treatment couch with 6 degrees of freedom is helpful to optimize patient alignment.

Additional Effects for CT Simulation in Particle Therapy

- *Organ motion*. The high selectivity of the dose of particle RT is critical in anatomical regions with organ motion. Breathing represents a limiting factor for particle therapy due to intra-fraction organ motion. The magnitude of internal motion resulting from breathing can range from a few millimeters to over 2 cm [37]. As the patient moves during treatment, the spacing between the beams may be no longer as planned, resulting in hot or cold spots in the delivered dose distribution. This effect can be particularly significant for small target volumes and has been shown to cause dose heterogeneity, potentially compromising treatment efficacy [38]. The solution to organ motion in particle beam therapy is to use image guidance techniques such as respiratory gating, breath-hold techniques, or rescanning. These techniques help adjust the particle beam delivery to the position of the tumor, minimizing the dose to surrounding healthy tissues and reducing the risk of treatment-related side effects. Respiratory gating involves monitoring the patient's breathing pattern and synchronizing the beam delivery with the respiratory cycle [39]. Breath-hold techniques involve instructing the patient to hold their breath during treatment delivery [40]. Rescanning aims to reduce the impact of interplay effects by repeating the delivery of the same scanning pattern multiple times, starting at a different breathing phase each time [41].
- *Artifacts*. For patients who have undergone surgery requiring reconstruction with traditional metal implants, which is most frequent in spine tumors, irradiation with particles should be avoided. Traditional metal implants differ significantly from normal tissues in terms of density and composition, leading to high perturbation effects on radiation beams and metal-related artifacts in both magnetic MRI and CT images. Especially for particle radiotherapy, the distortion of dose distribution and range uncertainties may compromise both disease local control and normal tissue sparing. It is suggested that reconstruction with allografts or, where available, carbon fiber when rods are necessary should be preferred at the time of surgery, and rod cross-links at the tumor level should be avoided. Carbon fiber stabilization devices lead to less image alteration and reduced contouring uncertainties, along with significantly higher dosimetric treatment planning accuracy (see Fig. 19.4) [42].
- *Spacer placement*. For patients with bowel strictly close to the tumor or area to irradiate, placement of a spacer prior to RT may allow full coverage of the tumor while reducing the

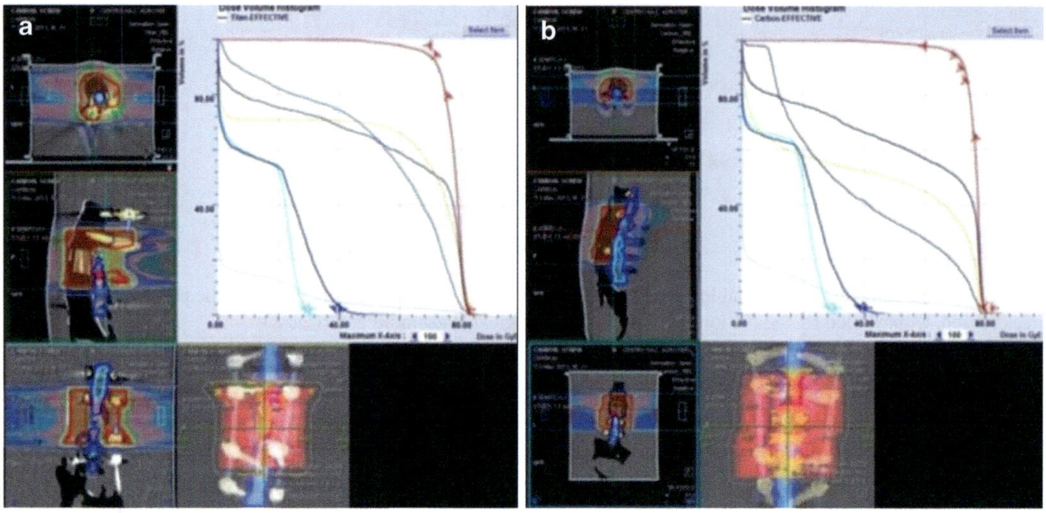

Fig. 19.4 Plan verification in water of irradiation of the vertebral body with a uniform radiation field with titanium (**a**) vs carbon fiber implant (**b**). Comparison of dosimetric calculation efficacy and interference effects. Carbon fiber devices did not determine artifacts on CT images improving image quality and dose calculation accuracy [42]

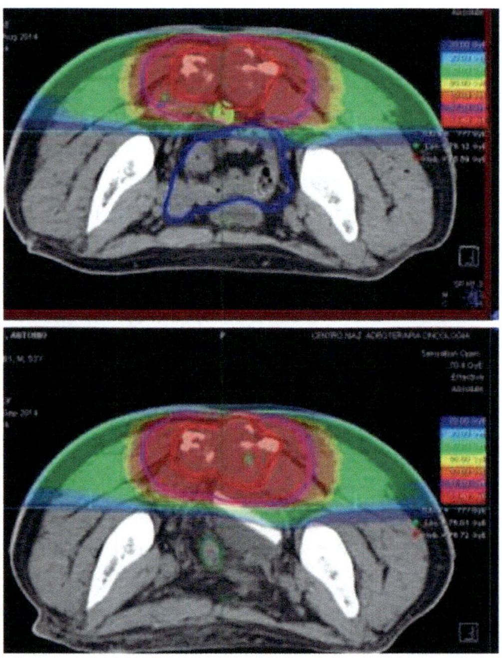

Fig. 19.5 Spacer implant for bowel displacement showing better coverage of the tumor with high dose while sparing bowel [43]

dose to the bowel. This approach may require an additional surgery if it is not previously planned during the therapeutic strategy definition (see Fig. 19.5) [43].

19.3.3 Target Volume and Organ-at-Risk Delineation

According to evidence-based recommendations for soft tissue sarcoma, the definition and delineation of target volumes for the treatment of MPNST is similar among different radiotherapy techniques. Target definition primarily depends on the treatment intent. Table 19.1 shows a summary of target volumes, suggested doses, and particle choices. Gross tumor volume (GTV) represents the detectable tumor mass and is delineated based on the planning CT scan registered with pre- and/or postoperative MRI images. The clinical target volume (CTV) represents the GTV plus an additional margin to capture areas of suspected microscopic spread. The planning target

Table 19.1 Target volumes, doses, and particle indication

Indication	Target	Definition	Dose protons	Dose carbon ions
Definitive unresectable tumors	GTV	Gross tumor delineated on planning CT and MRI		
	CTV-LR	CTV1 = GTV + 1.5 cm radial and 3–4 cm longitudinal with inclusion of peritumoral edema. Volumes should be adjusted to account for the anatomy, tumor spread way, natural barriers, etc.	54 Gy[RBE] 2 Gy[RBE]/fraction; 27 total fractions	39.6–43.2 Gy[RBE] 4.4–4.8 Gy[RBE]/9 fractions
	CTV-HR	CTV2 = GTV + 1.5 cm radial margin and 2 cm longitudinal. Volumes should be adjusted to account for the anatomy, natural barriers, etc.	16–24 Gy[RBE] 2 Gy[RBE]/8–12 fractions Total dose 70–78 Gy[RBE] 35–39 total fractions	30.8–33.6 Gy[RBE] 4.4–4.8 Gy[RBE]/7 fractions Total dose 70.4–76.8 Gy[RBE] 16 total fractions
Surgery negative margins (RO)	Preop-GTV	The position of the prior-to-surgery tumor delineated on preoperative CT/MR fused with planning CT		
	CTV-LR	CTV1 = preop-GTV + 1.5 cm radial and 3–4 cm longitudinal with inclusion of edema. Volumes should be adjusted to account for the anatomy, tumor spread way, natural barriers, etc.	50 Gy[RBE] 2 Gy [RBE]/25 fractions	
	CTV-HR	CTV2 = preop-GTV + 1.5 cm radial margin and 2 cm longitudinal. Volumes should be adjusted to account for the anatomy, natural barriers, etc.	10–16 Gy[RBE] 2 Gy [RBE]/5–8 fractions Total dose 60–66 Gy[RBE] 30–33 total fractions	

(continued)

Table 19.1 (continued)

Indication	Target	Definition	Dose protons	Dose carbon ions
Surgery positive margins (R1)	Preop-GTV	The position of the prior to surgery tumor delineated on preoperative CT/MR fused with planning CT		
	CTV-LR	CTV1 = preop-GTV + 1.5 cm radial and 3–4 cm longitudinal with inclusion of edema. Volumes should be adjusted to account for the anatomy, tumor spread way, natural barriers, etc.	50 Gy[RBE] 2 Gy [RBE] / fraction, 25 total fractions	
	CTV-HR	CTV2 = preop-GTV + 1.5 cm radial margin and 2 cm longitudinal. Volumes should be adjusted to account for the anatomy, natural barriers, etc.	18 Gy[RBE] 2 Gy [RBE]/9fractions Total dose 68 Gy[RBE] 34 total fractions	
Surgery macroscopic residual (R2)	GTV	Gross tumor residual delineated on planning CT and MRI + the position of the prior to surgery tumor delineated on preoperative CT/MR fused with planning CT		
	CTV-LR	CTV1 = GTV + 1.5 cm radial and 3–4 cm longitudinal with inclusion of peritumoral edema. Volumes should be adjusted to account for the anatomy, tumor spread way, natural barriers, etc.	54 Gy[RBE] 2 Gy[RBE]/fraction; 27 total fractions	39.6–43.2 Gy[RBE] 4.4–4.8 Gy[RBE]/9 fractions
	CTV-HR	CTV2 = GTV + 1.5 cm radial margin and 2 cm longitudinal. Volumes should be adjusted to account for the anatomy, natural barriers, etc.	16–24 Gy[RBE] 2 Gy[RBE]/8–12 fraction Total dose 70–78 Gy[RBE] 35–39 total fractions	30.8–33.6 Gy[RBE] 4.4–4.8 Gy[RBE]/7 fraction Total dose 70.4–76.8 Gy[RBE] 16 total fractions

(continued)

Table 19.1 (continued)

Indication	Target	Definition	Dose protons	Dose carbon ions
Relapse after surgery	GTV	Gross tumor relapse delineated on planning CT and MRI + the position of the prior-to-surgery tumor delineated on preoperative CT/MR fused with planning CT		
	CTV-LR	CTV1 = GTV + 1.5 cm radial and 3–4 cm longitudinal with inclusion of peritumoral edema. Volumes should be adjusted to account for the anatomy, tumor spread way, natural barriers, etc.	54 Gy[RBE] 2 Gy[RBE]/fraction; 27 total fractions	39.6–43.2 Gy[RBE] 4.4–4.8 Gy[RBE]/9 fractions
	CTV-HR	CTV2 = GTV + 1.5 cm radial margin and 2 cm longitudinal. Volumes should be adjusted to account for the anatomy, natural barriers, etc.	16–24 Gy[RBE] 2 Gy[RBE]/8–12 fraction Total dose 70–78 Gy[RBE] 35–39 total fractions	30.8–33.6 Gy[RBE] 4.4–4.8 Gy[RBE]/7 fraction Total dose 70.4–76.8 Gy[RBE] 16 total fractions

Abbreviations: *GTV* gross tumor volume, *CTV-LR* clinical target volume-low risk, *CTV-HR* clinical target volume-high risk, *RBE* relative biological effectiveness

volume (PTV) includes the CTV plus an expansion to account for daily variability in patient setup and uncertainty in treatment accuracy.

The majority of RT treatments are performed after complete surgical resection where no GTV is identified. Instead, the position of the primary GTV (preoperative GTV) must be recreated, as realistically as possible, using preoperative MR images rigidly registered with simulation CT. In postoperative treatment with macroscopic residual or disease recurrence, the GTV must be accurately delineated on CT and MRI acquired for treatment planning, taking into account the delineation of the preoperative GTV. The CTV is created from GTV by adding isotropic margins, based on the anatomical location of the primary tumor and further modified based on the possible anatomical spreading pathway. It is crucial to thoroughly examine the surgical and pathology reports to accurately define the CTV and appropriate doses. It is possible to distinguish two different CTVs: CTV-LR (low risk) and CTV-HR (high risk) corresponding to two dose levels. CTV-LR includes preoperative tumor extension and suspected microscopic spread. The margin can be modified manually to exclude natural anatomical barriers (such as fascia, bone, joints, skin, etc.) unless these structures are involved with the tumor. It is important that CTV-LR includes areas at risk of tumor spread, including the path of the affected nerve. CTV-HR is created from GTV in the presence of macroscopic residual after surgery or a volume involving microscopic disease in the tumor bed, receiving the higher dose level. CTV-HR is always included in CTV-LR (see Fig. 19.6). In the case of unresectable tumors, the definitive RT treatment can be planned with both CTV LR and CTV HR, starting with an initial larger volume followed by a boost to the tumor mass with smaller margins. However, clinical judgment may be used in cases of an extremely large volume.

PTV will include CTV. The most commonly used margins for PTV are 3–5 mm. The margins should be reduced at anatomical borders or to spare directly adjacent radiosensitive organs at risk at the discretion of the radiation oncologist. Generally, MPNST is not lymphophilic and does not require prophylactic irradiation of lymph node areas.

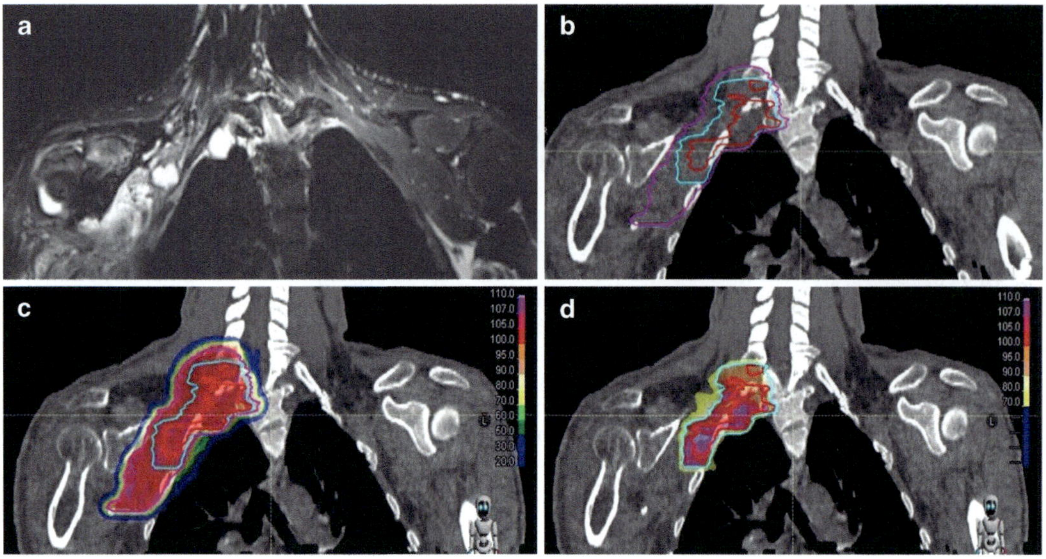

Fig. 19.6 Plan of MPNST treated with post-operative carbon ions RT. A 34-year-old male with high-grade MPNST, NF-1 associated, of brachial plexus with soft tissue component protruding on the right lung apex and perineural spread along the nerves. He underwent postoperative CIRT 76.8 Gy(RBE) in 16 fractions for macroscopic residual disease showing a fast increase after surgery. (**a**) MRI pretreatment showing residual mass. (**b**) Volume description: GTV is shown in red line, CTV HR light blue, CTV LR in pink. (**c**) CTV LR isodose 100% red color wash. (**d**) CTV HR isodose 95% orange color wash. Note the smaller area of low-dose spillage 20%. Case contributed by the National Center for Oncological Hadrontherapy (CNAO). Abbreviations. *CIRT* carbon ion radiotherapy, *GTV* gross tumor volume, *CTV-LR* clinical target volume-low risk, *CTV-HR* clinical target volume-high risk

Overall, the two-volume approach to radiotherapy for soft tissue sarcomas (including MPNST) has been proven effective and can improve outcomes for patients with these challenging diseases. However, the optimal treatment plan for each individual patient depends on their specific oncological history and previous treatments, tumor location, and proximity to critical organs. It is important to carefully balance the potential benefits and risks of RT and to closely monitor patients for possible side effects or complications [44–47].

19.3.4 Choice of Particle and Doses

If a facility has both proton and carbon ion beams, the choice of particle depends primarily on the presence of macroscopic disease. Hypofractionation high-dose carbon ion radiotherapy (CIRT) is recommended for unresectable tumors or macroscopic residual mass after surgery. Proton therapy is preferred in the absence of macroscopic tumor, after complete resection with positive margins, or without positive margins in cases with histopathological characteristics suggesting different profiles of aggressiveness. Other factors that are taken into account include the degree of extension/aggressiveness of surgical removal and related complications (e.g., presence of osteo/dural breach and occurrence of surgical fistula), the amount of tissues available for surgical reconstruction (e.g., autologous fat, preferably multilayered surgical flaps), and expected toxicities. Examples of three-dimensional dose distribution with carbon ion are shown in Figs. 19.6 and 19.7.

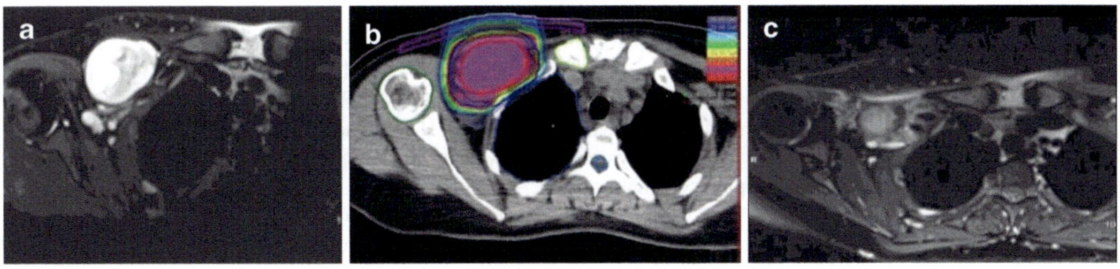

Fig. 19.7 Example plan of MPNST treated with postoperative carbon ions RT. A 21-year-old female with low-grade (G1) MPNST NF1 associated with brachial plexus. She underwent postoperative carbon ions RT for recurrence after surgery, total dose of 70.4 Gy (RBE) in 16 fractions. (**a**) Tumor volume pretreatment; (**b**) CIRT. (**c**) MRI after 16 months showing partial response. Follow-up after 8 years still showed partial response in stable disease. The low-grade and complete high-dose coverage was associated with long local control. Case contributed by the National Center for Oncological Hadrontherapy (CNAO)

19.4 Conclusion

MPNSTs are rare, aggressive soft tissue sarcomas that have a poor prognosis. The diagnosis and management of MPNSTs require a multidisciplinary approach, and treatment strategies should be tailored to the individual patient based on the extent of disease, location, and risk of functional impairment. Surgical resection with wide margins is the primary treatment for localized disease, but the extent of resection can be limited by the involvement of adjacent structures and the risk of functional impairment. Radiation therapy can be used as an adjuvant treatment to improve local control, but its efficacy in preventing distant metastases is limited. Advanced treatment modalities such as particle therapy should be considered for MPNSTs. Their radioresistance, requiring a higher dose or a more effective particle, the necessity to overcome the limited dose close to the organ at risk, the high risk of radioinduced secondary tumor in patients with NF1 make the particle therapy particularly suitable to lead to a significant improvement of the treatment of tumors characterized by poor prognosis and high mortality.

References

1. Bates JE, Peterson CR, Dhakal S, Giampoli EJ, Constine LS. Malignant peripheral nerve sheath tumors (MPNST): a SEER analysis of incidence across the age spectrum and therapeutic interventions in the pediatric population. Pediatr Blood Cancer. 2014;61(11):1955–60. https://doi.org/10.1002/pbc.25149.
2. Widemann BC. Current status of sporadic and neurofibromatosis type 1-associated malignant peripheral nerve sheath tumors. Curr Oncol Rep. 2009;11(4):322–8. https://doi.org/10.1007/s11912-009-0045-z.
3. Doorn PF, Molenaar WM, Buter J, Hoekstra HJ. Malignant peripheral nerve sheath tumors in patients with and without neurofibromatosis. Eur J Surg Oncol. 1995;21(1):78–82. https://doi.org/10.1016/s0748-7983(05)80073-3.
4. Amirian ES, Goodman JC, New P, Scheurer ME. Pediatric and adult malignant peripheral nerve sheath tumors: an analysis of data from the surveillance, epidemiology, and end results program. J Neuro-Oncol. 2014;116(3):609–16. https://doi.org/10.1007/s11060-013-1345-6.
5. Kattan MW, Leung DH, Brennan MF. Postoperative nomogram for 12-year sarcoma-specific death. J Clin Oncol. 2002;20(3):791–6. https://doi.org/10.1200/JCO.2002.20.3.791.
6. Ferrari A, Casanova M, Bisogno G, et al. Clear cell sarcoma of tendons and aponeuroses in pediatric patients: a report from the Italian and German Soft Tissue Sarcoma Cooperative Group. Cancer. 2002;94(12):3269–76. https://doi.org/10.1002/cncr.10597.
7. Baehring JM, Betensky RA, Batchelor TT. Malignant peripheral nerve sheath tumor: the clinical spectrum and outcome of treatment. Neurology. 2003;61(5):696–8. https://doi.org/10.1212/01.wnl.0000078813.05925.2c.
8. Cao Y, Wang YB, Bai Y, et al. Epidemiology, characteristic, and prognostic factors of primary sporadic intradural malignant peripheral nerve sheath tumor in the spinal canal: a systematic literature review. Front Oncol. 2022;12:911043. https://doi.org/10.3389/fonc.2022.911043.
9. Pervaiz N, Colterjohn N, Farrokhyar F, Tozer R, Figueredo A, Ghert M. A systematic meta-analysis of randomized controlled trials of adjuvant chemo-

therapy for localized resectable soft-tissue sarcoma. Cancer. 2008;113(3):573–81. https://doi.org/10.1002/cncr.23592.
10. Knight SWE, Knight TE, Santiago T, Murphy AJ, Abdelhafeez AH. Malignant peripheral nerve sheath tumors-a comprehensive review of pathophysiology, diagnosis, and multidisciplinary management. Children (Basel). 2022;9(1):38. https://doi.org/10.3390/children9010038.
11. Anghileri M, Miceli R, Fiore M, et al. Malignant peripheral nerve sheath tumors: prognostic factors and survival in a series of patients treated at a single institution. Cancer. 2006;107(5):1065–74. https://doi.org/10.1002/cncr.22098.
12. Wong WW, Hirose T, Scheithauer BW, Schild SE, Gunderson LL. Malignant peripheral nerve sheath tumor: analysis of treatment outcome. Int J Radiat Oncol Biol Phys. 1998;42(2):351–60. https://doi.org/10.1016/s0360-3016(98)00223-5.
13. Stucky CC, Johnson KN, Gray RJ, et al. Malignant peripheral nerve sheath tumors (MPNST): the Mayo Clinic experience. Ann Surg Oncol. 2012;19(3):878–85. https://doi.org/10.1245/s10434-011-1978-7.
14. Bishop AJ, Zagars GK, Torres KE, Bird JE, Feig BW, Guadagnolo BA. Malignant peripheral nerve sheath tumors: a single Institution's experience using combined surgery and radiation therapy. Am J Clin Oncol. 2018;41(5):465–70. https://doi.org/10.1097/COC.0000000000000303.
15. Miao R, Wang H, Jacobson A, et al. Radiation-induced and neurofibromatosis-associated malignant peripheral nerve sheath tumors (MPNST) have worse outcomes than sporadic MPNST. Radiother Oncol. 2019;137:61–70. https://doi.org/10.1016/j.radonc.2019.03.015.
16. Newell C, Chalil A, Langdon KD, et al. Cranial nerve and intramedullary spinal malignant peripheral nerve sheath tumor associated with neurofibromatosis-1. Surg Neurol Int. 2021;12:630. https://doi.org/10.25259/SNI_595_2021.
17. Chou D, Bilsky MH, Luzzati A, et al. Malignant peripheral nerve sheath tumors of the spine: results of surgical management from a multicenter study. J Neurosurg Spine. 2017;26(3):291–8. https://doi.org/10.3171/2016.8.SPINE151548.
18. Martin E, Muskens IS, Coert JH, Smith TR, Broekman MLD. Treatment and survival differences across tumor sites in malignant peripheral nerve sheath tumors: a SEER database analysis and review of the literature. Neurooncol Pract. 2019;6(2):134–43. https://doi.org/10.1093/nop/npy025.
19. Konieczkowski DJ, DeLaney TF, Yamada YJ. Radiation strategies for spine chordoma: proton beam, carbon ions, and stereotactic body radiation therapy. Neurosurg Clin N Am. 2020;31(2):263–88. https://doi.org/10.1016/j.nec.2019.12.002.
20. Tinganelli W, Durante M. Carbon ion radiobiology. Cancers (Basel). 2020;12(10):3022. https://doi.org/10.3390/cancers12103022.
21. Durante M, Paganetti H. Nuclear physics in particle therapy: a review. Rep Prog Phys. 2016;79(9):096702. https://doi.org/10.1088/0034-4885/79/9/096702.
22. Tsujii H, Kamada T, Shirai T, Noda K, Tsuji H, Karasawa K. Carbon-ion radiotherapy. Tokyo: Springer; 2014.
23. Kamada T, Tsujii H, Blakely EA, et al. Carbon ion radiotherapy in Japan: an assessment of 20 years of clinical experience. Lancet Oncol. 2015;16(2):e93–e100. https://doi.org/10.1016/S1470-2045(14)70412-7.
24. Imai R, Kamada T, Araki N, Working Group for Bone and Soft Tissue Sarcomas. Carbon ion radiation therapy for unresectable sacral chordoma: an analysis of 188 cases. Int J Radiat Oncol Biol Phys. 2016;95(1):322–7. https://doi.org/10.1016/j.ijrobp.2016.02.012.
25. Kepka L, DeLaney TF, Suit HD, Goldberg SI. Results of radiation therapy for unresected soft-tissue sarcomas. Int J Radiat Oncol Biol Phys. 2005;63(3):852–9. https://doi.org/10.1016/j.ijrobp.2005.03.004.
26. Matsumoto Y, Matsunobu A, Kawaguchi K, et al. Clinical results of carbon-ion radiotherapy with separation surgery for primary spine/paraspinal sarcomas. Int J Clin Oncol. 2019;24(11):1490–7. https://doi.org/10.1007/s10147-019-01505-y.
27. Bachmann N, Leiser D, Pica A, Bachtiary B, Weber DC. Clinical outcome after pencil beam scanning proton therapy of patients with non-metastatic malignant and benign peripheral nerve sheath tumors. Front Oncol. 2022;12:881665. https://doi.org/10.3389/fonc.2022.881665.
28. Sharif S, Ferner R, Birch JM, et al. Second primary tumors in neurofibromatosis 1 patients treated for optic glioma: substantial risks after radiotherapy. J Clin Oncol. 2006;24(16):2570–5. https://doi.org/10.1200/JCO.2005.03.8349.
29. Kahn J, Gillespie A, Tsokos M, et al. Radiation therapy in management of sporadic and neurofibromatosis type 1-associated malignant peripheral nerve sheath tumors. Front Oncol. 2014;4:324. https://doi.org/10.3389/fonc.2014.00324.
30. Ducatman BS, Scheithauer BW. Postirradiation neurofibrosarcoma. Cancer. 1983;51(6):1028–33. https://doi.org/10.1002/1097-0142(19830315)51:6<1028::aid-cncr2820510610>3.0.co;2-3.
31. Yamanaka R, Hayano A. Radiation-induced malignant peripheral nerve sheath tumors: a systematic review. World Neurosurg. 2017;105:961–970.e8. https://doi.org/10.1016/j.wneu.2017.06.010.
32. Imai R, Kamada T, Araki N, Working Group for Carbon Ion Radiotherapy for Bone and Soft Tissue Sarcomas. Carbon ion radiotherapy for unresectable localized axial soft tissue sarcoma. Cancer Med. 2018;7(9):4308–14. https://doi.org/10.1002/cam4.1679.
33. Jensen AD, Uhl M, Chaudhri N, Herfarth KK, Debus J, Roeder F. Carbon ion irradiation in the treatment of grossly incomplete or unresectable malignant periph-

eral nerve sheaths tumors: acute toxicity and preliminary outcome. Radiat Oncol. 2015;10:109. https://doi.org/10.1186/s13014-015-0414-8.
34. Fiore MR, Chalaszczyk A, Barcellini A, Vitolo V, Fontana G, Russo S, Rotondi M, Molinelli S, Mirandola A, Bazani A, Orlandi E. Clinical Outcomes of Carbon Ion Radiation Therapy for Malignant Peripheral Nerve Sheath Tumors. Adv Radiat Oncol. 2024;9(11):101619. https://doi.org/10.1016/j.adro.2024.101619. PMID: 39429980; PMCID: PMC11490661.
35. Doyle LA. Sarcoma classification: an update based on the 2013 World Health Organization classification of tumors of soft tissue and bone. Cancer. 2014;120(12):1763–74. https://doi.org/10.1002/cncr.28657.
36. Bertholet J, Knopf A, Eiben B, et al. Real-time intrafraction motion monitoring in external beam radiotherapy. Phys Med Biol. 2019;64(15):15TR01. https://doi.org/10.1088/1361-6560/ab2ba8.
37. Korreman SS. Motion in radiotherapy: photon therapy. Phys Med Biol. 2012;57(23):R161–91. https://doi.org/10.1088/0031-9155/57/23/R161.
38. Seco J, Robertson D, Trofimov A, Paganetti H. Breathing interplay effects during proton beam scanning: simulation and statistical analysis. Phys Med Biol. 2009;54:N283–94.
39. Minohara S, Kanai T, Endo M, Noda K, Kanazawa M. Respiratory gated irradiation system for heavy-ion radiotherapy. Int J Radiat Oncol Biol Phys. 2000;47(4):1097–103. https://doi.org/10.1016/s0360-3016(00)00524-1.
40. Patel CG, Peterson J, Aznar M, et al. Systematic review for deep inspiration breath hold in proton therapy for mediastinal lymphoma: a PTCOG Lymphoma Subcommittee report and recommendations. Radiother Oncol. 2022;177:21–32. https://doi.org/10.1016/j.radonc.2022.10.003.
41. Furukawa T, Inaniwa T, Sato S, et al. Design study of a raster scanning system for moving target irradiation in heavy-ion radiotherapy. Med Phys. 2007;34:1085–97.
42. Mastella E, Molinelli S, Magro G, et al. Dosimetric characterization of carbon fiber stabilization devices for post-operative particle therapy. Phys Med. 2017;44:18–25. https://doi.org/10.1016/j.ejmp.2017.11.008.
43. Cobianchi L, Peloso A, Vischioni B, et al. Erratum to: surgical spacer placement prior carbon ion radiotherapy (CIRT): an effective feasible strategy to improve the treatment for sacral chordoma. World J Surg Oncol. 2016;14(1):262. https://doi.org/10.1186/s12957-016-1020-4.
44. Salerno KE, Alektiar KM, Baldini EH, et al. Radiation therapy for treatment of soft tissue sarcoma in adults: executive summary of an ASTRO clinical practice guideline. Pract Radiat Oncol. 2021;11(5):339–51. https://doi.org/10.1016/j.prro.2021.04.005.
45. Le Péchoux C, Moureau-Zabotto L, Llacer C, et al. Radiothérapie des sarcomes des tissus mous de l'adulte [Radiotherapy of adult soft tissue sarcoma]. Cancer Radiother. 2016;20 Suppl:S235–43 (in French). https://doi.org/10.1016/j.canrad.2016.07.028.
46. Haas RL, Delaney TF, O'Sullivan B, et al. Radiotherapy for management of extremity soft tissue sarcomas: why, when, and where? Int J Radiat Oncol Biol Phys. 2012;84(3):572–80. https://doi.org/10.1016/j.ijrobp.2012.01.062.
47. Bahig H, Roberge D, Bosch W, et al. Agreement among RTOG sarcoma radiation oncologists in contouring suspicious peritumoral edema for preoperative radiation therapy of soft tissue sarcoma of the extremity. Int J Radiat Oncol Biol Phys. 2013;86(2):298–303. https://doi.org/10.1016/j.ijrobp.2013.01.032.

20 Radiomics and Machine Learning in PNST

Zhichao Wang, Chengjiang Wei, and Wei Wang

20.1 Imaging Method in PNSTs

Peripheral nerve sheath tumors (PNSTs) are one of the most common soft tissue neoplasms that arise from nerve sheaths outside of the central nervous system [1]. Most PNSTs are benign and are normally divided into four groups: schwannoma, neurofibroma, perineurioma, and hybrid nerve sheath tumors [1]. In addition to these benign tumors, some patients suffer from malignant peripheral nerve sheath tumors (MPNSTs), and most patients with MPNSTs have a poor prognosis [2]. Imaging methods, such as magnetic resonance imaging (MRI), computed tomography (CT), positron emission tomography (PET), and ultrasound, play an essential role in PNST management [3]. They offer valuable assistance for doctors, including early tumor screening, tumor evaluation, therapy efficacy prediction, etc. In this section, we introduce imaging methods that are commonly used in PNST management.

Z. Wang (✉) · C. Wei · W. Wang
Department of Plastic and Reconstructive Surgery, Shanghai Ninth People's Hospital, Shanghai Jiao Tong University School of Medicine, Shanghai, China
e-mail: wangzhichao@sh9hospital.org.cn

20.1.1 MRI Imaging

PNSTs are a group of soft tissue tumors that have better resolution on MRI than on CT. Consequently, MRI is normally the most preferred imaging method for PNSTs and plays an essential role in tumor evaluation, including determining the exact PNST location, characterizing its relationship to neighboring structures (intramuscular or intraosseous, neurovascular involvement), and therapy reaction assessment. Benign PNSTs are well-defined masses with low signal intensity on T1-weighted sequences and high signal intensity on T2-weighted sequences, and they show avid contrast enhancement on T1-enhanced sequences [3]. A central area of low intensity in PST lesions, called the "target sign," is sometimes observed, which indicates the presence of myxoid material peripherally and fibrous tissue centrally [4]. Among different types of benign PNSTs, the image features are similar and cannot be distinguished in many cases [5]. If the nerves from which the PNSTs were derived could be identified, a centrally located mass might suggest a neurofibroma, while an eccentrically positioned lesion suggests a schwannoma [5]. Compared with PNSTs, MRI imaging of MPNSTs sometimes reveals a perifocal edema reaction, which occurs due to the invasive growth of MPNSTs [6]. On T1-enhanced sequences, PNSTs normally show central focal enhancement, while MPNSTs show peripheral enhancement because of central

I. G. Vetrano, V. Nazzi (eds.), *Peripheral Nerve Sheath Tumors*,
https://doi.org/10.1007/978-3-031-76382-3_20

necrosis [6, 7]. Moreover, calcification and heterogeneity are more common in MPNSTs, but these can also be found in benign PNSTs such as ancient schwannomas [8]. Taken together, previous studies have proven the statistical significance of several morphological identifications based on MRI for distinguishing MPNSTs from benign PNF [9]. However, these features on MRI are not fully reliable because some patients do not have the typical signs mentioned above, and some patients even have opposite signs [9].

Recently, functional MRI (fMRI), such as diffusion-weighted image (DWI)-based fMRI, was further adopted for PST evaluation and provided information in addition to morphological sequences [4]. As the histological evaluation of MPNSTs commonly shows increased cellularity compared with benign PNSTs, DWI and DWI-based ADC are valuable for differential diagnosis because MPNSTs have lower ADC values than benign PNSTs [10, 11].

20.1.2 Computed Tomography

CT is an ideal radiology imaging method for observing bone, joint, and soft tissue lesions. In PNST evaluation, CT imaging provides valuable information for tumor location and local invasion [9]. Although MRI shows better resolution of PNST lesions, CT imaging is better than MRI for imaging bone, which is essential for PNST patients with bone deformities (such as patients with neurofibromatosis type 1). On CT imaging, benign PNSTs appear as well-defined hypodense masses relative to muscle, and PNST lesions can be enhanced under contrast administration [3]. Compared with benign PNSTs, MPNST lesions are normally much larger with poorly defined margins and high heterogeneity, but this can also be found in benign PNSTs [3]. In other words, there have been no definitive CT diagnostic features reported for distinguishing MPNST from benign PNSTs until now. Furthermore, no strong evidence currently suggests that the CT imaging characteristics could definitively distinguish the four benign neurofibromas. However, CT imaging can enable doctors to better observe the involvement of the adjacent orbital bone, and computed tomographic angiography (CTA) can further provide information on blood supply [12, 13]. CT imaging is also widely used in biopsy to access solid or enhancing components, avoid necrotic areas, and find the best path.

20.1.3 PET

By using radioactive medium such as 18F-FDG to reveal the metabolic activity of different tissues, PET scans have been proven reliable for distinguishing malignant and benign lesions [14]. Currently, the most popular PET imaging mode is PET/CT, which combines PET and CT imaging. In patients with PNSTs, PET/CT can provide at least two crucial pieces of information: the whole-body tumor loads and the metabolic activity of each tumor lesion. The former is of great importance to patients with diseases such as NF1, an autosomal dominant genetic disorder that has possible tumor occurrence all over the body [15, 16]. PET/CT examination could help the early discovery of PNSTs, which further contributes to early treatment [17]. Furthermore, PET/CT also provides tumor metastasis evidence for MPNST, which plays an essential role in the initial staging of MPNST [18]. Moreover, after a PET/CT examination, the maximum standardized uptake value (SUV_{max}) of each lesion is normally calculated. For MPNSTs, the SUV_{max} value is normally much higher than that for benign PNSTs, and a study showed that the SUV_{max} for all MPNST lesions was >3.15 [19].

Although PET/CT demonstrates high diagnostic value in PNSTs, one of the disadvantages is still the lower resolution of CT in soft tissue tumors compared with MRI. Another imaging method combining PET and MRI recently might exhibit better performance in PNSTs. Furthermore, as some PNSTs are characterized by a loss of a tumor suppression gene product that many children and adolescents need to be investigated, PET/MRI reduces both the number of necessary sedations in younger children and, compared to PET/CT, radiation exposure [20]. As a novel emerging method, further studies are needed to compare it with PET/CT in PNSTs.

20.1.4 Ultrasound

Ultrasound provides an economical, simple, and precise imaging modality for PNST evaluation compared with other imaging methods, such as CT and MRI [21]. Furthermore, ultrasound has the advantage of being a noninvasive and nonradioactive procedure that has no contraindications. On ultrasound images, most PNSTs appear as hypoechoic masses. The majority of benign PNSTs have well-defined boundaries and regular shapes, while most MPNSTs share features of irregular shape and an ill-defined boundary [22, 23]. Similar to the features on MRI, schwannoma is normally an eccentrically positioned lesion on ultrasound, while a concentric lesion might indicate a neurofibroma [21]. Plexiform neurofibromas are normally present as inhomogeneous nodular tumor bands, which are called "bags of worms" [21]. The target sign is considered a typical manifestation of benign PNSTs that is rarely exhibited in MPNSTs [22]. MPNSTs normally have richer blood flow signals than benign PNSTs [22]. Furthermore, ultrasound could be further applied to distinguish superficial masses such as PNSTs and enlarged lymph nodes because of its ability to find the hilum of the lymph gland. Compared with CT and MRI, ultrasound imaging is not well suited for tumor volume evaluation throughout the body. However, it provides a radiotoxicity-free and relatively economical method for certain tumor assessments and further follow-up evaluations.

20.2 Machine Learning and Radiomics in PNSTs

With the development of radiology imaging hardware, the resolution and quality of these imaging methods have increased significantly, which enables the potential discrimination of minute differences. These differences are sometimes difficult to identify by human eyes and even by some trained radiologists. The development of machine learning suggests a new direction to understand radiology imaging. With machine learning, it is now possible for computers to extract innumerable quantitative features from images, and the conversion of these medical images into high-dimensional data is called radiomics. The major advantage of machine learning is the ability to process massive radiomic data. In this section, we introduce the machine learning methods in PNSTs.

20.2.1 Machine Learning Models for PNST Diagnosis

Radiomics enables a quantitative approach for distinguishing among neurogenic soft tissue tumors [24, 25]. A study developed a radiomics model based on both T1-weighted MRI images and clinical features that outperformed human evaluators, even when more information from additional MRI sequences was provided to these human evaluators [26].

Accurate differentiation between benign peripheral nerve sheath tumors and MPNST is essential for patients because of the great differences in the treatment paradigms. As we introduced in Sect. 20.1, distinguishing benign PNSTs and MPNSTs with conventional imaging remains error-prone by subjectively defined features. Machine learning approaches can add quantitative insights to this qualitative interpretation, which offers other ways for early screening of MPNSTs from benign PNSTs. Michael Zhang et al. developed radiomic classifiers to differentiate between benign PNSTs and MPNSTs based on the features acquired by a machine learning approach, which showed an AUC of 0.845 on ROC curves on T1-weighted postgadolinium MRI images [27]. Another study further used the imaging features from T2w SPAIR MRI sequences and exhibited a higher mean AUC of 0.94 for the correct classification of benign PNST and MPNST (sensitivity: 94%, specificity: 97%) [28]. Moreover, this machine learning model further included atypical neurofibroma (ANF), an intermediate stage between PNF and MPNST, which showed lower sensitivity and specificity when combining ANF and MPNST [28]. This result was in accordance with the challenging histopathological differentiation, as traditional diagnosis of ANF is difficult for even experienced pathologists because of the nature of these tumors [29]. This study suggests that the machine learning approach could extend radiology imaging to more clinical applications that might not be possible by human radiologists. Another example is the success of building a deep learning model for head and neck MPNST differential diagnosis from PNF based on CT imaging, which has been reported to have no definitive diagnostic features that can be used to differentiate MPNST from PNF in NF1 patients [30]. Deep learning algorithms are another fundamentally different paradigm in machine learning without the need for a prior definition of features by human experts. However, deep learning models require much more data, called "big data," for automatic diagnosis, which has rarely been used in PNST evaluation. As more data are included in the future, deep learning models based on MRI might show better performance and could be expeditiously translated into the clinic. Furthermore, current explorations of these machine learning models, whether they are built on handcrafted engineered features or use the deep learning approach, only include limited modalities of images. Previously demonstrated effective sequences, such as DWI and PET, were not adopted, and the combination of these sequences in future models might further improve the performance of the models.

20.2.2 Machine Learning Models for Disease Progression and Therapy Response Prediction

In addition to differential diagnosis, radiology-based machine learning models have also been found to have potential roles in the prediction of therapy response and disease progression. For example, a study in local advanced rectal cancer found an MRI-based radiomic signature for predicting distant metastasis and chemotherapy benefit [31]. A study on vestibular schwannoma developed an MRI-based machine-learning model to predict the blood supply [32], and the success of this model suggested the correlation between radiomics features and disease conditions. Another radiomics analysis based on contrast-enhanced T1-weighted images also found that texture features were significantly associated with the rapid growth of vestibular schwannomas [33]. For therapy predictions, several studies used the machine learning approach and successfully built radiomics-based predictions of treatment response to radiosurgery of vestibular schwannomas [34, 35]. Furthermore, considering the recent emergence of targeted therapies for PNSTs, such as bevacizumab for schwannoma and selumetinib for plexiform neurofibroma, target therapy response prediction might also be an essential part of radiomics in PNSTs. A study on vestibular schwannoma in neurofibromatosis type 2 (NF2) proved several features from dynamic contrast-enhanced MRI as significant predictors for the volume response of the tumor [36]. This study indicated the potential of machine learning radiomics models in the target therapy response prediction of PNSTs, which merits further investigation.

20.2.3 Radiomics-Based Multiomics in PNSTs

Another direction for radiomics is the combination with other omics, such as genomics and pathomics. In this subsection, we introduce radiomics-based multiomics in PNSTs. The clarification of this indicated that radiology could provide doctors far more information than imaging itself with machine learning-based radiomics approaches.

Radiogenomics is the investigation of the association of quantitative radiology imaging features with genomic data. In total, radiogenomics is associated with two distinct scientific questions: one is the study of the correlation between germline genotypic variations and the large clinical variability observed in response to radiation therapy [37], and the other is the investigation of the link between specific imaging features and specific gene-expression patterns [38, 39]. In this section, we mainly discuss the latter, as the usage of radiation therapy is controversial and not widely adopted in PNSTs [40]. In neurofibromatosis type 1 (NF1), a study based on whole-body MRI revealed six major image feature patterns by unsupervised hierarchical clustering that were significantly correlated with predefined gene mutation domains [41]. Another study in NF1 found that a special MRI imaging feature of some patients, neurofibromatosis type 1 bright objects (NBOs), was more likely to appear in NF1 patients with frameshift variants. Considering that some PNSTs, such as neurofibromatosis type 1, are genetic disorders, studies with genotypic variations might be more essential than other tumor types.

Currently, studies for other radiomics-based multiomics are rare in PNSTs. However, the combination of radiomics with other "omics" might provide more information than single radiomics. For example, a study in advanced rectal cancer developed a radiopathomics model for predicting pathological complete response to neoadjuvant chemoradiotherapy and reached an AUC of 0.812 in the prospective validation cohort, which significantly outperformed single-modality prediction models (0.716 for the radiomics MRI model) [42]. With future studies of radiomics-based multiomics in PNSTs, the current performance of single radiomics-based models might be further improved.

20.3 Summary and Future Direction

Radiological imaging is essential for tumor diagnosis, evaluation, tumor staging, and therapy efficiency assessment in PNST patients over a long period. Recent advances in computer technology have facilitated the emergence of machine learning, which offers promise in numerous industries and applications, including radiology medical imaging [43, 44]. In PNSTs, many studies take advantage of machine learning approaches for analysis, and the adoption of these approaches in clinical scenarios might further improve the current time-consuming and substantial experience-needed manual work. The main directions of the current research are as follows: (a) differential diagnosis of PNSTs, (b) disease progression prediction, (c) therapy response prediction, and (d) the correlation between radiomics features and the genotype of tumors.

Simultaneously, we noticed some presently fundamental issues in machine learning-based radiology imaging research on PNSTs. First, the majority of current studies on PNSTs only applied a relatively small scale of data from a single center. The limitation of the scale of the data will restrict the adoption of deep learning-based models and inevitably lead to unsatisfactory accuracy of the machine learning-based model outcomes. Future efforts including collaboration among multiple centers to establish large open datasets might further assist the development of machine learning models for PNSTs. Second, the development of current machine learning models normally requires presegmentation of the region of interest (ROI). Traditional manual work is laborious and time-consuming, and one of the most popular research directions today is the automatic segmentation of tumors. In PNSTs, a full-automatic segmentation approach has not been reported. The development of this in the future will further assist the progress of machine-learning models in PNSTs. Third, most studies currently focus on the differential diagnosis, but the correlation with other clinical problems, such as tumor progression and therapy responses, remains unclear. Furthermore, studies correlating radiomics with other omics are also at the primary stage. This might partially be caused by the underlying mechanisms and clinical stages of PNSTs not being totally understood, and the scarcity of large cohorts might also contribute. As more data are included in the future and PNST development is further elucidated clinically, the study of radiomics will expand its application to these diseases. Finally, the adoption of deep learning models in studies on PNSTs is rare, and the main reason might be the scarcity of "big data," as mentioned above. Compared with most other machine learning-based models, deep learning models are neural network structures, which are loosely inspired by the human brain and give these models the ability to approximate complex nonlinear relationships. The future adoption of deep learning models with more data will further promote the application of radiology imaging methods in PNSTs.

In summary, radiology imaging is essential for PNST patients, but further large-scale cohort studies are needed in this area. This requires international cooperation because of the relative rarity and complexity of these diseases: every patient counts.

References

1. De Luca-Johnson J, Kalof AN. Peripheral nerve sheath tumors: an update and review of diagnostic challenges. Diagn Histopathol. 2016;22(11):447–57.
2. Valentin T, Le Cesne A, Ray-Coquard I, et al. Management and prognosis of malignant peripheral nerve sheath tumors: the experience of the French Sarcoma Group (GSF-GETO). Eur J Cancer. 2016;56:77–84.
3. Lin J, Martel W. Cross-sectional imaging of peripheral nerve sheath tumors: characteristic signs on CT, MR imaging, and sonography. AJR Am J Roentgenol. 2001;176(1):75–82.
4. Well L, Salamon J, Kaul MG, et al. Differentiation of peripheral nerve sheath tumors in patients with neurofibromatosis type 1 using diffusion-weighted magnetic resonance imaging. Neuro-Oncology. 2019;21(4):508–16.
5. Demehri S, Belzberg A, Blakeley J, Fayad LM. Conventional and functional MR imaging of peripheral nerve sheath tumors: initial experience. AJNR Am J Neuroradiol. 2014;35(8):1615–20.
6. Wasa J, Nishida Y, Tsukushi S, et al. MRI features in the differentiation of malignant peripheral nerve sheath tumors and neurofibromas. AJR Am J Roentgenol. 2010;194(6):1568–74.
7. Ogose A, Hotta T, Morita T, et al. Tumors of peripheral nerves: correlation of symptoms, clinical signs, imaging features, and histologic diagnosis. Skeletal Radiol. 1999;28(4):183–8.
8. Lee SK, Kim JY, Lee YS, Jeong HS. Intramuscular peripheral nerve sheath tumors: schwannoma, ancient schwannoma, and neurofibroma. Skeletal Radiol. 2020;49(6):967–75.
9. Liu J, Huang JN, Wang MH, et al. Image-based differentiation of benign and malignant peripheral nerve sheath tumors in neurofibromatosis type 1. Front Oncol. 2022;12:898971.
10. Ahlawat S, Blakeley JO, Rodriguez FJ, Fayad LM. Imaging biomarkers for malignant peripheral nerve sheath tumors in neurofibromatosis type 1. Neurology. 2019;93(11):e1076–84.
11. Yun JS, Lee MH, Lee SM, et al. Peripheral nerve sheath tumor: differentiation of malignant from benign tumors with conventional and diffusion-weighted MRI. Eur Radiol. 2021;31(3):1548–57.
12. Dai M, Wang T, Wang JM, et al. Imaging characteristics of orbital peripheral nerve sheath tumors: analysis of 34 cases. World J Clin Cases. 2022;10(21):7356–64.
13. Yang X, Liu J, Zhang Y, Richard SA. Tumor-associated hemorrhage in patients with vestibular schwannoma. Acta Neurochir. 2018;160(8):1563–9.
14. Fonti R, Conson M, Del Vecchio S. PET/CT in radiation oncology. Semin Oncol. 2019;46(3):202–9.
15. Cimino PJ, Gutmann DH. Neurofibromatosis type 1. Handb Clin Neurol. 2018;148:799–811.
16. Gutmann DH, Ferner RE, Listernick RH, Korf BR, Wolters PL, Johnson KJ. Neurofibromatosis type 1. Nat Rev Dis Primers. 2017;3:17004.
17. Filizoglu N, Ozguven S. Neurofibromatosis type 2: multiple meningiomatosis and vestibular schwannomas on 68 Ga-DOTATATE PET/CT. Clin Nucl Med. 2022;47(11):e710–2.
18. Xiao L, Li Y, Li L. Solitary lymph node metastasis from a malignant peripheral nerve sheath tumor of the thoracic vertebra revealed on FDG PET/CT. Clin Nucl Med. 2022;47(5):e419–20.
19. Tovmassian D, Abdul Razak M, London K. The role of [(18)F]FDG-PET/CT in predicting malignant transformation of plexiform neurofibromas in neurofibromatosis-1. Int J Surg Oncol. 2016;2016:6162182.
20. Reinert CP, Schuhmann MU, Bender B, et al. Comprehensive anatomical and functional imaging in patients with type I neurofibromatosis using simultaneous FDG-PET/MRI. Eur J Nucl Med Mol Imaging. 2019;46(3):776–87.
21. Gruber H, Glodny B, Bendix N, Tzankov A, Peer S. High-resolution ultrasound of peripheral neurogenic tumors. Eur Radiol. 2007;17(11):2880–8.
22. Jin Z, Zhao K, Guo W, Wang D, Deng Y, Chen T. Investigation of ultrasound parameters for the differential diagnosis of malignant and benign peripheral nerve sheath tumors. J Ultrasound Med. 2022;41:3091–101.
23. Chiou HJ, Chou YH, Chiou SY, Liu JB, Chang CY. Peripheral nerve lesions: role of high-resolution US. Radiographics. 2003;23(6):e15.
24. Ghosh A, Malla SR, Bhalla AS, Manchanda S, Kandasamy D, Kumar R. Texture analysis of routine T2 weighted fat-saturated images can identify head and neck paragangliomas - a pilot study. Eur J Radiol Open. 2020;7:100248.
25. Uthoff J, De Stefano FA, Panzer K, et al. Radiomic biomarkers informative of cancerous transformation in neurofibromatosis-1 plexiform tumors. J Neuroradiol. 2019;46(3):179–85.
26. Zhang M, Tong E, Wong S, et al. Machine learning approach to differentiation of peripheral schwannomas and neurofibromas: a multi-center study. Neuro-Oncology. 2022;24(4):601–9.
27. Zhang M, Tong E, Hamrick F, et al. Machine-learning approach to differentiation of benign and malignant peripheral nerve sheath tumors: a multicenter study. Neurosurgery. 2021;89(3):509–17.

28. Ristow I, Madesta F, Well L, et al. Evaluation of MRI-based radiomics characteristics for differentiation of benign and malignant peripheral nerve sheath tumors in neurofibromatosis type 1. Neuro-Oncology. 2022;24:1790–8.
29. Miettinen MM, Antonescu CR, Fletcher CDM, et al. Histopathologic evaluation of atypical neurofibromatous tumors and their transformation into malignant peripheral nerve sheath tumor in patients with neurofibromatosis 1-a consensus overview. Hum Pathol. 2017;67:1–10.
30. Wei CJ, Yan C, Tang Y, et al. Computed tomography-based differentiation of benign and malignant craniofacial lesions in neurofibromatosis type I patients: a machine learning approach. Front Oncol. 2020;10:1192.
31. Dong D, Fang MJ, Tang L, et al. Deep learning radiomic nomogram can predict the number of lymph node metastasis in locally advanced gastric cancer: an international multicenter study. Ann Oncol. 2020;31(7):912–20.
32. Song D, Zhai Y, Tao X, Zhao C, Wang M, Wei X. Prediction of blood supply in vestibular schwannomas using radiomics machine learning classifiers. Sci Rep. 2021;11(1):18872.
33. Itoyama T, Nakaura T, Hamasaki T, et al. Whole tumor radiomics analysis for risk factors associated with rapid growth of vestibular schwannoma in contrast-enhanced T1-weighted images. World Neurosurg. 2022;166:e572–82.
34. Langenhuizen P, Zinger S, Leenstra S, et al. Radiomics-based prediction of long-term treatment response of vestibular schwannomas following stereotactic radiosurgery. Otol Neurotol. 2020;41(10):e1321–7.
35. Yang HC, Wu CC, Lee CC, et al. Prediction of pseudoprogression and long-term outcome of vestibular schwannoma after Gamma Knife radiosurgery based on preradiosurgical MR radiomics. Radiother Oncol. 2021;155:123–30.
36. Li KL, Djoukhadar I, Zhu X, et al. Vascular biomarkers derived from dynamic contrast-enhanced MRI predict response of vestibular schwannoma to antiangiogenic therapy in type 2 neurofibromatosis. Neuro-Oncology. 2016;18(2):275–82.
37. Rosenstein BS, West CM, Bentzen SM, et al. Radiogenomics: radiobiology enters the era of big data and team science. Int J Radiat Oncol Biol Phys. 2014;89(4):709–13.
38. Rutman AM, Kuo MD. Radiogenomics: creating a link between molecular diagnostics and diagnostic imaging. Eur J Radiol. 2009;70(2):232–41.
39. Lambin P, Leijenaar RTH, Deist TM, et al. Radiomics: the bridge between medical imaging and personalized medicine. Nat Rev Clin Oncol. 2017;14(12):749–62.
40. Marjanska A, Galazka P, Wysocki M, Styczynski J. New frontiers in therapy of peripheral nerve sheath tumors in patients with neurofibromatosis type 1: latest evidence and clinical implications. Anticancer Res. 2020;40(4):1817–31.
41. Liu Y, Jordan JT, Bredella MA, et al. Correlation between NF1 genotype and imaging phenotype on whole-body MRI: NF1 radiogenomics. Neurology. 2020;94(24):e2521–31.
42. Feng L, Liu Z, Li C, et al. Development and validation of a radiopathomics model to predict pathological complete response to neoadjuvant chemoradiotherapy in locally advanced rectal cancer: a multicentre observational study. Lancet Digit Health. 2022;4(1):e8–e17.
43. Choy G, Khalilzadeh O, Michalski M, et al. Current applications and future impact of machine learning in radiology. Radiology. 2018;288(2):318–28.
44. Hosny A, Parmar C, Quackenbush J, Schwartz LH, Aerts H. Artificial intelligence in radiology. Nat Rev Cancer. 2018;18(8):500–10.

GPSR Compliance

The European Union's (EU) General Product Safety Regulation (GPSR) is a set of rules that requires consumer products to be safe and our obligations to ensure this.

If you have any concerns about our products, you can contact us on ProductSafety@springernature.com

In case Publisher is established outside the EU, the EU authorized representative is:

Springer Nature Customer Service Center GmbH
Europaplatz 3
69115 Heidelberg, Germany

Batch number: 10732801

Printed by Printforce, the Netherlands